D0953452

FUZZ

FUZZ

When Nature Breaks the Law

MARY ROACH

W. W. NORTON & COMPANY
Independent Publishers Since 1923

For information about permission to reproduce selections from this book, write to Permissions, W. W. Norton & Company, Inc., 500 Fifth Avenue, New York, NY 10110

For information about special discounts for bulk purchases, please contact W. W. Norton Special Sales at specialsales@wwnorton.com or 800-233-4830

Photograph credits: Frontispiece: EyeEm / Alamy Stock Photo; Chapter 1: Frederick Bass / Getty Images; Chapter 2: Everett Collection; Chapter 3: Sueddeutsche Zeitung Photo / Alamy Stock Photo; Chapter 4: Aditya Singh / Getty Images; Chapter 5: Collection of Mary Roach; Chapter 6: "Scats of the Cat Family" from *A Field Guide to Animal Tracks*, 2nd edition, by Olaus J. Murie. Copyright © 1974 by Margaret E. Murie. Reprinted by permission of Houghton Mifflin Harcourt Publishing Company. All rights reserved; Chapter 7: Sourced Collection / Alamy Stock Photo; Chapter 8: Izel Photography / Alamy Stock Photo; Chapter 9: The Picture Art Collection / Alamy Stock Photo; Chapter 10: Album / Alamy Stock Photo; Chapter 11: Library of Congress; Chapter 12: Christian Musat / Alamy Stock Photo; Chapter 13: Christian Musat / Alamy Stock Photo; Chapter 14: US Patent Office; Chapter 15: BIOSPHOTO / Alamy Stock Photo.

Manufacturing by LSC Communications, Harrisonburg
Book design by Chris Welch
Production manager: Beth Steidle

Library of Congress Cataloging-in-Publication Data

Names: Roach, Mary, author.
Title: Fuzz : when nature breaks the law / Mary Roach.
Description: First edition. | New York : W. W. Norton & Company, [2021] | Includes bibliographical references and index.
Identifiers: LCCN 2021006594 | ISBN 9781324001935 (hardcover) | ISBN 9781324001843 (epub)
Subjects: LCSH: Human-animal relationships. | Animals and civilization. | Animal behavior. | Wildlife management.
Classification: LCC QL85 .R623 2021 | DDC 591.5—dc23
LC record available at https://lccn.loc.gov/2021006594

W. W. Norton & Company, Inc., 500 Fifth Avenue, New York, N.Y. 10110
www.wwnorton.com

W. W. Norton & Company Ltd., 15 Carlisle Street, London W1D 3BS

1 2 3 4 5 6 7 8 9 0

For Gus, Bean, and Winnie. To the farthest star.

Contents

FUZZ

A Quick Word of Introduction

On June 26, 1659, a representative from five towns in a province of northern Italy initiated legal proceedings against caterpillars. The local specimens, went the complaint, were trespassing and pilfering from people's gardens and orchards. A summons was issued and five copies made and nailed to trees in forests adjacent to each town. The caterpillars were ordered to appear in court on the twenty-eighth of June, at a specified hour, where they would be assigned legal representation.

Of course no caterpillars appeared at the appointed time, but the case went forward anyway. In a surviving document, the court recognizes the rights of caterpillars to live freely and happily, provided this does not "impair the happiness of man . . ." The judge decreed that the caterpillars be assigned a plot of alternate land for their sustenance and enjoyment. By the time the details were worked out, the defendants, having pupated, were surely through with their devastations, and all parties no doubt left the proceedings satisfied.

The case is detailed in an unusual 1906 book, *The Criminal Prosecution and Capital Punishment of Animals*. When I first paged through it, I wondered if it might be an ambitious hoax. Here were bears formally excommunicated from the Church. Slugs given three warnings to stop nettling farmers, under penalty of "smiting." But the author, a respected historian and linguist, quickly wore me down with a depth of detail gleaned from original documents, nineteen of which are reproduced in their original languages in a series of appendices. We have the itemized expense report of a French bailiff, submitted in 1403 following the murder trial of a pig ("cost of keeping her in jail, six sols parisis"). We have writs of ejectment issued to rats and thrust into their burrows. From a 1545 complaint brought by vintners against a species of greenish weevil, we have not only the names of the lawyers but early examples of that time-honored legal tactic, the stall. As far as I could tell, the proceedings dragged on eight or nine months—in any case, longer than the life span of a weevil.

I present all this not as evidence of the silliness of bygone legal systems but as evidence of the intractable nature of human-wildlife conflict—as it is known today by those who grapple with it professionally. The question has defied satisfactory resolution for centuries: What is the proper course when nature breaks laws intended for people?

The actions of the magistrates and prelates made no rational sense, of course, for rats and weevils cannot understand property law or be expected to conform to the moral principles of human civilizations. The aim was to cow and impress the populace: *look here, even nature must bend to our rule!* And it was, in its way, impressive. The sixteenth-century judge who granted leniency to moles with young offspring made a show not only of his authority but of his temperance and compassion.

Wandering through the Middle Ages and the centuries just

beyond, I began to wonder what the modern epoch had brought to bear on these matters. Having sampled the esoteric solutions of law and religion, I set out to see what science has been bringing to the table, and what answers it might offer for the future. So began more wandering. My guides were people with titles unfamiliar to me: Human-Elephant Conflict Specialist, Bear Manager, Danger-Tree Faller-Blaster. I spent time with predator attack specialists and attack forensics investigators, builders of laser scarecrows and testers of kinder poisons. I traveled to some of the "hot spots"—back alleys in Aspen, Colorado; leopard-terrorized hamlets in the Indian Himalaya; St. Peter's Square the night before the pope's Easter Mass. I considered the contributions of bygone professionals—the economic ornithologists and the rat searchers—as well as the stewards of the future, the conservation geneticists. I taste-tested rat bait. I was mugged by a macaque.

The book is far from comprehensive. Two thousand species in two hundred countries regularly commit acts that put them at odds with humans. Each conflict needs a resolution unique to the setting, the species, the stakes, the stakeholders. What you have here is the highlights of a two-year exploration, a journey through a world I had not known existed.

The first half of the book considers the felony crimes. Murder and manslaughter, serial killing, aggravated assault. Robbery and home invasion. Body snatching. Grand theft, sunflower seed. The perpetrators include the usual suspects, the bears and the big cats, and some less usual—monkeys, blackbirds, Douglas firs. The later pages explore acts less grievous but more widespread. We consider the jaywalking ungulates. The vultures and gulls that vandalize property for no discernible reason. The littering geese and the trespassing rodents.

Of course, these are not literal criminal acts. Animals don't follow laws, they follow instincts. Almost without exception, the wildlife in these pages are simply animals doing what animals do: feeding, shitting, setting up a home, defending themselves or their young. They just happen to be doing these things to, or on, a human, or that human's home or crops. Nonetheless the conflicts exist, creating dilemmas for people and municipalities, hardships for wildlife, and material for someone else's unusual book.

1

MAUL COPS

Crime Scene Forensics When the Killer Isn't Human

For most of the past century, your odds of being killed by a cougar were about the same as your odds of being killed by a filing cabinet. Snowplows kill twice as many Canadians as grizzly bears do. In the extremely uncommon instance when a North American human is killed by a wild North American mammal, the investigation falls to officers and wardens with state or provincial departments of fish and game (or fish and wildlife, as less hunty states like mine have rebranded themselves). Because the incidents are so rare, few of these men and women have much experience with them. They're more accustomed to poaching cases. When the tables turn and the animal is the suspect, a different kind of forensics and crime-scene know-how is called for.

Without it, mistakes are made. In 1995, a cougar was presumed to have killed a young man found dead on a trail with puncture wounds to the neck, while the true murderer, a human being, walked free. In 2015, a wolf was wrongfully accused of pulling a man from his sleeping bag and killing him. Cases like these are one reason there is WHART: Wildlife-Human Attack

Response Training (and by its founders' admission, "a horrible acronym"). WHART is a five-day course—part lecture and part field training—taught by members of the British Columbia Conservation Officer Service.*

Because they have the experience. British Columbia has more cougar attacks than any other North American state or province. It has 150,000 black bears—to Alaska's 100,000—17,000 grizzlies, and 60 predator attack specialists, 14 of whom (the specialists but not the bears) have driven down from Canada to serve as WHART instructors this week. WHART 2018 is being hosted by the Nevada Department of Wildlife, which has offices in Reno. This fact helps explain why a training course for wilderness professionals would be held in a casino complex, where the resident wildlife amounts to the furry hominid on the Betti the Yetti slot machine and an unspecified "biohazard" that closed down the pool for a day. WHART seems to be the only booking at the Boomtown Casino event and conference center this week. Management has a bingo game going on in the next room.

The WHART student body, some eighty of us in all, has been split into small groups, each led by one of the predator attack specialists. Like many Canadians, they are distinguishable from white Americans mainly by sound. I'm referring to that uniquely far-northern habit of ending statements with folksy interrogatives. It's an endearing custom thrown somewhat off-kilter by the present subject matter. "Quite a bit of consumption and feedin' and whatnot, eh?" "Holdin' on by two, three tendons, right, ya know?"

Our conference room, the Ponderosa, is a standard offering with a podium and a screen for slides and videos. Less standard are the five large animal skulls sitting in a row on a long table at the front of the room, like participants in a panel discussion. On the screen, a grizzly bear is attacking Wilf Lloyd of Cranbrook, British

---

* Canadian for "Fish and Game Department."

Columbia. The footage is part of a presentation entitled "Tactical Killing of a Predator on a Person." The instructor sums up the challenge that Wilf's son-in-law faced in trying to shoot the bear but not the man: "All you could see was the body of the bear and a limb of Wilf once in a while." The son-in-law saved Wilf's life but also shot him in the leg.

Another challenge: Marksmanship deteriorates under the influence of adrenaline. Fine motor skills are out the window. The thing to do, we are told, is to "run directly up to that animal, plant the barrel and shoot upward" to avoid hitting the victim. Though you then run the risk of "attack redirection." That's a calm, technical way to say that the animal has dropped its victim and now it's coming after you.

A second video illustrates the importance of order and discipline in the face of animal-attack mayhem. In it, a male lion charges a safari hunter. The other members of the hunting party wheel and scatter. The video is paused at various moments when a rifle is pointing both at the lion and at a hunter directly behind it. "Stay tight and communicate," is the advice here. We will be practicing this kind of thing later, in an immersive field scenario out in the scrub near the Truckee River, below the casino.

The cursor glides to the Play arrow again, and the lion resumes its charge. I used to work at a zoo, and the roaring in the Lion House at feeding time was God-like. It twisted my viscera. And that was just their mealtime conversation. The lion in this video means to intimidate and destroy. The bingo party has to be wondering what the hell is going on in the Ponderosa Room.

After one more presentation, we break for lunch. Preordered sandwiches are waiting for us to pick up at a small deli over in the casino. We stand in line, attracting curious glances. It's unusual, I suppose, to see so many uniformed law enforcement professionals inside a gambling establishment. I collect my lunch sack and follow along behind a small group of conservation officers heading to the lawn outside. Their leather hiking boots squeak as they walk. "So she looks in her rearview mirror," one is saying, "and there's a bear in the back seat, eating popcorn." When wildlife officers gather at a conference, the shop talk is outstanding. Last night I stepped onto the elevator as a man was saying, "Ever tase an elk?"

While we were off on lunch break, the instructors stacked the chairs against the walls and laid out soft-touch male and female training manikins on the tables, one per group. Working from photographs, some of the more artistically inclined instructors have used paint and, apparently, hacksaws to create convincing facsimiles of actual wounds from attacks. *Wounds* is a tepid word for what teeth and claws can do.

My group's manikin is a female, though it would be difficult to know this from what remains of her face, or from the sign attached to the table, which reads BUD. Later, walking to the bathroom, I pass a badly mauled LABATT and a decapitated MOLSON. Instead of being numbered, the manikin workstations have been beered. I take this to be an effort, a very Canadian-dude effort, to lighten the mood.

Our first task is to apply our newly acquired forensics savvy and determine what species it was that did the mauling. We're looking at what's known in attack forensics as "victim evidence": injuries and clothing. The worst of the visible damage is above our manikin's shoulder. (Only one remains.) Part of her neck is flayed, and a flap of scalp hangs loosened, like peeling stucco. Missing eyelid, nose, lips. We all agree it doesn't seem like the work of *Homo sapiens*. Humans rarely eat their victims. If a murderer removes body parts, it's likely to be hands or head—to stymie matches with fingerprints or dental records. Murderers occasionally take a trophy, but a shoulder or lip would be an unusual choice.

The consensus is that she was killed by a bear. Bears' teeth are their main weapon, and their lightly furred face is their weak spot. When bears attack humans, they apply the tactics they use in fights with other bears. "They go teeth to teeth, right? So their instinct is to go right for your face." Joel Kline, our youthful, forthright instructor, has been an investigator on ten cases of bear attack. "They come right at you and you have all these massive injuries right to the face." Joel's own face—our focus as we take in his words—is blue-eyed, unblemished, peachy clear. I work hard not to picture it in that state.

Bears are inelegant killers partly because they're omnivores. They don't regularly kill to eat, and evolution has equipped them accordingly. They feed on nuts, berries, fruit, grasses. They scavenge trash and carrion. A cougar, by contrast, is a true carnivore. It lives by the flesh of animals it kills, and thus it kills efficiently. Cougars stalk, well hidden, and then pounce from behind and deliver a "killing bite" to the back of the neck. Their molars close like scissors blades, cutting flesh cleanly. A bear's mouth evolved for crushing and grinding, with flat molar surfaces and jaws that move side to side as well as up and down. Wounds made by bears' teeth are cruder.

And more numerous. "Bears are more bite bite bite bite." Our manikin, says Joel, is how it usually goes. "It's a big mess."

Looking around at the manikins, I see not just bites and scratches but broad scalpings and skinnings. Joel explains the mechanics of this. A human skull is too large and round for a bear or cougar to position between its jaws and get the leverage it would need to crush or bite into it. So when it brings its teeth together, they may skid off the skull and tear away skin. Think of biting into a very ripe plum, how the skin pulls away.

Deer, a popular entrée among cougars, have longer, more muscled necks than we have. When a cougar tries to make its trademark killing bite on a human, its teeth may encounter bone where normally there would be muscle. "They try to dig their canines in and they bring their teeth together and they take the flesh and remove it," said WHART co-founder Kevin Van Damme, in a talk called "Cougar Attack Behavior." Van Damme has astronaut looks and a voice that carries to the back of the Ponderosa Room without a microphone. I opened a decibel meter app on my phone at one point and was impressed to see him hit 79, about the level of a garbage disposal.

The claw marks on our simulated victim rule out a cougar. Cats' claws, unlike dogs', create a cluster of triangular punctures as they sink in to grip their prey. With a bear attack, you're more likely to see what we have here in front of us, the parallel rakings of a swipe.

Joel takes a step closer to the manikin's head. " 'Kay, what else do we have here? Missing nose, lips, right? So later we're going to think of looking for those in . . . ?"

"The bear's stomach," a few of my group mates call out.

"Stomach contents,* right on." Joel says "Right on" a lot. Writing the chapter later, I would recall "bingo"s, too, but that may be a memory that seeped in from the other side of the wall.

None of the manikin torsos in the room are laid open. There's none of what Van Damme calls "feeding on innards." I'm initially surprised by this. I know from research for a previous book that predatory carnivores tend to tear into the abdomen of their prey straightaway to get to the organs—the most nutritious parts. One

---

* Scientists with the long-ago Division of Economic Ornithology used stomach contents as evidence in cases of birds accused of raiding farms, hunting stock, and commercial fishing operations. A 1936 U.S. Department of Agriculture report provides examples: eiders accused of decimating scallop beds, yellow-crowned night herons shot by froggers when in fact the birds had been eating crayfish, hunters killing marsh hawks because they thought they were preying on quail. In each case, the birds were exonerated by their stomach contents, a happy outcome for all except of course the individuals examined, who gave their stomachs that others might live. Maryland's Patuxent Wildlife Research Center housed a collection of thousands of glass jars of bird stomach contents, until pressing storage needs triggered a massive emesis into a Patuxent dumpster.

possible reason you don't see this as much on human victims, say our instructors, is that humans wear clothing. Both bears and cougars avoid clothed areas when they're feeding or scavenging. Perhaps they don't like how the cloth feels or tastes, or they don't realize there is meat underneath.

Joel indicates a suite of wounds on the neck and shoulder. "Are we thinking perimortem or postmortem?" In other words, was our victim alive or dead as these wounds were inflicted? It's important to know this, because otherwise a bear that was just scavenging could take the fall for a killing. Based on the bruising around the puncture wounds, we judge them to be perimortem. Dead people don't bleed or bruise, a bruise being essentially a bleed beneath the surface of the skin. If blood is not being pumped, it doesn't flow.

Joel tells us the story of a gnawed-upon corpse that was found near its car in the woods, partially buried under leaves. The bites appeared to have come from a bear, and a bear was trapped nearby, but there was little blood on and around the man's body. Investigators found needle marks between the toes and a used syringe on the car floor. An autopsy confirmed that the man had died of an overdose. The bear, as Joel says, "just saw an opportunity to get some good, high fat and calorie content" and pulled him from the car and ate some of him and cached the body to come back to later. The bear was released.

Joel rolls our manikin onto its front side, revealing one or two additional perimortem gashes on the back. I point out two small divets along the spine, which exhibit no purpling or blood. I hazard a guess, based on a slide from yesterday showing postmortem rodent damage, that a small woodland creature might have been gnawing on our corpse. Joel exchanges a look with one of my group mates, a wildlife biologist from Colorado.

"Mary, those are marks from the injection molding." Part of the manufacturing process of the manikin, he means. This would be less embarrassing for me had I not, as group notetaker in an earlier exercise, transcribed teeth-wound measurements using the abbreviation for centimeters instead of millimeters, entering into evidence a tip-to-tip canine-tooth span not seen since the Jurassic period.

We move on now from victim evidence to animal evidence: evidence on or in a "suspect" that has been shot or captured near the scene of the attack. For instance, Joel is saying, you can look for the victim's flesh up in the pockets of the gums of the (immobilized) animal. It's odd to think of a bear getting human stuck between its teeth, but there you go.

With cougars, Joel adds, it's sometimes possible to recover the victim's blood or flesh from the crevice on the interior of a claw. "So you need to push those out, those retractable claws, and you might have evidence under there, right?"

Claws can be misleading as indicators of the size of an attacker's paw. When the animal steps down and transfers its weight onto a foot, the toes splay, making the foot appear larger. Investigators have to be cautious with measurements of claw or tooth holes in clothing as well, because the cloth could have been wrinkled or folded over as it was pierced.

" 'Kay, what else are we looking for?"

"Victim's blood on the fur?" someone offers.

"Yup, right on." Joel cautions that if the bear had been shot at the scene of the attack (rather than trapped afterward), its blood could mingle with the victim's blood and muddy the DNA tests. "And how do we prevent that?"

"Plug the wound!" And that is why men with the British Columbia Conservation Officer Service keep a box of tampons in the truck.

What we're seeking, the end point of all this, is linkage: crime-scene evidence that connects the killer to the victim. Joel goes over to get one of the skulls from the table at the front of the room. He brings the upper teeth down onto a row of wounds in the manikin's shoulder. This is the glass-slipper moment. Do the upper canines and incisors fit into bite marks on the manikin's shoulder? And if so, do the lower teeth match a corresponding set of marks on the other side of the body?

It's a match. "Pressure and . . ." Joel positions the lower jawbone into the wounds on the manikin's backside. "Counterpressure. There's your smoking gun."

At the outset of this chapter, I mentioned a man found dead on a hiking trail with puncture marks on his neck. Investigators deemed it a cougar attack, even though there were no marks to suggest a set of matching upper and lower teeth. The wounds, it turned out, weren't made by anyone's teeth but by an ice pick. The murderer got away with the crime until twelve years on, when he bragged about it to a fellow inmate while serving time for something else.

Every so often, the opposite happens. A human is found guilty of a killing that was in fact committed by a wild animal. Most famously, there is Lindy Chamberlain, the Australian woman who screamed that she'd seen a dingo run off with her baby while the family was camping near Ayers Rock in 1980. We heard a presentation on the case from one of our instructors, predator attack specialist (and—stay tuned—survivor) Ben Beetlestone. Because the Australian investigators had no body and no dingo in custody, they could not do what we're doing today. They could not link the victim evidence to the animal evidence. Without linkage, the trial turned on assumptions (for instance, that a dingo could not or would not carry off a ten-pound baby), human error, and a media frenzy that swayed public opinion. About three years after Chamberlain was convicted, a search party looking for the remains of a rock climber found a dingo lair with remnants of the baby's clothes. Chamberlain was released and acquitted, and her conviction was overturned. The dingo really did eat her baby.

These days linkage often takes the form of a DNA match. Does DNA from the captured (or killed) suspect match DNA from hair or skin under the victim's fingernails? Does the animal's DNA match DNA from saliva on the victim? With animal attack cases, scavengers can complicate these efforts. While animal saliva near tooth marks on, say, a jacket has likely come from the attacking animal, saliva swabbed from the victim's skin could have come from an animal that fed on the corpse later.

Up in the Canadian wilderness there tend to be a lot of bears around, so good linkage is vital. Van Damme shared a story about a woman killed by a bear in her yard in Lillooet, British Columbia. His team set traps and ran DNA on two "bears of interest" before they scored a match with the third. The innocent bears were released.

It's beer o'clock (Canadian for 5:00 p.m.). Instructors are straightening tables and carrying manikins to the back of the conference room and piling them on the floor near the refreshment table. You need to straddle a corpse to get a last refill on your coffee. I waylay one of my group mates, Aaron Koss-Young, of Yukon Conservation Officer Services, for a quick overview of something that isn't covered in WHART: what people should do in an attack situation, or even just a surprise encounter. Aaron says sure. He's of the same vintage as Joel, with similar fair features and good manners.

You may have heard the ditty "If it's black, fight back. If it's brown, lie down." The idea being that brown bears, of which grizzlies are a subspecies, may lose interest in a person who appears to be dead. Right away, a problem: brown bears' fur can be black, and some black bears look brown. A more reliable way to distinguish the two is by the length and curvature of their claws, but by the time you're in a position to make that call, the knowledge will be of limited practical use. The most important thing to consider, Aaron says, is not what kind of bear you are facing, but what kind of attack. Is it predatory or is it defensive? Most bear charges are defensive. They're not really attacks, they're bluffs. You've startled the bear, or you're too close, and it would like you to back off. "It's going to come across as big and scary. Its ears are upright, not back." Aaron pauses to blow his nose. He has a miserable summer cold. "It may be swatting the ground. Huffing." Popping or clacking its jaws. (But not roaring or growling. That's mainly a movie thing.)

Aaron stuffs the Kleenex in the pocket of his fleece. "It just wants to scare the crap out of you." Grizzlies evolved in more open, less forested terrain than black bears. They often can't just disappear into the trees as a startled black bear can and typically does. So they make *you* run instead.

The recommended response to a bluff is to be as nonthreatening as you can. Back away slowly. Talk to the animal in a calm voice. You'll probably be fine—even if the bear is a sow with cubs. For all British Columbia's bears and bear encounters, and for all the hype you hear about the danger posed by protective mother bears, the province has seen only one fatal attack of that nature. (It was a grizzly. No black bear sow with cubs has ever killed a person in British Columbia.)

With a predatory attack, the survival strategy is the opposite. The rare predatory bear attack begins quietly, with focused intent. Counter to common assumption, it's more often a black bear than a grizzly. (Though with both species, predatory attacks are rare.) The bear may be following at a distance, circling around, disappearing and reappearing. If a bear starts to charge with its ears laid flat, you're the one who needs to look scary. Open your jacket to make yourself look larger. If you're in a group, get together and yell, so you look like one big, loud creature. "Try to give the message, 'I am not going to give up without a fight.'" Aaron says. "Stomp your feet, throw rocks."

The same holds true for an attacking cougar. Take inspiration from the Kansas pioneer N. C. Fancher, who in the spring of 1871 noticed a cougar eyeing him as he stood inspecting a buffalo skeleton. As recounted in *Pioneer History of Kansas*, Fancher shoved his feet inside the dead buffalo's horns, banged its femurs over his head while jumping up and down, and "bellowed desperately." The cougar, and really who wouldn't, took off.

And if the animal goes ahead and attacks anyway? "Do whatever you can to fight back," Aaron says. If it's a bear, go for the face. Aaron points in the direction of his nose, a red chapped thing. "Don't play dead." If you play dead at that point, there's a good chance you shortly won't be playing.

The worst thing you can do in any situation where a predator seems bent on attack is to turn and run. This is especially true with a carnivorous hunter like a cougar, because running (or mountain-biking) away triggers the predator-prey response. It's like a switch, and once it's flipped on, it stays on for a surprisingly long time unless a kill is made.

WHART instructor Ben Beetlestone experienced firsthand the determination and persistence of a cougar in attack mode. As a Conservation Officer in the mountainous West Kootenay region of British Columbia, he handles a fair number of predator attack calls—most involving bears and minor injuries. A few years ago, he responded to an unusual call. An emaciated cougar was skulking around a couple's property. Beetlestone shared the experience during a presentation yesterday. He told us he got out of his truck, unarmed, and went up to knock on the door, not realizing the cougar was stalking the couple *at that moment*, through their windows. "If the guy left one room and went into another," he told us, "the cougar went to that window." The windows had paw prints.

Suddenly the man slams the door. Beetlestone turns to see the cougar, five feet away, crouched, with its ears flat to its head and its tail swishing. "I'm yelling and screaming and kicking at it, all that stuff we tell the public to do. None of it is working." The cougar jumps him. He tries to choke it, but it pulls away, turns, and sinks its teeth into his work boot. He grabs a broom that's leaning against the house and hits the cougar, but it won't let go. He manages to push the broom handle down the animal's throat. Meanwhile, the couple in the house are just watching through the window. Beetlestone is holding off a cougar with a cheap tin broom, yelling, "Hey! *HEY!*"

"Finally the old guy opens the door and goes, '*What?*' And I'm like, 'I need a knife!'" The man goes to the kitchen to look for a specific knife that turns out to be in the dishwasher. Finally he finds the knife and gives it to Beetlestone, who "Bates–Motels" the cougar. (A necropsy revealed a piece of a running shoe wedged in the opening of the cougar's stomach, blocking it and starving the animal.)

The bingo game is letting out as Aaron and I collect our things and leave the conference room. One of the players, spry but slightly stooped, is making his way toward the men's room as Kevin Van Damme sets off to cross the hall with a bloody, half-naked manikin under one arm. Van Damme is an imposing figure, a purposeful strider. The bingo player halts. "Excuse me," says Van Damme, offering no explanation.

Very few cars drive the quarter-mile road from the Boomtown Casino parking lot down to the Truckee River. Today would be a diverting day to travel this road, because multiple crime scenes are cordoned off with yellow police tape. Uniformed men and women with neon-green Predator Response Team vests come and go with rifles and body bags. It's WHART field-scenario day.

My group's crime scene lies between the guardrail and the bottom of a steep, rubbly embankment. Last night we received a pretend text about an attack. Following a fight with his fiancée, we were told, a young man left the couple's Winnebago to sleep outside in his sleeping bag. At 4:00 a.m., the sheriff got a missing-persons call from the fiancée and drove out to have a look. He found the empty sleeping bag and saw a wolf, which he shot and killed. Then he turned the investigation over to a Predator Response Team. That's us.

Our first task is to secure the area, to be sure no large animals are lurking. Cougars and bears sometimes cache the bodies of their victims, burying them lightly with leaves and brush and coming back later to feed some more. This makes the "crime scene" potentially dangerous for the response team.

A young woman walks up to the man in our group who has taken the role of Incident Operations Chief. "Where's my brother?" she says. "What's going on?" It takes me a moment to realize she's a role-player. She delivers the line with no trace of agitation. More of a *Hey, what's up.* Meanwhile, at the scenario up the road, we have some N. C. Fancher–style desperate bellowing: *"YOU HAVE TO FIND HIM! HE'S A TWELVE-YEAR-OLD BOY!"* This is how it goes with these scenarios. You have one Al Pacino and everyone else is channeling C-SPAN2.

Our Ops Chief puts his hand on the sister's shoulder. "Well, we got a report there's an animal in the area."

"What kind of animal?" Like she might go back and get her binoculars. She lifts one foot to step over the police tape. "I need to be down there looking for him."

Ops Chief takes her arm gently. "Now, we don't want you to go down there and get hurt. We've got a strategic team down there, doing a diamond-shaped security sweep."

We practiced the diamond-shaped sweep earlier. Four people move along back to back to back to back, weapons ready. It's a human octopus with guns. Each person scans the quadrant in front of her (named for hours on a clock face: 12, 3, 6, and 9) and calls "Clear" if she sees no danger. Whereupon the person to her right calls "Clear." Et cetera, around and around. Not only can the surroundings be monitored in all directions, but it's safe in that no one can inadvertently point a weapon at anyone else. Should someone spot a threat, she calls it out, whereupon the people on either side move into position beside her. Now three rifles are aimed and ready, while one person watches the rear. When we practiced this earlier, Joel played the dangerous animal. I had hoped for some pantomime, maybe even a costume, but he'd just step in front of us and say, "I'm a bear."

Four of my teammates move through the brush in the diamond formation. Aaron climbs onto a boulder to assume "lethal overwatch," his appearance of lethalness dimmed somewhat by the Kleenex wadded in the palm that supports his rifle. I'm on paperwork detail again (because "you're a writer").

"Bear, three o'clock!" It's not Joel this time. It's a lifelike bear model, one of those hard-foam target practice items used by bowhunters. Six o'clock and twelve o'clock glide into position beside three o'clock, sliding their feet along the rough ground without looking down. They raise their weapons in unison. It's kind of balletic. It's like synchronized swimming with rifle shooting, and can we please make that an Olympic event?

On a quick count of three, pretend shots are fired at the polyethylene bear. Someone calls for tampons, and the excitement is over.

Was the wolf that the sheriff shot last night a red herring, an innocent bystander? It's our job now to figure that out. It's a wildlife whodunit.

The victim—played by one of yesterday's manikins—is shortly found down the hill from the empty sleeping bag, under a bush. A team member pretends to photograph the body, quickly, because an affable coroner, played by Joel, wants to remove it before the midday heat sets in. We'll have a chance to examine it later, at the morgue/Ponderosa Room.

Once the scene is secured, it's time for evidence collection. Items of evidence, as we know from TV police procedurals, are called exhibits. Bodies, sleeping bag, footprints, paw prints, drag marks—these are all exhibit items. Items destined for the lab are assigned numbers and put into evidence bags after they're photographed in place. A corresponding evidence flag is stuck in the ground where the item was found. My role is to note all of this—a short description of the item, its number and location—on an Exhibit Report, illegibly and probably in the wrong place.

The animal tracks in the dirt are from a bear. This is good, because we didn't learn about wolf attacks in class. (Because they almost never occur.)

The team is on hands and knees now, searching for animal hairs and blood. It's uncomfortable, hot, tedious work, but important. Much can be learned from blood at a crime scene. Round drops on the ground suggest a "gravity pattern": blood falling by its own weight from a wound. Oblong gravity drops suggest a victim running as he dripped. A "force-related pattern"—blood ejected by the force of, say, a paw swipe or the pressure of a major artery—is elongated, with a tail like a comet. It's a spatter, not a drip.

Someone finds a trail of drips. Joel tells us to look closely at their size. When drips of blood grow smaller as the trail progresses, they're probably not coming from a wound. They might be dripping from the animal's fur, or a murderer's blade. If the size of the drips remains constant—a "replenishing trail"—they are likely coming from an "active bleeder." A smear of blood is a "contact pattern," perhaps a place where the victim fell or placed a bloody hand.

When we're sure we've found everything there is to be found, Joel reaches down and flips over a leaf, revealing a tiny drop of blood on the underside. We missed this. We missed a lot—blood on rocks, plants, on the ground. "Splatter pattern," someone says knowingly.

Joel nods, but adds quietly, "S*patter*, not *splatter*."

Together the blood and the marks in the dirt tell the narrative

of the attack. Drips and blood on the sleeping bag from the initial bites. A drag mark and replenishing drips as the man is pulled
from the sleeping bag into the brush. Scuffle marks and blood
in the dirt as the man tries to escape, and then a spatter pattern
on the plants and rocks, perhaps caused by the bear shaking the
man to stop the struggle. Had the body lain dead for any length
of time, the chemicals of its decay would have left a final piece of
evidence, a stain or area of blackened vegetation, called a "decomposition island." No pretty beaches there.

Our victim's injuries, Joel tells us, have been recreated on one
of the manikins. It's not here at the scene, but we'll examine it in
class tomorrow morning when we try to establish linkage.

And that brings us to beer o'clock again. Joel collects the props
and the evidence flags and the polyethylene bear, and we all troop
back along the road and up to our hotel rooms to change. By the
time I come back downstairs, my group has gathered at a small
sports bar behind the blackjack tables. They're intent on hockey,
Oilers versus the Toronto Maple Leafs.

"Hey," I try. "Shouldn't that be Toronto Maple *Leaves*?" I can't
compete with hockey, so I go for a walk. I end up at a Cabela's
outfitters. I don't hunt, but I enjoy the taxidermy. This outlet has
an outstanding mountain diorama and a musk ox on top of the
dressing rooms. Also a Gun Library, which, I discover, contains
used guns, not books.

The man behind the counter waits for me to say something. I
ask about getting a library card. "You can't borrow these guns,"
he says. "They're for sale."

"Then it's not much of a library, is it?" Seems like I should
probably call it a night.

The manikin from our crime scene comes with some extras. Joel
has just emptied onto a tabletop a bag of realistic moulage bear
stomach contents: an ear and an eye and a strip of scalp with part

of a mohawk haircut. These are passed around among our group. It's early in the morning for such things. Doughnuts sit untouched.

The stomach contents are a match for what's missing on our manikin's head, suggesting that indeed the bear, not the wolf, was behind the attack. The mohawk seems like a fanciful touch, but turns out not to be. Joel reveals that our scenario from yesterday was based on an actual attack—real bear, real man, real mohawk. Joel investigated this case in 2015. All the WHART manikins, in fact, represent not just real wounds but real attack victims.

Joel brought along photographs from the actual attack scene. One shows the victim's backside. The largest wound, a raw, gaping, messy chomp, is to the buttocks. The man had been sleeping in one-piece long johns, and the flap, Joel says, must have opened while the bear was dragging him. "So that's why there's feeding right there." After a moment, Joel adds, "You know the one with the bear paw prints on it? On the butt flap?" This is apparently a common item in Canada, because several of my group mates nod. "That's what he was wearing."

There's a clean set of bite marks on the manikin's shoulder. From the position of the upper and lower canine marks there, we can tell that the man had been sleeping on his back. The bear, Joel surmises, came upon the sleeping figure, maybe licked the salts from his skin. The man woke up and probably made some noise. "So the bear figures, *Well, I either finish this or I run away.* He chose to finish it."

Meanwhile, what was inside the stomach of our other suspect, the wolf shot by the sheriff when he arrived on the scene? Gum wrappers and tinfoil. No human tissue or clothing. Case closed. No DNA analysis was needed.

Once the forensics is completed and the perpetrator known, what happens next? If this bear hadn't been shot near the scene of the attack, what would have been its fate? Kevin Van Damme talked about this after a lecture. Prison isn't an option. Canadian zoos won't take bears older than three months, because they tend to pace and because zoos generally have enough bears. Capital punishment is what happens. "If a bear treats a person as food, it will do it again," Van Damme said. "I have spent twenty-six years as a predator attack specialist. I know some of you disagree with me, but if it hurts a person, it's going to die."

As any criminologist can tell you, prevention is better than punishment. The safest thing for both species is to keep them apart. Don't let bears learn to associate humans with easy meals. Require that people in bear country secure their garbage. Tell them to stop feeding birds and leaving dogfood on the porch. The man in the long johns lived in the woods, where there was no garbage pickup. Trash likely piled up outside the trailer. The tinfoil and gum wrappers in the wolf's stomach suggest that this was a place wild animals had become comfortable scavenging for scraps. Garbage is a killer.

2

BREAKING AND ENTERING
AND EATING

How Do You Handle a Hungry Bear?

Stewart Breck is a tall, narrow plank of a man. His arms don't stray far from his sides as he walks, and he carries no backpack or bag to break up the long vertical plane he occupies in space. You notice this when you walk behind him, which I've been doing a lot of because his stride covers several city blocks. Though personable, his demeanor, too, tilts to restraint. Over the course of the day I've spent with him, he has not raised his voice or gesticulated memorably or used a bleepable word. He's composed, considerate, reasonable. I'm telling you this so you'll understand how I was a little shaken when Stewart Breck, a moment ago, went, "Are you *FUCKING KIDDING ME?*" and his arms shot out from his sides, where they remain, palms up, the universal gesture of exasperation.

Because I am, again, lagging behind, I don't at first see what Breck sees. Now I do: two fat trash bags ripped open, with food scraps spilling out onto the pavement. It is 3:30 a.m., bear time in the back alleys of compact, restaurant-dense downtown Aspen,

Colorado. The sound of Breck's approaching SUV must have scared off a bear mid-scavenge. Compost and garbage are known in the parlance of human-bear conflict as "attractants." Aspen municipal code requires both to be secured in bear-resistant containers.

"Give me a break." Quieter now, the hands back at his sides. "We spent hundreds of thousands of dollars on this." *This* equals: multiyear, multicity research into how best to get people in the midst of bear country to properly lock up attractants, and how much difference it makes when they do. The work was funded by Colorado Parks and Wildlife (CPW), who get the calls when bears damage property while looting unsecured human food; Colorado State University, where Breck teaches a course in human-wildlife conflict; and Breck's employer, the National Wildlife Research Center (NWRC), headquartered in Fort Collins, Colorado.

NWRC is the research arm of Wildlife Services, which is part of the United States Department of Agriculture (USDA). The "services" are provided mainly to ranchers and farmers who are having problems with wildlife cutting into their livelihood, and often they take the form of killing that wildlife. Breck was hired by NWRC to research nonlethal alternatives. His job gives him lots of opportunity to deploy his admirable composure. There are old-schoolers at Wildlife Services who hate him for rocking the boat, and there are animal welfare activists who hate him for not rocking it hard enough. I like him because he's trying to stand on the impossible middle ground.

What the garbage studies showed is that reinforced, locking bear-resistant containers make a solid difference—provided people take the time to latch them properly. In an area where 80 percent of the containers were used as they're meant to be, there were 45 human-bear conflicts over the course of the study. A similar area with only 10 percent compliance had 272 conflicts.

What this says is that containers aren't enough. You also need laws requiring people to use them, and fines for people who ignore those laws. Aspen has all of this, but there has been a reluctance to follow through with the fines. Especially here, downtown. Breck has been told that in the intervening years the situation has improved.

Just now, it's not seeming that way. Coming down the alley in that unhurried, endearingly pigeon-toed way is a full-grown black bear. Breck and I are standing near his vehicle, which is parked twenty feet back from the mess. The bear nears the garbage, which has been its focus until this moment, and then it looks over at us. It clacks its jaws, an indication that it's uneasy. For here are two staring humans, one with some good height to him, at a time of night when humans are rarely about. On the other hand: kitchen scraps from Campo de Fiori! The bear considers the situation a moment longer, then lowers its head to eat.

Because there's a lot of eating to be done. It's early fall, the time of year when black bears eat with purpose and abandon, to construct the fat they will live off in their dens over the winter.* A hyperphagic black bear doubles or even triples its daily calorie count, taking in as much as 20,000 calories. As omnivores, bears happily eat a variety of foods; during hyperphagia, what they are

---

* You may be wondering: When you live off your own fat, do you need to use the toilet? If you are a bear, you do not. Hibernating bears reabsorb their urine and form a "fecal plug." Cubs, on the other hand, let it go inside the den. Not a problem, because the mother bear eats it—partly as cleanup, but mostly as food. She is nursing, after all. While hibernating. Black bear hibernation isn't the same as sleep. They're just sort of slowed down and out of it. Surreally, black bear sows give birth halfway through their hibernation. They deliver a couple of cubs, snack on the placenta, then go back into hibernation, nursing and tending their cubs in a state of semi-alertness until spring. According to a scientist who has taken blood samples from hibernating black bears, they do not have sleep breath and their dens don't stink. They smell like roots and earth and that is all.

drawn to most powerfully is a concentrated source. They want to take in lots of calories without having to burn lots of calories wandering around looking for calories. The mountains around Aspen have always supplied that: acorn-dropping oak brush, fruiting serviceberry and chokecherry trees, the outrageous fecundity of crabapple trees. Come the 1950s and '60s, the skiers began to move in. Bears looked up from their nuts and berries and went, *Hurunh? Birdseed hanging on a tree? Bag of kibble sitting on a deck? Yes, please.* Soon they ventured into town, following the humans, because the humans provide. The alleys behind Aspen's multitudinous restaurants are concentrated-food-source nirvana.

Breck nudges me. Another bear is coming down the alley, this one darker and slightly smaller. The lighter, dominant bear turns its attention to the newcomer and makes a low, rumbly sound. *You may have those hearts of romaine and that spinach gnocchi, but do not come near my grilled sustainable Skuna Bay salmon.*

Breck raises his phone to take a picture, which surprises me. This is a man who uses the word *routine* to describe the act of hand-darting a hibernating black bear to replace its tracking collar. It turns out he's not photographing bears. He's photographing irony. "Look at the lid." He aims his flashlight at the wheeled compost cart lying, open, on its side. The molded-plastic lid features a bear face, and inches from this decorative bear face is the face of the actual bear now enjoying the contents of this certified bear-resistant container that has failed to resist it.

"They jump on them," Breck says, "and they pop open."

Or the locking mechanism may be broken. This was the case with another of this same model of compost cart farther along the alley, which we saw earlier in the day. Breck walked up and lifted the lid on fifty reeking bananas. "Be sure to latch," chided a sticker. "A bear's life depends on it." In the next alley over, Breck led me to an uncovered vat of used cooking grease. It was as tall and big around as a drinking fountain, and bears sometimes use it like one. Breck has seen paw prints in grease leading away down the alley.

Chapter 12.08 of Aspen's solid-waste code, entitled "Wildlife Protection," was modeled on that of the neighboring ski and mountain-bike resort village Snowmass. There the similarity fizzles out. Snowmass Animal Services/Traffic Control consists of Tina White and Lauren Martenson, and they are *on it*. "We ticket everyone," White told me when we met yesterday. She recently put together a slide presentation in Spanish for restaurant kitchen staff, many of whom hadn't realized what happens to bears that start raiding dumpsters when people neglect to lock them. Her efforts have been working. It's been several years since a bear causing problems in Snowmass was, as White put it, "pulled out of the mix." At the time of my visit, Aspen was up to nine for the year. Then again, Aspen is three times as populous, with four times the number of restaurants.

Aspen's garbage violations are handled by community response officers, five in all. Breck and I met with their representative, Charlie Martin, yesterday morning in a conference room at the Aspen Police Department. Charlie wore a black and yellow uniform, and a pair of socks on which rainbows alternated with unicorns. "It's not Friday and I wasn't on bike patrol," he said, mysteriously, when I commented on them. Charlie listed some of the things his team was already struggling to keep up with when bear-related garbage infractions were added to the list: traffic violations, barking dogs, idling construction vehicles, 911 calls, rabid bats, lost and found, sidewalk snow, jump starts, vehicle lockouts, community picnics, and removing dead deer from roadways.

Charlie was a trifle defensive about the alley situation. "We've passed out almost ten thousand dollars in tickets this year." The fines for leaving garbage or compost unsecured range from $250 to $1,000. Breck and I could have matched the year's total in one day. Except that, as Charlie pointed out, the fines wouldn't stick. "You've got multiple parties sharing one dumpster," Charlie said, referring to containers in both condo developments and restaurant back alleys. "You write someone a ticket, and they'll say, 'It was someone else. We left at ten p.m., and we locked it. Prove to me that it wasn't locked when we left.'"

Aspen's waste management companies are, by law, required to assign a number to each compost and trash container, and to keep a database that links these numbers to the person or company responsible for keeping the contents secured and paying a fine if they are not. Aspen contracts with five of these companies, and none appears to have set up such a system. (Snowmass does its own pickup. Also, Tina White will happily climb into a dumpster and rummage through a trash bag for mail with a name and address. She has heard people refer to her and Lauren as "the bear bitches.")

You read about this kind of thing over and over in communities that have tried to switch to bear-resistant containers. Generally speaking, waste management companies are fiercely concerned with their bottom line, and not so fiercely concerned with the welfare of bears. The containers need to fit the lifts on the trucks, which means that on top of the expense of the bins, there will be the expense of new trucks or retrofitted trucks, and either way it's money the companies would prefer not to spend. And the people who respond to the bear calls are not the people who draft the ordinances or the people who run the garbage companies. It's a stinking mess.

While wandering the alleys this afternoon, Breck peered over the lip of a dumpster marked CARDBOARD ONLY. At the bottom were French fries, an olive, and some squeezed lemon halves. As it's written, city code doesn't require recycling dumpsters to be bear-resistant or locked or even covered, and people often toss in bags of trash. On the residential side, problems arise when homeowners rent out their property and the vacationers either aren't told about, or don't remember or don't care about, the garbage laws.

Charlie agrees with Breck that Aspen needs an overhaul. It needs to replace the busted bear-resistant carts for the downtown compost and trash. It needs to resolve the shared dumpster loophole. Most importantly, it needs to hire enough staff to stay on top of things.

That would not, Breck added, be a heavy lift for Aspen. The county is home to about as many billionaires as bears. The Koch brothers have property here. The Bezos parents. The Lauder siblings. There is oil money, hedge fund money, cosmetics money, tech money, lingerie money, tinfoil money, chewing gum money. Breck believes that may be part of why the Aspen enforcement effort stumbled, that the city's council members kowtow to its alpha residents.

Of course, the billionaires don't manage the restaurants. That part may be Charlie's fault. "I've got to live in this town too," he said to us at one point. "And I'd like to go out to a restaurant and eat. I just gave them a thousand-dollar ticket, and I'm gonna go in their restaurant?" Aspen needs bear bitches.

The lighter-colored bear is working a crab leg, while its colleague noses through cabbage leaves. "What have these bears just learned?" Breck is saying. "*I can eat garbage with people standing and watching me and nothing bad happens.*" When Breck first joined the National Wildlife Research Center, he did some human-bear conflict research in Yosemite National Park. In the park's early days, he says, staff would set up bleachers and lights around the garbage dump and charge visitors for the show: twenty or more black bears gorging and pushing each other around.

Right now we're the people in the bleachers. We've just given these two a little less reason to worry about humans. As a result, they may start coming into the alley earlier or standing their ground longer. Odds are they'll end up like the bear that dined out at the dumpster behind Steakhouse No. 316. One night not long ago, the restaurant's manager, Roy, came out to roust the animal. Because the dumpster was set in an alcove, the bear's escape was blocked on three sides. On the fourth side was Roy. With only one way out, the bear lunged and, quoting Charlie, "bit Roy in the ass." According to University of Calgary professor emeritus and bear attack researcher Stephen Herrero, 90 percent of black bears that injure humans are bears that have habituated to them—that is, accustomed to their presence and lost their fear— and developed a taste for their foods.

Based on a description of the bear that Roy provided, the animal was found, trapped, and, because it had injured someone, put down. (What the description said beyond "dark hair" and "heavy-set" I can't imagine; however, DNA from saliva on Roy's pants was a match with the bear's.)

Roy and his staff could have been more careful about keeping the dumpster locked and that, too, bit him in the ass. Townspeople picketed the steakhouse following the bear's death. People don't want bears destroyed because of other people's neglect. If anything, they want them hazed or relocated—the two nonlethal approaches you hear about most with "conflict bears." (There's also electric fencing, but the prison-camp look doesn't play well in residential areas.)

Hazing refers to the practice of supplying a frightening or painful experience such that the animal associates the unpleasantness with the location or the behavior underway when it began, and then avoids such in the future. In the case of these two bears, you'd need to station someone here in the alley during the wee hours with an implement of less lethal unpleasantness,* most likely a gun that shoots rubber ammunition or bean bags. If you are a law enforcement ignoramus like me, you may be picturing the colorful handsewn item tossed at holes from a distance or juggled by clowns. These bean bags are smaller, about the size of a walnut. They don't penetrate skin, or hide, but they smart.

"Hazing is never going to solve this," says Breck. The bigger bear rips deeper into his garbage bag. "There's too much to be gained." How well hazing works depends on the push and pull of risk and benefit. These bears have learned that a visit to this alley is likely to offer a caloric windfall. Weighed against those calories, the risk of another smack to the flank would be a risk worth taking. "And there's too much other stuff nearby," says Breck. "If you were to haze these bears right here, they'd just go over to the next alley."

When hazing does work, it generally does not do so for long. In

---

* Taser International briefly sold a wildlife taser, the X3W, that some thought held promise as a hazing tool. The devices were purchased mainly to be used as they are on humans, to gain control in a threatening situation without necessitating the firing of a lethal weapon. The item sold poorly, a company representative told me, because it was costly and because it only worked on very tall mammals—a moose, or a bear on its hind legs—and at a distance of less than 25 feet. (Otherwise one of its two probes, the downwardly aimed one, would hit the ground.) The impetus for the X3W was an agitated moose whose calves were stuck in an open house foundation. The moose had chased Larry Lewis, of the Alaska Department of Fish and Game, and a state trooper three times around the patrol car, when the trooper pulled out his Taser. The moose was stunned, recovered, and ran off, allowing Lewis to safely extract the calves. Impressed, he contacted the Taser people and worked with them to design a wildlife version that was safety-tested at the Kenai Moose Research Center ("a world leader in moose science"). Tasing was found to be less stressful to the animal than tranquilization, and safer in that there's no risk of killing it by overdose. (Darts are filled according to an estimate of the animal's weight.) So it seemed like a promising alternative when a situation develops quickly and lives are at stake, or when a moose has "a chicken feeder stuck on its head"—the example Lewis provided for *Alaska Fish & Wildlife News.*

2004, a team of Nevada wildlife biologists assessed the effectiveness of hazing black bears in an urban locale. One group of bears
was hazed with rubber bullets, pepper spray, and loud noises, and
another group got all that plus a barking Karelian bear dog to run
it off. A control group was not hazed. In terms of how much time
passed before the bears returned, neither group stayed away significantly longer than the group that hadn't been hazed. All but five
bears out of the sixty-two followed in the study showed up again
eventually, and 70 percent were back in fewer than forty days.

Breck spent many a late night trying to haze bears during an
epidemic of car break-ins at Yosemite campgrounds. Between
2001 and 2007, eleven hundred automobiles were broken into
by bears. (Minivans were hit most often. While it's possible some
structural weakness contributed, Breck believes it had more to
do with what minivans typically hold: kids, lots of them, spilling juice, dropping crumbs, grinding chips into the footwells.
He guesses the bears were keying in to the smells of this "micro-
trash.") Hazing efforts proved futile. "Once they learn what's
inside . . . forget it." The bears quickly came to recognize the
sound of Breck's truck. They'd take off when they heard it coming, and go back when they heard it drive away.

It turned out that fewer than five bears—sows and their cubs—
were behind the break-ins. This is typical. From the start of the
year to the September of my visit, bears in Snowmass have broken into houses through unlocked doors or windows sixty times.
Wildlife-camera images have implicated just four bears. A bear
research scientist with the Minnesota Department of Natural
Resources, Dave Garshelis, told me about a call he got from a
National Guard camp where bears were raiding pallets of military
rations called MREs, which bears apparently enjoy more than soldiers do. He was told that around a hundred bears were raiding
the supplies. "The guy said, 'I'll bring you to this place where you
look across to this ridge that is completely pockmarked with bear

dens.' I was like, 'This sounds cool.' " The "dens" turned out to be natural landscape features, and "a hundred bears" in fact was three.

Great, so just trap the few brigands and deliver them deep into the forest, and your troubles are over, no? Say hello to the disappointing reality of translocation. Adult black bears rarely stay put where they're released. They have made their way home in journeys as lengthy as 142 miles—in one case including a 6-mile ocean swim. It is a remarkable achievement given that, unlike migrating birds, they can't rely on internal magneto-gadgetry to help them navigate. Whether they are picking up sensory cues—the smell of the ocean, say, or the sound of an airport—or just trying out different directions until something feels familiar, is not known, but they are motivated and they are good at it.

In a 2014 study, sixty-six conflict bears were radio-collared and translocated by Colorado Parks and Wildlife. Thirty-three percent of the adults made their way back to the spot where they'd been captured, and none of the subadult bears did. Those read as fairly optimistic statistics; however, if you define success not as the failure to return but rather as surviving a year in the new home, the picture is less rosy. Translocated bears often wander into a new town close to where they've been released and start getting into the same kind of trouble. More than 40 percent of translocated bears in Yellowstone National Park and 66 percent in Montana were involved in another "nuisance event" within two years. Yosemite rangers tried translocating the bears that were breaking into cars, moving them to the other side of the park. The result: car break-ins on the other side of the park.

Another factor is at play in the decision. Were a person to be seriously harmed by a translocated animal in its new location, the agency that brought it there could be held partly liable. The Arizona Game and Fish Department settled out of court for $4.5 million after a bear they'd translocated mauled a young girl at a campsite.

Dave Garshelis has worked with humans and bears for almost forty years. I asked him, by phone, how he felt about translocation. "People think this is a kind thing to do, but I'm not sure it is all that kind," he said. Often it's sows with cubs that get into trouble, because they need the most food. "Here she is living in her home range, teaching her cubs where the foods are. Now all of a sudden you plop her down somewhere else that she's completely unfamiliar with. With a whole bunch of other bears which she's competing with for food. You're injecting them into a social system they're not familiar with." When bear biologists from the state of Washington surveyed forty-eight U.S. wildlife agencies, 75 percent said they sometimes translocate problem bears, but only 15 percent believed it was an effective way to resolve the problem. It's more often done in high-profile cases, when media attention has put the animal and the agency in the spotlight. Generally speaking, translocation is a better tool for managing the public than it is for managing bears.

The most promising candidates are young bears translocated early in their "criminal" careers. This is partly because yearlings are less inclined, or less able, to find their way back, but mainly because dumpster diving is a gateway crime. Next comes breaking and entering, burglary, home invasion. As garbage-eaters become habituated to humans, as they start to associate them with jackpots of food, the risk-benefit ratio shifts. Less perceived risk, dependable benefits. Why stop with the metal boxes in the restaurant back alleys? Why not get inside the big boxes in the hills with the enticing cooking smells? Since the end of hibernation, in April, Colorado Parks and Wildlife has had 421 calls about Pitkin County bears damaging property while going after people's food. Most of these calls go to District Wildlife Manager Kurtis Tesch, whom Breck and I are meeting up with tomorrow.

The darker bear, perhaps weary of being harried by its dominant associate, has snagged a bag and run up a short set of steps.

We follow it up and around a corner, to the upper level of a swank mini-mall. Ordinarily, I would take delight in the optical non sequitur of a bear standing in front of a Louis Vuitton boutique. This poor goober with the burrata on its snout, innocent and utterly unaware of its likely fate, makes me want to cry.

Kurtis Tesch has bear stories, but maybe not the kind you expect. The things that stay with him are not the displays of strength or violence but rather the intelligence and occasional unexpected lightness of touch. The bear that unwrapped the foil on a Hershey's Kiss. A bear that stood up, grasped a door on either side and pulled it from its frame, then carefully leaned it up against the house.

"They'll reach in and take things out of the fridge, like eggs, and set them aside without breaking any." We are on the way to the scene of a break-in high on a ridge road, Kurtis and Breck and myself, careening around switchbacks in Kurtis's cluttered, thrumming CPW truck. An egg wouldn't last long in here.

Black bears are keeping Kurtis unusually busy this year. This was unexpected, because the spring was wet; human-bear conflicts are typically thought to intensify with drought, not with plentiful rain. But the year before was very dry, and Kurtis says he's heard that drought spurs some plants to produce an excess of reproductive material, or "mast"—fruit, seeds, berries, acorns—and then less of it the following year. "They're trying to spread their seed, thinking that they're about to die off. And then when a wet year comes, they're more concerned about growing." I don't know if that's what has happened here, but I like this worldview of trees that worry and prioritize and plan for their demise.

From the back seat, Breck volunteers that the general trend toward warmer temperatures also contributes, by shortening the length of the bears' hibernation. In a 2017 study, he and six CPW biologists radio-collared 51 adult black bears and monitored the timing and duration of their hibernation, along with environmental factors. For every 1.8-degree Fahrenheit increase in temperature, hibernation shortened by about a week. Based on current climate change projections, black bears of the year 2050 will be hibernating 15 to 40 days less than they are now. That's 15 to 40 more days out on the landscape looking for food. Add "more bear break-ins" to the list of possible consequences of climate change.

Food supply also affects hibernation. In a year of plentiful food, bears hibernate for shorter periods. For a bear that starts relying on human-sourced foods, every year is a plentiful year. Breck found that bears that foraged mostly urban areas hibernated a full month less than bears that foraged the natural landscape. Another concerning consequence of plentiful food is that reproduction rates rise. Black bear sows have a reproductive option called delayed implantation. Fertilized eggs become clusters of cells, called blastocysts, that loiter in the uterus over the summer. Whether they implant in it come fall—and how many of them do so—depends on the mother's health and how well she's been eating.

We've arrived at the driveway of our destination. From here, the house looks to be of average size. Turns out this is because most of what we're looking at is garage. The house pours down the mountainside two, three, I don't know how many stories. Breck steps down from the truck and walks to the edge of the blacktop. I assume he's marveling at the view, but as I walk over I hear him calling out the names of bushes and trees growing wild around the house, the ones black bears feed on: serviceberry, chokecherry, oak.

"Yup," says Kurtis. "This is some of the best bear habitat in Colorado. We moved into their habitat. You know?" Kurtis wears reflective orange-tinted sunglasses that stay on his face the whole time we're with him. He's light-haired and fit, with a good jawline, and that's as far as I can take you.

The owners of the house have been out of town. The housekeeper, Carmen, discovered the break-in and called the police, who in turn called Kurtis. Carmen lets us in and takes us downstairs to the entry point: a floor-to-ceiling window in a bedroom with its own deck. She says it was locked, but bears can wedge their claws into any small gap in a window frame and pry the unit out. An interior screen window lies on the carpet. The wall-to-wall is white, but the bear left no tracks. Nor, Carmen says, did it knock anything over on its way upstairs to the refrigerator. You get the sense that if there'd been a mop handy, it would have cleaned up the kitchen floor.

This bear reminds Breck of one that was breaking into Aspen homes back when his study was underway here. They called him Fat Albert. "He was just kinda laid-back. He'd gently open a door of a cabin, go in, eat some food, and leave. People would go, 'Wow, he didn't destroy my place at all.'" That's why he was fat, and that's why he was alive. There's more tolerance for a bear like that. An aggressive bear that trashes the place or otherwise makes homeowners feel violated and in danger is very quickly going to be, to use Breck's word, whacked. The upside, if it can be said there is one, is that natural selection favors the Fat Alberts. Aggressive bears are likely to be put down before they have much opportunity to pass on their genes.

With a growing percentage of Fat Alberts, will coexistence eventually become a possibility? Or even a policy? Could we live with bears in the backyard the way we live with raccoons and skunks? I posed this question to Mario Klip, a bear specialist with the California Department of Fish and Wildlife (CDFW) in the Lake Tahoe region. Many people in his area already do, he said. Say a couple of homeowners find a bear under the deck. Rather than call Fish and Wildlife, they may call the Bear League, a local advocacy group. "They'll send someone out to crawl under and poke it with a stick and get it to run off, and then help board up the space for you."

Klip practices coexistence with the BEAR League. "They are," he points out, "filling a vacuum." More and more people want nonlethal options for bears that trespass or break into houses. And not just Californians. Dave Garshelis works in rural northern Minnesota, where most people have guns and are allowed—encouraged, even—to solve their bear problems themselves. "I've been here thirty-six years," Garshelis told me. "I can sense a sea change in attitudes to bears."

What would happen if wildlife managers did nothing, if they stopped destroying the recidivist bears? The fear is this: Those bears' cubs will learn to break into homes, and ditto those cubs' cubs. As break-ins escalate, tolerance erodes. As Garshelis put it, "It's hard to be tolerant when there's a bear in your kitchen."

Back upstairs, Carmen describes the scene as she found it. The bear appears to have gone straight to the refrigerator. It opened the door, pulled out and scarfed a tub of cottage cheese, broke a bottle of maple syrup and a jar of honey and lapped those up, and then moved on to a pint of Häagen-Dazs in the freezer. (Pitkin County bears consistently prefer premium brands. "They will not touch Western Family ice cream," Tina White reports.)

Behind us a set of French doors leads to another deck. Carmen found these doors open and assumes this is where the intruder left the house. French door handles, locked or unlocked, are so easy for black bears to open that they're known as "bear handles" and are prohibited by local building code. But people like them, and do-it-yourselfers either don't know or don't care about the finer points of building code, and Kurtis sees them everywhere. Hollow doorknobs are likewise prohibited; bears crush and grip them in their teeth and easily turn them. (Some businesses make things even easier. Automatic doors open for bears, too.)

Kurtis thinks we may be looking at the work of two different bears. The first one entered and exited through the downstairs bedroom window, and a different bear came up to the French doors on the kitchen deck and smelled or saw the aftermath of the first pillage. His reasoning is based on the position of the doors as Carmen found them: opened inward. It would be unusual, he says, for a bear to pull a door inward in order to pass through. It's also possible the same bear returned to the scene a second time. Kurtis says they often come back at least once.

Like human burglars, bears typically break in when the homeowners are away. Given the large percentage of Aspen properties that are let as vacation rentals part of the year, empty homes are easy for bears to find. With bolder bears, burglary may escalate to home invasion. Often the bear comes in while people are asleep, especially, Kurtis says, when it's hot and someone has left the windows open. Or a sliding door is left unlocked. Sometimes the residents are not asleep. "We've had people eating dinner at their table and the bear walks in, grabs some food off the table, and runs back outside. We've had bears ripping doors or windows out while people are in there, hiding in their bedroom or bathroom."

Kurtis gives Carmen his card and tells her to have the owners call him if they want a live trap set. She doesn't ask him what would happen to a bear that ends up in the trap. Colorado Parks and Wildlife, like many state wildlife agencies, has a two-strike policy. If Kurtis gets a call about a bear nosing around someone's trash or hanging out in a back yard, say, he will attempt to trap it, and if he succeeds, he'll ear-tag it and take it into the woods and release it and hope it doesn't come back. (A trap is left in place no more than three days, to lower the odds of trapping the wrong bear.) Often the trap stays empty. "We're not catching them like we used to," Kurtis confided later. "I don't know if they've just gotten smarter, or what the deal is."

The bear that broke in here would not be granted a second strike. Because it's breaking into locked windows, and will likely continue to do so—and, if it's a sow, teach its cubs to do so—the agency considers it a threat to public safety. Kurtis says people often decline to report break-ins, because they know their call may set in motion a death sentence for the bear. The black bear is a ridiculously lovable species. There's a reason kids have teddy bears, not teddy goats or teddy eels.

"So what would happen if you were to trap this bear?" We're climbing back into the truck now, to head back to town. I notice a lint roller in the door caddy, as if sometimes bears sat up front in the cab.

"When you do trap the right one and take it out," Kurtis says, "you notice a slight decrease in break-ins in the area. For a short period of time. And eventually another bear comes in and takes over. So."

"It's a temporary solution," says Breck. "You're just mowing the grass."

That wasn't exactly what I was asking about. I was more asking about the "taking out." I'm going to have to be more direct. "And it can't be fun to have to put down a bear." All these euphemisms.* Taking out. Putting down. Are we killing an animal or unloading a truck?

"No, it's not," Kurtis says flatly. "Last week I had to put down a sow and a cub." The pair were repeatedly breaking into houses. "And that is not fun. At all." We drive along in a grim quiet broken intermittently by the walkie-talkie clearing its throat.

"In that instance," Kurtis adds, "I was struggling on how to do

---

* I understand the desire on the part of those whose job sometimes requires them to kill an animal to avoid that verb. *Kill* has a taint of murder. The sheer number of euphemisms suggests a long-running struggle to find something right. I collected them for a while: *cull, take, dispatch, remove, lethally remove.* As a word person, I balk at *euthanize*, which implies the relief of suffering, and *harvest*, which makes animals sound like corn. I heard one person say "*use lethal force on*," which seems better suited to SWAT operations and Gary Busey movies.

it exactly. I didn't want to put the cub down and have the mom watch that. I didn't want to put the mom down and have the baby see that. I ended up darting the baby so it went to sleep. I put the mom down, and then while the baby was asleep, I put it down. That way neither one had to see the other. So."

When Kurtis says "so" it is shorthand for the many frustrations the job holds. Uncaring homeowners who don't bother to follow the laws. Who then blame and hound him when a bear crosses a line and is destroyed. Government agencies that would rather pass the buck than spend a buck.

I try to imagine how I would feel if I lived in the house we just left and I had seen how effortless it was for a bear to get in. I ask Kurtis how people usually react. "Some are terrified," he answers. "Some are nonchalant." So far, no one in the area has been killed during one of these break-ins. Black bears, by and large, are not aggressive animals. Still, I'm surprised that what sometimes happens when human burglars break into homes hasn't happened here: homeowner or homeowner's dog surprises burglar, homeowner and/or dog goes after burglar, burglar panics and kills homeowner.

"Oh, it's coming," Kurtis says. Black bears may be no more aggressive than raccoons, but they're a lot larger.

What if we accepted that risk? What if we chose to live not only with the occasional bear in the kitchen but with the likelihood that someone at some point will be killed by one of those bears? Planes are allowed to operate even though every now and then they crash and people die. One difference is that with airlines, sales revenue covers the expense of lawsuits and insurance. When a bear harms or kills a person, the state wildlife agency may be held liable, and bears, unlike planes, aren't generating the revenue it would take to cover the costs. There have been a couple of lawsuits recently, one in Utah and another in Arizona, with

large payouts to the families of the victims. The agencies had been
aware of the bear's presence in the area but had opted to monitor
the situation rather than set a trap.

Breck rolls down his window. "So that's going to be your lim-
iting factor on that idea."

Away from the back alleys, Aspen is picturesque and pris-
tine. There are almost as many window boxes as windows, and
although we're closing in on October, I have yet to see anything
in them that's starting to wither or turn brown. It's like this town
has so much money and power that even the laws of nature shrug
and surrender. Flowers bloom in fall, and women's hair goes ash-
blond as they age.

Where I see prettiness, Breck sees attractants. "These right
here?" Breck points over our heads, to one of the small trees that
line the pedestrian walkway where we've been questing for an
affordable lunch. "Crabapples. The city *planted crabapple trees*." People enjoy the profusion of pink spring blossoms. That then turn
into bite-sized apples that bears mouth straight off the branch,
like cartoon emperors with their clusters of grapes. Black bears
show up midday in downtown Aspen regularly enough that the
city passed a law making it a ticketable offense to ignore the CPW
officer standing guard and go right up to a bear and snap a selfie.
Kurtis Tesch's predecessor tried, unsuccessfully, to convince the
city council to replace the crabapple trees. When I got home I
stumbled onto an online Aspen Arbor Guide to residential tree
selection and planting. Among the recommended species: crabapple, oak, chokecherry, and serviceberry trees. I hesitated to tell
Breck. I thought his head might explode.

We locate a moderately priced restaurant that is not one of the
eighteen fined for having unsecured garbage and publicly shamed
in this week's *Aspen Times*. I dig out a list of questions, questions
that basically boil down to: What is happening here, and is there

an answer? I bring up something Kurtis Tesch said when we were driving back to town. He shared a theory about how the ballot measure that banned Colorado's spring bear hunt (because it orphaned cubs) had caused the rise in the number of bear conflicts. Breck says he often hears this argument. "There's a sentiment carried forward by a lot of hunting communities and parks and wildlife people that one way out of this is to hunt our way out of it. But there's no good science to say that lowering the bear population will lower the number of conflicts."

For one thing, he says, there's a mismatch between where the hunters go and where the conflicts are. "Hunting quotas are set according to game management units." I don't catch all the details because I'm distracted by some intense celebrity name-dropping at a large table next to ours. I hear: "This game unit has Aspen, Snowmass, Carbondale—" ". . . Reese Witherspoon—" "They'll say, 'Okay, within this unit, harvest x number—'" "So then Reese . . ."

Hunting does alter the behavior of the hunted somewhat, Breck allows, in that it perpetuates fear and avoidance of humans. But Colorado still has a fall bear hunt, so he doesn't buy less hunting as a reason for the rise in bear conflicts in this case.

It's worth mentioning that the money for Kurtis's salary—like most items in state fish and wildlife agency budgets—comes partly from hunting and fishing license fees and taxes on equipment. "I'm not here to criticize that model," Breck says, "but you have to be aware that that's underlying all this stuff."

I am. Somewhat uncomfortably so. In the course of researching this book, I met a lot of good, intelligent people at these agencies, professionals who saw their job as protecting people and animals both. But because of the financial model, it can be hard to set aside the nagging sense that institutional priorities are at play. The money's coming from hunters, to a large degree—and that makes it hard for agencies to win the trust of everyone else. (And creates perplexing mottoes like "Support Nevada's Wildlife . . . Buy a Hunting and Fishing License.")

Breck shakes open his napkin. There's legislation making its way through Congress that would add a billion-plus dollars in federal funds to wildlife agency coffers. The money would be earmarked for conservation-oriented projects. "Which would alter that dynamic." We skim our menus. At the next table, they're on Miley Cyrus. ("She's amazing." "She *is* amazing.") Breck is constitutionally immune to the distraction. On the drive over, I had asked about Aspen celebrities. I got: "Jack . . . Nicholson. Nicklaus? Which one is the golfer?" He knew that Kevin Costner has a place, because Kevin Costner once had a bear problem.

Breck puts his menu down. "The thing that's not being talked about enough is this. You have a recovering bear population that was suppressed mightily at the beginning of the last century." The prevailing attitude toward America's wildlife in the early part of the twentieth century had changed little since settlers first crossed the divide. The first to push west were ranchers, subsistence farmers, cattlemen, fur trappers. Wild animals were either a commodity or they were varmints. Bounties were widespread. Bears were routinely poisoned, up through the 1970s. "We wiped out everything," Breck says.

The government was there to help. Breck's employer, the National Wildlife Research Center, has had many incarnations and names over the past hundred and fifty years but always one goal: effective, cost-efficient wildlife damage control. Whether the wild animals were predators taking livestock or birds and rodents helping themselves to crops; whether the name on the door was Division of Economic Ornithology and Mammalogy or Eradication Methods Laboratory or Division of Predatory Animal and Rodent Control, the goal was to help the rancher and farmer. What looked like pure wildlife biology—studies of animal behavior, food habits, migratory patterns—was biology in service of prosperity.

With the birth of environmental and animal welfare movements in the 1960s and '70s, a national conscience began to emerge. Activists pushed back against practices like den-shooting and airdrops of strychnine-laced bait. In 1971, Defenders of Wildlife, the Sierra Club, and the Humane Society of the United States sued to end the use of poisons in predator control. The following year, the Environmental Protection Agency (EPA) canceled registrations for strychnine and two other predacides. Advocacy groups sparked a shift in public sentiment, which over time has become impossible—and impolitic—to ignore.

A growing percentage of Americans feel a strong emotional connection to wild animals and disapprove of their destruction for economic reasons. In 1978, three thousand Americans were asked to rate their like or dislike of twenty-six different animal and insect species. In 2016, researchers at Ohio State University readministered that same survey. Compared to the first go-round, the proportion of respondents who reported liking wolves and coyotes was up by 42 and 47 percent, respectively. (The cockroach had also risen in popularity, from most despised—that honor now going to the mosquito—to second most-despised.)

Which is a gassy way of saying: bears are back. To the point where they're starting to get all up in people's worlds. "This is new territory for wildlife biologists," Breck says, forking salad greens. "And we're not very good at it. When I was an undergrad it was all about, How do we bring these populations back? How do we count them, manage them? Now it's all about human-wildlife interactions. How do we manage *this*? We're seeing wildlife biologists going . . ." Breck mimes banging his head against the table. "The game has changed."

At the moment, it feels unwinnable, this game. There are more bears, more wolves and coyotes, and ever more humans moving into their ranges. And no cultural consensus on what should be done when one of those animals ransacks a kitchen or kills some sheep or nips a steakhouse owner's backside. We have human-wildlife conflict and human-human conflict. We have ranchers and farmers and animal lovers hating on each other in a cultural clash that can feel as dug-in as the politics in this country. *Kill them all! Don't harm a single one!*

Breck and other specialists in human-wildlife conflict are starting to move their focus from animal biology and behavior over to human behavior. As a science, it's called human dimensions. The goal, stated unscientifically, is to find pathways to compromise and resolution. It often begins by bringing together people who aren't normally in the same room and getting them to listen to and even empathize with one another. Breck recently co-founded the Center for Human-Carnivore Coexistence. In early 2020, the center organized a two-day gathering of hunters, trappers, ranchers, and representatives of conservation and animal welfare groups to talk about reintroducing wolves in Colorado.

Breck came away hopeful. By the end of the second day, he was hearing people talk without hostility and in ways that struck him as productive. "The question now is, What happens when everyone goes back to their own corners?" Breck's hope is that whatever path the state pursues, it will be a decision reached by a group like this, rather than a few legislators in a back room.

Lately Breck has been spending time with Zach Strong, director of Carnivore Conservation at the Natural Resources Defense Council. The NRDC's more standard approach to Wildlife Services has been to sue them. Breck encouraged Strong to forge a relationship with the Wildlife Services director in Montana. Partly as a result of this unlikely pairing, three nonlethal—or "wildlife conflict-prevention"—specialist positions were created within Wildlife Services, two in Montana and one in Oregon. By showcasing the effectiveness of these hires, NRDC and Defenders of Wildlife were able to secure federal funds for hiring and evaluating similar positions in ten other states. Breck is hopeful that the new developments at Wildlife Services signal a shift within the culture there. Meanwhile, the Fish and Game Department in Idaho still funds a hunter- and rancher-friendly nonprofit that offers bounties on wolves.

All the government agencies agree on one thing: the fate of a wild animal that kills a human. Might even this someday change? Are there places in this world where the consensus favors sparing the animal's life, particularly in the case of a defensive attack? Bear biologist Dave Garshelis has been spending time on the Tibetan plateau, where brown bears routinely break into the homes of herders who have left for the summer to graze their stock. "They'd come back and the house would be a wreck, completely demolished. But these people are strict Buddhists and they don't want any retribution." Garshelis told me about a conversation he had with the local officer who responds in cases of animal attacks. "I asked him, 'What if you were called to a situation where you saw a bear on top of a person, mauling a person? Would you shoot the bear?' And he said, 'I don't have the right to decide whose life is more important, the person or the bear.'"

In India, around five hundred people are killed every year in encounters with wild elephants. Government policy is to compensate the families but not, with few exceptions, to destroy the elephant. The state with the highest numbers—403 deaths in the past five years—is West Bengal. Maybe some answers are there, too.

3

THE ELEPHANT IN THE ROOM

Manslaughter by the Pound

In India there's a thing called an "awareness camp." I first heard the term from a researcher at the Wildlife Institute of India who runs elephant and leopard awareness camps. I imagined a camp in the American sense, with bunk beds and marshmallows, and I tried to meld this in my head with large, dangerous animals. Naturally I wanted to go. As it turns out, an awareness camp is closer to a national awareness day. I've seen listings for dengue fever awareness camps, diabetes awareness camps, traffic safety awareness camps, and at least one Snoring and Sleep Apnea Awareness Camp, which sounds like our bedroom. They're informational gatherings with the aim of making people aware of dangers they may not know much about or prefer not to think about and offering guidance in how best to avoid or survive them.

Every December that researcher, Dipanjan Naha, locks up his office in Dehradun, slaps a Government of India ON DUTY placard on a hired four-wheel drive, and heads out on a sort of awareness camp road trip. This year his cousin Aritra is coming along as his assistant, and I'm in the car, too. We're starting in North

Bengal—confusingly, a region of West Bengal—where each year, wild elephants kill, on average, 47 people and injure another 164. Forty-seven people per year, in an area the size of Connecticut. India's forest department has wildlife rangers that get involved in these cases, but they don't kill the elephant. A few rangers will be on hand at Naha's first awareness camp, in the village of Baman-pokhri. I'm eager to learn what it is they do.

Outside the car window, agriculture rolls by: tea planta-tions, marigold farms, neat rows of rice plants set in the earth like toothbrush bristles. Small villages punctuate the paddies and plots—homes of corrugated metal and thatch, a temple, a few open-front bodegas. Cows loiter in the roads and small black goats are parallel-parked along the sides, but I see no other animals, no logical setting for elephants.

Elephants! Naha assures me they're not far off. It's winter, the time of year when herds are on the move. They forage at night and sleep by day in patches of teak and lal—remnants of the for-ests that once stretched from the Indian state of Assam through North Bengal all the way to the eastern border of Nepal. This "elephant corridor" has since been fractured and diminished, first by the many sprawling tea estates planted by the British and more recently by military bases and settlements of refugees and immigrants from Nepal and Bangladesh. Ever more humans are coming into these forests to cut wood and graze their livestock, turning elephant habitat into human. In their attempts to cross the land, the elephants encounter barriers, dangers, dead ends. The corridor is now a pinball game. Herds can become isolated in a single pocket of forest. They become "pocketed elephants." Keeping elephants in a pocket is as ill-advised as it sounds. The gene pool stagnates and the population density spikes. Soon there's not enough food to support them. They wander into villages to eat what they can find, and what they find is people's crops and granaries. Hello, human-elephant conflict.

Aritra points out his window as we pass a turnoff. "Two kilometers up that road a man was killed by an elephant. A few days ago. Three people were working on the roadway. They ran when they saw the elephant, and one man was separated, and the elephant followed him."

This is hard for me to imagine. I grew up with Babar and *National Geographic.* Elephants were gentle and slow-moving. They wore spats and bright green suits. They were never something to fear. This has created a minor disconnect between myself and my hosts. Our first night on the road, we stayed at a government-owned bungalow in a teak forest just down the road from a sign marking an elephant crossing zone. The bungalow's cook said he saw one near the gate the night before we arrived. My reaction to this piece of information was to announce my intention to go for a walk. It was around 7:00 p.m., two hours past the time when elephants head out in search of food and Dipanjan Naha and his cousin stop going near the forest.

"Don't go very far, Mary," said Aritra. We were sitting on the porch, drinking tea under the geckos and moths. Aritra has a round head and a friendly, slightly giggly disposition. He slips easily from his role as Naha's assistant to the more familiar role of younger cousin.

Naha didn't like my plan, either. "Please be very careful."

They looked at each other, and then they set down their cups and got up to accompany me.

We got as far as the railroad tracks at the end of the driveway. Naha said a few words about the history of the narrow-gauge railway in India. We stood for a few minutes, as though waiting for a train. Aritra toed the gravel between the ties. "Let's go back inside."

To better appreciate the danger of an unplanned elephant encounter, sit down with someone who investigates the deaths. Saroj Raj is the range officer for the Bamanpokhri Beat of the local forest division, where every year since 2016, someone has been killed by an elephant.

Officer Raj has come to Bamanpokhri's blue-walled, one-room community hall—the site of today's awareness camp—to talk to people and answer questions. So far it's just me asking. The people who show up on time for an awareness camp, it seems, are the people who have to. Today that's a group of schoolchildren in plaid uniforms and a half dozen rangers from local wildlife squads. Naha isn't bothered. It's Diwali—a holiday week—and it's after lunch. "So they are sluggish to come."

Officer Raj gives me the particulars of the most recent fatalities. He begins, each time, with the exact date. You get the feeling there's a lot of paperwork. "Thirty-one October, 2018. Three workers in the road." The spot we passed earlier. "Suddenly one elephant appeared."

One can be scarier than many. Herds comprise females and young elephants, the peace-loving jumbos of my childhood. A loner is typically male, and males can be trouble. Bull elephants go through a periodic hormonal tumult called musth, during which their testosterone levels are as much as ten times higher than at other times. This gives them a competitive edge with other males and with the herd's dominant matriarch, but also a degree of random volatility. The musth mood state ranges, in the words of Asian elephant specialist Jayantha Jayewardene, from "hyper-irritability" to "a charging or destructive tendency" toward other elephants, human beings, "and even inanimate objects." Villagers know this. "The men tried to escape into the bushes," says Officer Raj. "One person unfortunately fell down."

Officer Raj supplies the detail Aritra was careful to omit. The elephant stepped on the man's head. When you weigh 6,000 pounds, simply stepping or kneeling on a person—or, in the 1992 case of a riled circus elephant named Janet, performing a headstand upon the person—is an effective means of killing. But Officer Raj's determination in this case, based on footprints and disturbances of the surrounding vegetation, was that the death was accidental.

"Sixteen October, 2016. Also an accident." A man was climbing up a riverbank when he encountered an elephant. "It was slippery," recalls Officer Raj. "Both slid down the bank. The elephant just rolled over him." Elephants sometimes kill the way cars kill: by being large and running into—or over—something much smaller. (Elephant keepers try not to get between a wall and an elephant.)*

"These elephants were not in the intention to kill," Officer Raj says. How does he know? Because the bodies were in one piece. "If an elephant is in an angry mode, the body will not be intact. It will get in pieces." A book by Jayantha Jayewardene includes a list of the nine recorded methods by which an angry or musth-addled elephant has killed a human being. "Placing a fore foot on one limb of the victim and ripping off the other with the trunk" is number 3. (Elephants use a similar anchor-and-pull technique to strip an uprooted shrub of limbs and leaves to eat.) Rulers of Ceylon (now Sri Lanka) in the 1600s were said to have taken advantage of this natural behavior, training elephants to serve as executioners. An engraving from *An Historical Relation of the Island of Ceylon* depicts an elephant so engaged. One forefoot rests on the malefactor's torso, and the trunk is wrapped around the man's raised left leg. Were it not for the caption ("An Execution by an

---

* At the zoo where I worked in my twenties, the pay rate for the elephant keeper was slightly higher than that of workers who cared for other animals—though not because of the risks. She was paid a "shit differential" because she had to shovel excessive amounts of it. Well earned: according to a 1973 *Smithsonian Contributions to Zoology* document, an Asian elephant defecates 18 to 20 times a day, dropping "4 to 7 boluses" of around 4 pounds each, for a daily output of 400-plus pounds.

Eliphant") and the raggedly torn-off arm in the foreground, you could get the impression that Ceylonese monarchs trained elephants to serve as bodyworkers.

Naha's awareness presentation will stress the importance of not upsetting an elephant. This is summed up by item 20 on the Best Practices for Mitigating Human-Elephant Conflict poster Aritra has just hung up on the wall of the community hall: "No Rambo-Style Action!" Shooting an elephant is not only illegal but, depending on the caliber of weapon, pointless. Our Janet withstood fifty-five shots from Palm Bay, Florida, police officers' 9mm duty revolvers as well as the first round fired by an off-duty SWAT officer with a stash of ammo designed to penetrate armored personnel carriers. (The second round stilled her.)

The safe thing for a villager to do upon sighting an elephant or a herd in the area is to call the 24/7 hotline and let Officer Raj's elephant squad handle it. The squad knows that elephants are social animals, so they stay calmer if they're herded off the premises in a group. The rangers converge on them from the sides, like cowhands, and move the whole group in the direction of the forest out of which they've come. By now, the animals recognize the sound of the elephant squad vehicles. "We drive into the area, and they go." Officer Raj smiles slightly. He's not a smiley guy. "This is a convenience for us."

Officer Raj makes elephant patrolling sound like mall security, but apparently it's quite risky. A beat officer sitting with us has been charged four times. "They tell you not to run," he says. "And I tell you, this is very hard to do when an elephant is coming straight at you!" My request for a ride-along is turned down, as is a second request. The expression on Officer Raj's face suggests I've become a "nuisance event," so I leave it.

Deaths, when they happen, tend to occur in the half hour or more it takes for the squad to arrive. Upon discovering elephants raiding their crops, villagers rush out of their homes, yelling, throwing stones, lighting torches and firecrackers.* A village may have freelance "elephant chasers" wielding spikes and carrying out other non–Best Practices. Bulls and dominant matriarchs may charge in defense, and normally placid females and calves may panic and stampede. In the dark of unlit fields and paddies, people stumble and fall and elephants are running blind and, as my mother liked to say, *somebody's* going to get hurt.

"The elephant we can guide easily," says Officer Raj. "To guide the people is the hard part. They are not in the condition to listen." They're upset, and that is understandable. Village farmers work hard and have little to show for it. A single Asian elephant may consume three hundred pounds of vegetation in a day. Between the raiding and the trampling, a small herd can quickly torpedo a season's labor and livelihood.

An elephant among the crops is a powerful impetus to unwise action. Throw in the wobbly judgment and dimmed impulse control of inebriation, and the results can be dire, Naha says. He squats in front of a speaker, untangling a spaghetti heap of wires. "This is what we see. A group of people are drunk. Someone wants to be the hero, so he goes in front of the animal, harasses it, and that animal in self-defense . . ." Naha, too, avoids the verb *kill*, with its undertone of intent. "There is an accident." By his own data, 36 percent of the people killed by elephants in North Bengal between 2006 and 2016 were drunk. Later, I would see this head-

---

* That elephants are reliably spooked by fire and sudden loud noises limited their usefulness in war. Though the sight of armor-sided "elephantry" with swords lashed to their trunks conferred, from a distance, a psychological advantage, this quickly evaporated as the two sides drew closer. Records exist of elephants turning and breaking ranks at the sound of musket fire or the sight of flaming arrows. A fleeing, sword-waving elephant storming its own battalion likely racked up as many causalities as it would have inflicted on the enemy.

line in the *Hindustan Times*: "Drunken Man Challenges Elephants' Herd, Trampled to Death in Jharkhand." (Jharkhand borders West Bengal.) "He tried to fight with them," a forest ranger told the reporter. "Them" was eighteen elephants.

Dangerously, an elephant also enjoys a snort. In North Bengal, elephants drink what the villagers drink: *haaria*, a home brew fermented and stored in quantities sufficient to inebriate an elephant. (Because elephants lack the main enzyme that breaks down ethanol, it takes less than you'd think.) According to Officer Raj, two things happen when elephants liquor up. Most just stumble away from the herd and sleep it off. But every herd seems to have an aggressive drunk—the matriarch, often, or a bull in musth. Whatever you do in this life, stay away from an inebriated bull elephant in musth.

There is data to support Officer Raj's observations. In 1984, as part of a study at the UCLA Department of Psychiatry and Biobehavioral Sciences, three Asian elephants "with no known history of alcohol use" and seven African elephants from Lion Country Safari were served a "large calibrated drum" of grain alcohol-laced water. The animals tended to wander off, away from the herd. They stood or leaned with their eyes closed or "wrapped their trunks about themselves." They skipped meals. They didn't bother to bathe. The matriarch became louder and more aggressive, as did a bull named Congo.

To discourage elephants from tippling, villagers may drag the hooch indoors. A terrible idea. Because now it is not drunk elephants they need to worry about, but elephants determined to become drunk—that is to say, elephants who smell booze inside the house and see no reason not to knock down a wall to get to it. In Naha's survey, 8 percent of North Bengalis killed by elephants were sleeping inside their homes when it happened.

The chairs are mostly full now, and Naha takes up the microphone to begin his talk. He speaks in Hindi, of course, so Aritra translates, intermittently leaning over and cupping a hand to my ear to deliver a stand-alone sentence in a decisive, oracular cadence. "Being *drunk*, you should not go in front of the elephant." "Always drive by the *back* side of the elephant."

Naha is an expressive speaker with a lexicon of fluid hand gestures. I didn't necessarily expect this. Offstage, there's an economy of movement and words. He stands with shoulders squared and his feet angled outward as if to create a stable base. He's like the man I keep seeing in the Ambuja Cement billboards, standing firm and unperturbed with an entire thundering hydroelectric dam in his arms. I've heard him say "I was once chased by a tiger" in the same workaday tone in which you or I might say "I was once in Omaha."

The philosophy behind an awareness camp is simple. If you want to get through to people, talk to them while they're relaxed and clearheaded. Sit them down and let Aritra serve them a cup of tea and a samosa. The more people understand about the biology of elephants and the behavior of herds, the safer the encounters. It mostly boils down to staying calm and giving the animals space. Especially mothers with calves, and even more especially, lone males, and extra-strength especially, males in musth. (Some hallmarks of musth, courtesy of Jayantha Jayewardene: profuse oozing from glands on the temples, frequent erections, and "fully opened ogling eyes with roving eyeballs.")

There is another point worth emphasizing with people. Officer Raj mentioned it earlier, when we were chatting: "We the human beings, *we* are disturbing *them*."

Awkwardly, this includes the actions of professionals like Officer Raj. Herding elephants into the nearest forest patch provides immediate benefits for the local villagers, but in the long run, Naha told me later, it aggravates the problem. Because it aggravates the elephants. They begin to associate humans with the anxiety and privation of being chased off when they're trying to eat.

They start to stand their ground. There are reports coming out of conflict regions in the neighboring state of Assam that say female elephants are starting to become as aggressive as bulls.

A better system, Naha believes, would incorporate sensors to detect the approach of a herd. A warning would go out to village heads and trained local response teams who would monitor the situation and try to intervene before crops are trampled and chaos erupts. Naha doesn't mean motion sensors or heat sensors, both of which would be triggered by other mammals. He means seismic sensors: sensors triggered by vibrations so powerful only the footfall of an elephant (or a small earthquake) would have created them. Meanwhile, you do what you can to reduce the heavy footfall of humanity: keep working to restore forests and set aside preserves.

The assistant manager of the Gopalpur Tea Estate wears long, trim shorts of suit fabric and bulbous neon athletic shoes. His head is set at a backward tilt that makes him seem aloof, or maybe just in need of a new glasses prescription. He came to greet us as we drove in. An awareness camp for the workers, our second stop this week, is scheduled to begin in half an hour. But first, a cup of tea.

The assistant manager is a numbers man. The estate covers 1,200 acres, he says, placing cups in front of us. Two thousand one hundred workers collect the tea leaves. We listen and sip, and then he leads us back out the door and across the way, to an open-walled pavilion where Naha will be speaking.

The workers have already arrived. They look through the handouts that were placed on each chair inside a reclosable clear plastic bag (brand name: My Clear Bag). The women sit on one side, segregated from the men. Naha fiddles with the sound system, a holdover from the days before movie streaming, when the estate brought in live bands for weekend entertainment.

It's hot and thickly humid. The tea estate managers are late.

The workers fan themselves with Their Clear Bags. Time passes. Someone from the assistant manager's house brings a tray. More tea! Ours is served in china cups with saucers. The workers sip from paper—tiny cups no bigger than a shot glass. *WTF*, I want to say, *you've got twelve hundred acres of tea.*

Here come the managers! SUVs arrive like squad cars, pulling up fast and braking hard. Doors open and slam, and the managers, five total, stride onto the scene. The workers rise from their seats in unison. Rather than joining Aritra and me in the audience chairs, the managers ascend to a row of tables on the stage. They crack open the water bottles that have been set at their places along with notepads and pens. There is a spectacular variety of mustaches up there on the stage.

The managers take turns with the microphone. Aritra yells his translations over the blast of dance hall speakers. The first manager urges the workers to pay attention, because we are near a river and a forest, and so we have elephant problems. He hands off the mic to the manager beside him, who outlines the estate's present elephant deterrence strategy: a team that patrols on a tractor, lighting firecrackers as needed. The microphone continues its travels. The next manager is another numbers guy. In the twelve years he has worked here, seven or eight people have been killed by elephants.

At last the microphone is handed to Naha. The workers do in fact pay attention, so closely that I wonder whether there might be a penalty for looking away. The managers whisper with each other and peek at their phones in their laps. One takes a call, holding a hand over the device as if, like a burp, the behavior could thereby be made less rude.

When the presentation is over, Naha invites the workers to join in with questions and comments. One tea collector immediately rises from her chair. She is older than many, maybe fifty, and is dressed, like all the women, in the colorful, patterned sari she wears to work the tea plots. Aritra jumps up to bring her the mic, but she doesn't need it. Her anger is an amplifier. The mustache showroom shifts in its seats.

Aritra resumes translating. "You tell us to change crops," the woman says, referring to the small gardens kept by the workers. "From corn or rice to something like ginger or chilis, which elephants don't like. But we grow corn and rice to feed ourselves. Also, once the tractor passes, the elephant is coming, coming, coming again. Elephants need a lot of food." The woman sits down again. "We need different measures."

She's right, but good solutions are elusive. Measures that seem intuitively obvious are, in practical fact, limited—by their expense and by the new problems they create. Electric fencing is a for-instance. You need enough of it to keep the herds out, but not so much that it blocks their migrations. Maintaining and repairing long stretches of fencing is time-consuming and costly and often doesn't get done. Or is done wrong. The voltage has to be high enough to discourage an elephant but not so high as to electrocute it. On average, fifty elephants a year are electrocuted in India.

And there is the considerable challenge of elephant intelligence. An Indian elephant, confronted with an electric fence, will likely soon figure out how to get past it without getting shocked. It will figure out that wood doesn't conduct electricity. It will push down the posts or it will pick up a log and use it to press down the wires to let others step through.

The elephant's intelligence has not always served it well. Elephants have been put to work—by the Indian military, historically, and more recently by the timber industry. They're treated like employees, in that the forest department keeps a log of their hours on a work register. These "duty elephants" are of course not paid a salary but, Naha told me, at age fifty they receive a "pension," in the form of retirement lodgings at a *pilkhana* with meals and a daily bath followed by a rubdown with oil.

As people get up from their chairs to leave, I ask Aritra to introduce me to the woman who dared to speak out. Her name is Padma. She has reason to be upset. A week ago, she was awakened at 4:30 in the morning and saw that an elephant had smashed through a wall and eaten the grain she sells in a small bodega in the worker colony. She has yet to receive the compensation owed to her from the tea estate.

A subset of managers hovers within earshot. One sidles up to derail the conversation. "Hello, you are from America, my son works in Memphis, in the best hotel, do you know the Memphis hotel ducks?" I actually do know this hotel, these ducks that descend the lobby staircase for no discernable duck or hotel reason. I keep my focus on Padma.

Aritra continues translating. "This is the second time it has happened to her."

Still playing on the manager channel: "At five o'clock the ducks come down . . ."

Naha joins us. He suggests we drive over to the workers' colony to see the wreckage of Padma's shop. The managers exchange urgent looks, but it's too late. We squeeze into our car with Padma and back away.

The tiny bodega isn't so much ransacked as flattened. A wall of corrugated metal lies crumpled beneath a concrete support beam. On another occasion, an elephant broke into Padma's home while she slept. This is a place where "the elephant in the room" is not a metaphor, where elephant jokes are no joke. What time is it when an elephant sits on your fence? Probably around 11:00 p.m.

Elephants are vegetarians, but they are not picky eaters. They'll eat most parts of a plant—grains, grass and leaves, stems, twigs, bark. On a tea estate in the Sonitpur district of Assam in 2017, three wild elephants broke into a workers' shop at 2:00 a.m. and helped themselves to the cotton fiber product known as rupees. They broke open the cash box and consumed 26,000 rupees in large denominations.

One thing Indian elephants won't eat is tea leaves. Everyone here likes to drink tea, but few, human or beast, like to eat it. The leaves are too bitter. Small crop losses result when elephants trample plants as they pass through an estate, but, by and large, it is the workers, not the owners and managers, who suffer.

Nonetheless, Padma and her neighbors say they feel no anger for the elephants. Seventy-five percent of the people Naha interviewed about elephants that come into their villages report positive feelings for them. Given the amount of death and destruction caused by elephants in North Bengal, retaliatory killings are surprisingly rare. Naha says just three to five elephants a year.

With Aritra's help, I tell Padma what happens to large mammals in the United States that injure humans or break into homes. I ask her whether people she knows ever express a desire to kill an elephant that has broken into their home or shop. "Why would you kill a god?" she says, referring to the elephant-headed Hindu god Ganesh. "We just say, 'Namaste and please go away.'"

Padma leads us out to the tea field currently being harvested. Leaf collectors are spread out along the rows, working the waist-high bushes. Only the new, tender bright green leaves are plucked. The workers remind me of steel pan drummers, standing still with their arms moving crazy quick. They are fast because they have to be. If they don't meet a set quota, their pay is docked.

Naha bends down to show me the empty space below the tea bushes. The leaf pluckers sometimes surprise a female leopard resting in the shade with cubs. The animal may awake in a panic and, if it feels cornered or threatened, jump at the worker. Deaths are rare, but injuries happen. Ninety percent of leopard attacks in North Bengal happen on tea estates.

We watch the women work. As soon as their hands are full, they reach behind their heads to stuff the leaves in a cloth sack slung over their foreheads. Now I understand why Padma, earlier, had shown intense interest in my backpack. You'd think the managers could spring for bags more ergonomically suited to their task. I say this to Naha.

He rolls up his sleeves. "They are paid one hundred fifty rupees a day." This is less than the cost of a cappuccino at the Delhi airport. "It's a colonial mentality. They are the same tribal laborers that the British brought in from central India. They use them because they are considered hardworking and obedient."

When Naha told me about the work register for animals, about pensioner elephants and their bathing regimens, I was initially impressed. Here is a government guaranteeing working animals some of the same benefits normally accorded human workers. Seeing, now, how the tea collectors are treated, are legally permitted to be treated, it's less uplifting than it had seemed. Depending on your species, religion, gender, and caste, India may be a better place to be an animal than it is to be a human. In 2019, the government of Delhi announced plans to revamp one of the five sanctuaries it maintains for the city's free-ranging, traffic-snarling, sacred cows. Perhaps in response to criticism that the city provided better care for its cows than for its citizens, the city's Minister of Animal Husbandry announced, "We are planning a unique coexistence programme where elderly will be allowed to stay with the cows."

Things have lately grown more extreme. The current prime minister, Narendra Modi, rose to power on a surge of Hindu nationalism. This is a man who bestowed personhood status on the Ganges. A river enjoys human rights protections, while women like Padma earn 150 rupees a day and Muslims are lynched for selling beef.

As we get in the car to go, the assistant manager trots over with an armful of Mylar bags: a pound of tea for each of us. Aritra thanks him, then turns to Naha as the car gets underway. "It's CTC." (The acronym describes a processing method.) To me he explains, "The cheapest kind."

Wildlife is the theme at the government-owned Jaldapara Tourist Lodge, where I'm staying tonight. The grounds feature life-size plaster models of local wildlife. A number of the figures have been knocked over or have broken-off appendages that lie where they've fallen, on the incongruously well-manicured lawn. The whole courtyard has the feel of a miniature golf course after a party of drunken golfers played through, beaning the statuary with errant balls or maybe their golf clubs.

I love this kind of place, love the surreal decay of it, love the clerk who does not know where breakfast is served or even *if* breakfast is served, love everything, really, except the rat turds on my balcony. I tried to imagine what could be the draw, for a rat, of a balcony with no food and nothing in which to nest and not even much of a view. It appeared to be simply the place rats go to have a dump. The Jaldapara rat toilet.

The lodge is scheduled for a remodel, because the West Bengal forest minister is working with the West Bengal Tourism Development Corporation to create a rhino preserve out of the neighboring forest. This happens to be the forest to which Naha, Aritra, and I are shortly headed. We're going to track a radio-collared "conflict leopard"—leopard 26279—that was relocated here a year and a half back, when the land was earmarked to become a wildlife preserve. The animal was not full-grown then, and Naha wants to check to be sure it hasn't outgrown its radio collar.

While he's here, Naha will talk to villagers who live on the edges of the forest. Radio-collaring provides points on a map, but it doesn't answer the questions that ought to be asked when a predator is released into people's backyards. Has the leopard been taking your goats? Are people okay with it being here? Naha has been following up, like a social worker on a foster care placement. He's been tracking this leopard remotely and phoning alerts if he notices it moving toward their homes.

This morning we have a new driver, Ashok. He rarely joins the conversation, preferring to give his full focus to the driving. This is a switch from the last driver, who, to keep himself from nodding off, suction-cupped his smartphone to the inside of the windshield and began streaming a sitcom. (Naha was unalarmed. "One eye is on the phone, and one eye is on the road.")

We turn off the paved roadway onto a dirt track through increasingly thick underbrush. Branches scrape the sides of the car. Ashok has grown quiet. He seems tense. Did I say something wrong? Is he worried about scratching his paint job?

Just past a cluster of homes, we pull up alongside a man dousing his cauliflower patch with pesticide from a backpack tank. Naha gets down from the car, and the man shuts off the nozzle. One eye is cloudy. Three other men wander over. Aritra listens in. Nothing to report. They haven't seen the leopard in a while.

We drive on. We pass a watchtower used by anti-poaching squads. Naha asks Ashok to stop so he can climb up and get a better signal. Aritra tells me to stay inside the car with him.

Naha descends the tower stairs and gets back in the car. He reports that we're now just a thousand feet from the leopard.

We drive on. The road dead-ends at a wide river. Naha gets out again. He walks along the sandy riverbank holding an antenna above him like a torch. On the other side, a group of men are waist-deep in the water, clearing an overgrowth of water hyacinth.

Naha leans in the car window to let us know we are now 500 feet from the leopard. *Dut-dut-dut-dut*, goes the receiver. We can't get closer, because the cat is on the other side of the river. He points. "Just beyond where those men are."

There's no bridge near, so we turn back. On the last few miles of the drive, Ashok breaks his silence. He and Naha speak, in Hindi. After Ashok drops us off, I ask Naha what they were talking about.

"His father was killed by a leopard."

The father had gone to collect firewood. When he didn't return home, Ashok, who was twelve then, went with some friends to look for him. They found him, still alive, in a riverbed. The kind of place where the men were working today, a few dozen feet from a leopard. "They brought him to the hospital," Naha says, "but he couldn't recover. He sustained a lot of injuries. His eyes, and all that. He must have been very badly mauled."

Ashok will not be our driver for the next leg of our trip, and this is for the best. We are headed for Pauri Garhwal, a hot spot for leopard attacks. These are not the kind of encounters that take place on tea estates when a worker surprises a leopard sleeping under the plants. These are stalkings, killings undertaken with intent.

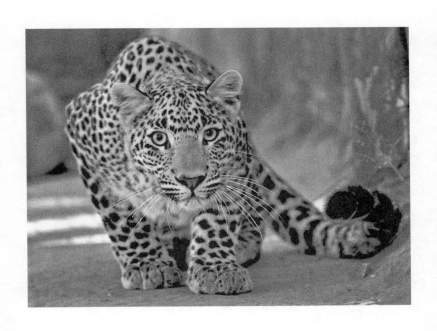

4

A SPOT OF TROUBLE

What Makes a Leopard a Man-Eater?

The road to Pauri Garhwal carries travelers through the Middle Himalaya—the sparsely settled, mildly mountainous realm between the low foothills and the airless white monsters of the Greater Himalaya. It's a lovely drive but, as road signs will say to you, HIGHLY ACCIDENTAL PRONE. Landslides happen so often that the sides of some mountains, at a distance, look like ski areas. The higher you drive, the steeper the slopes and sharper the curves, to the point where you're forced to take each bend blind, honking and bracing for impact.

The road traces the old pilgrimage trail that links the Hindu shrines along the Ganges river. In past centuries, devotees walked the trail barefoot and slept in simple thatch shelters. The danger then was not crashes but leopards, which had a documented proclivity for slipping through an unsecured doorway. Between 1918 and 1926, government recorders ascribed 125 kills to a single leopard, known in the global media of the day as the Man-Eating Leopard of Rudraprayag.

The shrines remain and the people still come, but now they drive and book hotels. Modest accommodations line the road: Hotel Nirvana, Om Hotel, the unrelaxing-sounding Shiv Hotel. Our driver today, Sohan, is friendly and seemingly imperturbable, steady in the face of all that India drops in our path: the cows, the rockslide rubble, speeding motorcyclists, a dilapidated loom. I've seen the serenity lapse only once, at the sight of a man urinating over a guardrail. Sohan drives so competently that I've stopped fretting about the nonfunctional seat belts in the backs of Indian cars, a preoccupation Naha finds amusing. (Naha on airbags: "You mean that balloon that comes up?")

We'll be visiting three mountain hamlets today and tomorrow, all of them leopard attack hot spots. If there's time, we'll go to Rudraprayag, which has grown by now to a small city. A monument stands on the spot where the famous man-eater was killed by the famous man-eater hunter Jim Corbett. Naha made a pilgrimage of his own last year, to look for descendants of Rudraprayag villagers who may have been alive then. He leans into the back seat to show me photographs. The Corbett monument needs repair. The pedestal is cracked, and the great hunter's mustache is chipped. Naha managed to track down a grandson of the village priest Corbett worked with. One of the things the man told him was that when a leopard stalks and kills more than three or four people, villagers consider it a demon.

I don't buy in to demons, but I do wonder what's going on with these leopards. In North Bengal, where we've just come from, leopard attacks are inadvertent encounters. After a quick scuffle, the surprised cat takes off. There may be injuries, but rarely are there fatalities. Here in Uttarakhand's Pauri Garhwal district, leopards stalk humans as prey. Every year, in an area smaller than Delaware, leopards typically kill three or four humans. Between 2000 and 2016, Naha reports, leopards attacked humans 159 times. A large majority of these attacks were predatory, he says.

What causes a species to update the menu? What happened in Pauri Garhwal?

Corbett blamed the flu pandemic of 1918. So many people died so quickly, he wrote, in *The Man-Eating Leopard of Rudraprayag*, that the Hindu funerary custom of carrying bodies to the Ganges for cremation was for a time replaced with a more expedient rite. A burning coal was placed in the mouth and the corpse cast down the hillside in the general direction of the river. A leopard will readily scavenge a meal, and Corbett surmised it was these bodies that gave the Garhwal carnivores their taste for human meat. Likewise, the man-eater of Panar—another Jim Corbett project— began its killing spree in the wake of a cholera outbreak. Corbett claimed this animal had killed four hundred humans by the time he stepped in. (The figure is questioned in *Leopards in the Changing Landscapes*, by H. S. Singh, formerly of the Gujarat state forest department, and by others, including Naha. Corbett stalked book sales as skillfully as he stalked big cats.)

One thing that may work in favor of the North Bengalis is that their leopards were heavily hunted by the British Raj and their royal Indian cronies. Singh wrote that between 1875 and 1925, hunting parties killed 150,000 leopards. "They may," Naha says, "still regard humans with fear." Now that my state's mountain lions (aka cougars)* are no longer bounty-hunted, are they

---

* "Mountain lion," "cougar," and "puma" are regional names for the same species. Florida calls them "panthers," and in South Carolina they're "catamounts." The name "Rowdy" applied to just one, a cub captured by Clark Gable on a 1937 hunting expedition. Rowdy was one of two cubs Gable intended to bring back to surprise paramour Carole Lombard, who had jokingly told him to bring her back "a wildcat or two." According to Stanley P. Young, co-author of *The Puma,* Rowdy escaped the first night, in his new collar with the engraved nameplate (only to be bagged a year later, collar intact, by a mystified hunter). Rowdy's sibling was presented to Lombard and soon thereafter donated to the MGM Studios zoo. Lombard had earlier given Gable an enormous ham with Gable's face pasted to the wrapper, so the cougar appears to have been the victim of misguided gift-giving one-upsmanship.

becoming less wary of humans? I posed the question by email to California mountain lion researcher Justin Dellinger (whom you'll shortly meet). He didn't think so. Mountain lions are not so much wary as stealthy—a trait that evolved over eons, probably because it made them successful hunters.

The hills of Pauri Garhwal have a terraced, wedding-cake appearance. Terracing creates level spaces for hillside farming, but I see no farmers and, when I look closer, no crops. Naha explains that the region has undergone a sizable out-migration. Villagers have left to look for work in cities, because almost anything is easier and more profitable than farming the hills. Terraces are difficult to irrigate, and when the crops mature, monkeys and wild boars show up to raid them. In the decade between 2001 and 2011, 122 villages were abandoned. You see it in the scenery: mile after mile of fallow terracing. It's like driving through a topographical map. In places, the contour lines have begun to soften as the native vegetation returns. This "rewilding" creates scrubland that serves as cover for leopards on the hunt. Nearly 99 percent of the villagers Naha interviewed for a paper published in PLOS ONE reported a belief that this has brought leopards closer to people's homes. Seventy-six percent of leopard attacks on residents of Pauri Garhwal happened in a spot with medium or dense shrub cover.

As the people have moved out, more livestock have been left to graze unguarded: easy pickings for leopards. Naha makes the point that chasing prey on steep slopes is, like farming on them, a challenge. Goats and calves make dinner a snap. Compared with deer and other natural prey, domesticated animals are slower and less wary. ·

As are human children. Forty-one percent of the Pauri Garhwal villagers killed in leopard attacks, Naha's survey shows, were between one and ten years old. An additional 24 percent of the fatalities were youth between eleven and twenty.

Here Sohan joins the conversation. Aritra is dozing, so Naha translates for me. "He has seen this." It was 1997. A thirteen-year-old girl was alone in a field, cutting grass with a sickle. It was around 4:00 p.m. Sohan was resting in a car a few yards away when a leopard appeared. "He saw everything in front of his eyes," Naha says. "The leopard attacked from behind. It jumped on her back and pierced her jugular. There was blood coming out. It was very harsh."

I ask Sohan about the physical condition of the leopard. Did it limp? Was it old or underweight? Jim Corbett, in another of his memoirs, put forward a theory about man-eating Bengal tigers that held that they were sick or injured, and that they went after humans because that's all they could bring down. (Like the cougar that attacked Ben Beetlestone.) "With the leopards of Pauri Garhwal," Naha says, "this has not been the case." He scratches the side of his face. His once-precise beard line is rewilding.

Sohan agrees. After the killing, the girl's family was persuaded to leave her body, briefly, where it lay. The leopard would return to its kill, and local hunters had been called. The body was chained to an iron stake, and the hunters waited. It was Sohan who drove the leopard's body to the forest department for necropsy. Other than the gunshot wounds, the animal had no injuries, and no teeth were broken or missing. "It was in perfect condition."

More often than not, Naha says, man-eaters are females with cubs to feed. In his book, Singh objects to the term "man-eater," because it suggests an animal "gone 'mad.'" It blames the leopard rather than the changes humankind has brought—the rapid disappearance of forests and the prey that lived in them. Besides, he points out, to a carnivore, meat is meat. "Big cats eat all varieties of flesh. . . . So why not human beings? Who gave these magnificent big cats this derogatory tag to begin with?" he asks. And then answers: Jim Corbett.

Naha's wife, Shweta Singh, also a wildlife biologist, has joined us for this leg of the trip. The couple met at the Wildlife Institute of India, where both are employed, but Shweta did not come along for her work. She came because it's beautiful here and the air is cleaner, and because this week is Diwali and she'd like to spend it with her husband. Shweta is a little younger than Naha and lighter of spirit. She and he share a passion for research and the wild places it brings them to. She and I, for the colorfully packaged snack foods of India, bags of which you find at every roadside bodega, hanging in strips like linked sausages.

Sohan has pulled over in a small town. We're stopping for tea (and two bags of Masala Munch). Below the cafe window, a steep incline bottoms out at the Ganges. The river's glacial origins are evident in the water's chalky aqua hue. It's noticeably colder up here. The women wear cardigans under their saris.

Naha is filling me in on forest department policy for leopards that harm or kill humans. Unlike in the United States and Canada, where the animal would be, as they say, "lethally removed," here a distinction is made between defensive and predatory behavior—or as Naha puts it, provoked versus unprovoked. If a particular leopard is known to have killed and fed on three or more humans, only then will the chief wildlife warden of the state formally declare it a "man-eater," whereupon hunters or forest department staff may shoot it. How do they know it's the work of one leopard? They set up wildlife cameras in the area and learn to recognize the local cats and their turf. (They keep them straight by their spots. The pattern of a leopard's rosettes is unique, like a set of human fingerprints.)

Declared man-eaters are no longer translocated. Here the thinking mirrors that of North American wildlife agencies. If you move a leopard and it kills a person in the new location, now you, the forest department, may be held liable. The human-leopard conflict researcher Vidya Athreya reports that translocation itself makes attacks more likely. After forty leopards were moved to a forested area in Maharashtra in 2001, the average number of attacks per year jumped from four to seventeen—and not just because there were more leopards in the area. Athreya attributed the rise to two factors. First, the leopards had lost their fear of humans during the period of captivity leading up to the translocation, and second, the stresses of being captured and released into unfamiliar territory had made them more aggressive.

On top of everything else, translocation is, as with black bears, likely to be a temporary solution. Remove one leopard from its turf, and another soon takes it over. Often the new arrival is a subadult recently dispersed from its mother. That can mean trouble, in that inexperienced hunters are more inclined to go after easy prey.

If not translocation, what, then? What does a forest department do with a leopard that has killed just once, or is preying on livestock? Naha pulls his attention from a nature program on a TV mounted on the cafe wall above our heads. A narrator is sharing facts about shrews.

"It can be trapped and kept in captivity," Naha says. When I ask where, he uses the word *zoos*. I ask about the possibility of visiting one. "They are not open to the public."

So, not really a zoo. I'm trying to picture it. "Open spaces, like a fenced preserve? Or cages?"

Naha runs his hands through his hair. The side of his head has been styled by the four-hour blowout of an open car window. "Both. They are enclosed for some time, and then released into a bigger area for exercise."

"Like a prison." Naha doesn't argue with the comparison. The leopards do time.

Later, tooling around the internet, I found a master plan for an upgrade to what appeared to be a similar place—the South Khairbari Rescue Centre in West Bengal, not far from where we were last week. Twenty-five "night shelters" opening onto a paddock area. It was not for conflict leopards per se but for rescued circus tigers and orphaned tiger and leopard cubs. This facility, too, was closed to the public.

The heading "Disposal of Solid and Liquid Waste" caught my eye. "Faecal matters" were treated differently than carcasses, which went to a Bone Collection Unit. "The bones so collected are disposed of by way of selling." Was the government of West Bengal taking part in the illicit trade in "medicinal" wildlife parts? Given the high price of tiger and leopard bone, I wondered about the temptation to change a life sentence to a death sentence.

As in the United States, there are people who object to the government's policies and take matters into their own hands. But where we have, say, Californians freeing bears from Fish and Wildlife culvert traps, here in Pauri Garhwal the anger runs in the other direction. Villagers want "man-eaters" killed, and they don't want to wait for a second or third victim. A mob mentality can quickly take hold. In one of the hamlets on our route, a leopard had recently killed two people. Residents did not contact the forest department but set a cage trap themselves. A young man who spoke some English brought us to the spot where the animal had been trapped. "There was too much angerness," he said. "So due to this angerness, he was burned, the leopard. In the cage itself." After a quiet moment that I mistook for melancholy, he held up his smartphone. "Can I selfie?"*

Shweta knows this case. Her forensics lab at the Wildlife Institute of India received the evidence and the remains—in this case, "ashes, and stones with blood." Forest department staff will

---

* Proximity to the border with Tibet/China means a heavy military presence, and that means military-sponsored cell towers and one of the many curious contrasts of modern India: villagers who have smartphones before they have indoor kitchens.

endeavor to be sure they've caught the correct leopard, but villagers will kill whatever steps into the trap. "They don't know if they have the right animal or not," Shweta says. "They just want to take their justice." Had things been done properly, officials would have first tried to match the DNA of the trapped leopard with DNA found on the victim's skin or under the nails. (Linkage!)

Naha glances around the cafe. "This is not a good place to talk about this." Though few here speak English, they know the L-word. This morning I noticed that the government ID placard is gone from the windshield.

A similar conundrum faces wildlife managers in American states that have three-strike or two-strike laws. If the state were to adopt a never-kill policy, ranchers would likely go back to killing the predators themselves—a practice known among wildlife professionals as "shoot, shovel, and shut up."

Again: too much angerness. "If you're a rancher, your life is sheep," Stewart Breck said, back in Aspen. "There's a strong emotional component." Breck has six llamas. You would not say that his life is llamas, but he will not forget the day he walked behind his house and saw two of the neighbor's dogs at the neck of one of his llamas. Here in the Himalaya, it's not just people's livestock being killed but also their family members.

Naha takes his jacket from the back of his chair. "Let's go."

By noon, we're deep in leopard country. Naha is pointing out the car window at the place where an eleven-year-old was attacked and killed walking home from school. The last five miles have been a stop-and-start narration of death, in Naha's streamlined monotone.

At a bus shed on a lonely stretch of road near the village of Kolkhandi: "One elderly man was sitting there, and he gets attacked."

On the road to Ekeshwar, our next destination: "Here were

two attacks. An old lady, at five in the morning. And in the same spot, three years back, a person, thirty-eight years old. Returning from the fields."

Across from a field that abuts a forest in Maletha, near Ekeshwar: "Fifteen or sixteen people were there, all cutting grass. It was a most audacious attack. One lady was taken. Right in the daylight."

Sohan pulls onto the shoulder. The road to Ekeshwar is too narrow for our vehicle, so we'll park here. Naha pulls his bag from the back and shuts the door. He points to the hillside that slopes down to the village. "Here a lady was attacked and eaten. Late evening. 2015."

Along the half-mile walk, we pass a woman walking with a sickle in her hand. "See that lady," Naha says, tipping his head in her direction but not pointing, as though he's about to share some juicy item of village gossip. "She's going to the forest. She is taking a risk, going alone to cut grass." She has no choice. Winter snows are coming, and the cows will need hay to eat.

Local government in Pauri Garhwal takes the form of a "head system." Win the trust and support of the village head, and/or the priest, and your job is vastly easier. Naha has been back and forth to this area many times, creating relationships with both, and it's been paying off. We stop first at the home of Ekeshwar's village head. He is out, but we are welcomed by his brother Narender, a tall, gap-toothed man wearing flip-flops, one maroon and one gray, despite the chill. He invites us in, or rather up. It rains so infrequently this time of year that rooftops become living spaces. Red chiles have been spread in the sun to dry. A satellite dish is held upright by an arrangement of rocks.

Shweta translates for me. "He likes leopards even though they sometimes take his livestock. He says this is its natural prey, and he is acceptable to that. He does not support anyone who does killing." In the words of an American rancher I met last year who is also, improbably, a mountain lion activist, "When you have livestock, there's going to be some deadstock."

Naha has asked Narender and his brother to nominate villagers for a wildlife response team. This would be the same sort of early responder team Naha has been helping to establish in North Bengal, but here it's humans, not animals, that the team tries to control. Most of the nominees are army veterans, because they are respected in the community and because they are, Naha explained, "capable of controlling a mob." I've seen the equipment list for the team members. It includes "polycarbonate riot shield of 3–5 mm thickness" and "fibre stick used by Police Department."

Shweta points out that people's ire is for the government as much as for the leopards. If there were school buses, children wouldn't have to walk two miles at dusk, when the risk of a leopard attack is greatest. If there were hospitals and ambulances, an attack might not mean a life lost. But there are not. A leopard is an expedient outlet for their anger.

Naha has held awareness camps at many of these villages. He encourages parents to have their children walk home from school in groups. He tries to discourage people from dragging their dead livestock onto the road for the vultures, because the carcasses also attract leopards. Attitudes and behavior change slowly in a small village like this. Twenty years ago, Naha recalls, there were cases of Pauri women being nabbed by leopards as they squatted in the brush to relieve themselves at night. Indoor toilets were eventually built, but people wouldn't use them at first. "Slowly they are understanding it's okay to shit indoors."

Naha goes to check on a light he installed on an earlier visit. It's part of a controlled study to assess whether "fox lights" help keep leopards away from people's houses. These are solar-powered lights that turn on and off randomly to mimic the sight, at a distance, of humans patrolling with flashlights. It is shaping up to be an effective, if temporary, solution. To forestall habituation, the lights should be used intermittently. To get people to understand this has been difficult, Naha says. They want to leave them on continuously. Stewart Breck has had the same challenge with some of the ranchers who try fladry—fluttering ribbons tied to livestock fencing wire to spook coyotes and wolves. When they see that it works, they leave it up, rather than restricting its use to calving season and other periods of heavy predation.

This year Naha has been encouraging village heads to apply for funds from the Mahatma Gandhi National Rural Employment Guarantee Scheme. The money would allow the village to hire someone to keep the brush trimmed back from around the houses, and to build secure nighttime enclosures for livestock. These are the same suggestions that more progressive USDA Wildlife Services operators give to property owners who call because they want a mountain lion killed for preying on their livestock or pets. What if Wildlife Services made these things a requirement rather than a suggestion? Better yet, what if they arranged and paid for the brush-clearing, or for the enclosures to be built? What if non-progressive operators had to start being progressive? I ran this by Stewart Breck, by phone, when I got home. He did not see it becoming policy. "It's more of a philosophy."

I exhaled in a snorty, dismissive way. "That Wildlife Services pays lip service to?"

"Let's just say the ship is slow to turn. But it is turning."

The day ends high on a hill in the village of Khirsu, where the wildlife institute leases a house. It's unfurnished and unheated, but the view across the valley is a balm after the hours in a car. The slope directly behind the house is forested, and people have tied bunches of grass around the tree trunks to dry, out of reach of free-ranging cows. Aritra stands beside me on a balcony, intent on the hula-skirted trees. He hears something.

Naha watches his cousin. "Aritra is scared it's a leopard about to jump down from the hill." He points to a patch of sandy ground on one side of the house. "If there were a leopard around, there would be pugmarks there."

A large langur, black-faced and beefy as a baboon, drops from a tree and bolts across the slope. Scares the stuffing out of Aritra and me. "And there," deadpans Naha, "is Aritra's 'leopard.'" He goes inside to help Shweta fix dinner.

To enjoy the view (and because we have no table or chairs) we eat outside on a patch of concrete in front of the house. Shweta has built a cook fire and rigged it with a spit made from branches. Someone opens a quart bottle of Godfather Super Strong. By the time we finish eating, it is past ten. Shweta keeps the fire going. Naha is talking about the Aghori, Hindu monks who practice ritual cannibalism. His words are cut off by a bellowing outcry, louder and more jagged than a scream, human but barely. It's the sort of thing horror movie sound effects people aspire to but seldom attain.

"Leopard!" says Aritra. *"Leopard!"* He doesn't mean that a leopard is making this sound. He means a leopard is attacking someone. I know this is what he means, because it is absolutely the sound I, too, would imagine coming out of a person being killed by a leopard—terror plus pain plus the squeezing of vocal cords by clamping jaws. It's unfolding directly below us, at the bottom of the hill where the path from the village shops meets a cluster of houses. "Get up, get up!" We do, we get up, and we stand there, freaked and listening, trying to understand what's happening. Leopard? Insane person? Drunk? Aghori? Down below, other voices join, but they don't sound like people watching a leopard kill a neighbor or trying to stop it. Soon the voices trail off, as the afflicted entity is led or carried away.

It's late and the path is steep and unlit. We'll inquire in the morning.

We pack after breakfast and begin the precarious walk down the trail to the car with our stuff. Cook smoke rises from the houses below, and Himalaya morning sounds—women sweeping, men coughing, cowbells. At the bottom, Naha stops to talk to a woman standing in her doorway, to find out what was going on down here the night before.

He catches up with us at the car. It wasn't a leopard or a drunk. "It was a case of demonic possession," I hear him say in his tossed-off Naha way, as if it were no more remarkable than a sprained ankle.

Does this sort of thing happen often around here?

Aritra pushes a heap of sleeping bags deeper into the rear of Sohan's hatchback. "At least monthly," he answers after some thought.

I think back to what the priest from Rudraprayag said to Naha about leopards that kill three times—that they are demons. So maybe a leopard after all.

Our destination today is the city of Dehradun, home of Naha and Shweta and the Wildlife Institute of India. We are leaving behind the demons, but not necessarily the leopards. In 2009, an emaciated leopard showed up in Dehradun. Nineteen people were injured before the animal was shot.

Because most Indian leopards that visit a city do so at night, nabbing a stray dog or scavenging garbage and returning to the forest before daybreak, the excursions go largely unnoticed. It's when the big cat lingers past dawn that trouble sets in. H. S. Singh's book includes summaries of forty-three cases of "leopards straying in cities." They do the things people do. They visit temples and stroll college campuses and go to the hospital. One afternoon a leopard showed up at the Central Institute for Cotton Research. In a township outside Chandigarh, a woman came home to find a leopard asleep on her bed in front of a television program. In 2007, a leopard that had been seen for several days in and around the city of Guwahati was eventually trapped inside an upscale shopping complex, where it had been observed "prowling near an ATM," as though out of cash.

When things go well, the leopard is quickly tranquilized and released in nearby forest. More often, the story unfolds as it did in the case of the leopard who jumped the wall of the gated compound belonging to Bollywood star Hema Malini:

1. The person who first encounters the animal runs or hides. ("The gardener and the watchman locked themselves in a room.") In a common variation, it's the leopard that is locked in a bedroom or bathroom.
2. Police are summoned and, lacking tranquilizing darts and training, prove to be of minimal help. ("When one of the police teams tried to enter the house, the leopard growled at them, after which they waited for the forest department to arrive.") Though to be fair, some officers have proved to be skillful improvisers. Police responding to a call about a leopard in a suburban Delhi plywood factory hoisted a netted cricket batting-practice cage and threw it over the cat.
3. Long before the forest department arrives (a four-hour wait, in the Hema Malini case), the leopard makes a run for it. Malini's leopard escaped unharmed.

The Wildlife Institute of India sits on the periphery of a small forest, but Naha doesn't recall the Dehradun leopard. What they do have roaming around the institute are rhesus macaques, dozens of them. And a troop of macaque researchers, some of whom I'm planning to see tomorrow.

We're halfway back to Dehradun now. Shweta has her earbuds in, nodding to her tunes. Aritra is trying to explain Hinduism to me. As we drive down out of the hills, the black-faced blond-gray langurs we'd been seeing on rooftops and in trees are gone, and in their place are smaller, pink-faced monkeys with light red fur—rhesus macaques. These monkeys come right down to the road. They sit on the cement slabs that serve as guardrails, waiting for handouts or trash thrown from car windows.

Everyone in northern India has macaque stories. Naha awoke one morning to find one sitting on his chest. After a brief, intense stare-down, he yanked the blanket up between them and the monkey ran off. Another time, a macaque came up the back stairs of their apartment building and leapt onto the kitchen counter. "It could have just eaten some things and left, but no. He took the induction cooker and threw it on the floor. Just came in and did that and left. That made his day."

I've heard they'll snatch a person's sunglasses. "Yes," Naha says. "Or their cellphone. And then they drop it from the tree. Their main purpose in life is to harass people."

Shweta pulls out her earbuds. "They do this because this behavior is rewarded. When a monkey takes a phone, the person comes back and offers it food, because they know the monkey will take the food and give up the phone."

Naha isn't having it. "Shweta, remember they came to the terrace that time and flipped the flowerpot upside down?" He turns to me. "And then they'll go and shit right there. They feel good when you are harassed."

Shweta puts her earbuds back in.

Naha looks out the car window. "Definitely they do."

5

THE MONKEY FIX

Birth Control for Marauding Macaques

The Wildlife Institute of India is a thrown-down cluster of concrete buildings connected by outdoor walkways. Because these corridors have no walls, rhesus monkeys from the neighboring forest can occasionally be seen walking along behind or beside the humans. Neither species pays the other much mind, as if the monkeys, too, have meetings to get to and photocopies to make. This nonchalant coexistence is in contrast to the state of human-monkey relations elsewhere in India.

"Simians Lay Siege to Agra," blares a headline in the *Times of India* the week I arrive. It's from a multi-feature full-spread special section, complete with signature two-color MONKEY MENACE graphic, the letter O taking the form of a fang-flashing monkey head. The lead story describes a baby fatally wounded after a rhesus macaque snatched him from his mother's breast. "Earlier this month," another *Times* piece states, "a bunch of monkeys had stoned a 72-year-old to death." The *National Herald* has had Agra's monkeys "in armies marching from one area to the other." During the eleven months I've had Google on alert for monkey news in Delhi and Agra, Indian newspapers have reported eight lethal macaque "attacks."

The past decade has seen a minor epidemic of people plummeting from balconies because of monkeys. I found news accounts of six deaths in the last three years alone. Most famously, there was the 2007 fall of Delhi deputy mayor S. S. Bajwa. While taking the air, Bajwa was startled by a group of macaques set on storming the house to look for food. As he tried to stop them—or get away from them (there were no eyewitnesses)—he lost his footing and hurtled over the railing.

While I question the hostile intent that the word *attack* suggests, a monkey home invasion is surely an unnerving experience. Each evening on a recent visit to Udaipur, I would sit in one of the city's many rooftop restaurants and watch langurs and macaques appear out of the dusk to begin their evening depredations. They sprinted up fire escapes and leaped from building to building like Tom Cruise in righteous pursuit. One night, making my way through an unmemorable dal, I looked up to see a langur vault onto a decorative beam above my table. Had the waiters not kept a stick close at hand for scaring off monkeys, the meal would quickly have become more memorable. When a forty-pound monkey abruptly drops in, you move without thinking. If you happen to be on a balcony or a roof, you too may shortly be dropping in from above.

The Wildlife Institute of India, according to several of these articles, is working on a contraceptive vaccine. The *Times of India* wrote of a shot that would " 'sterilise' the animal within minutes." Here is the dream! Easily administered, long-lasting birth control for overpopulous, problematic wildlife. Before leaving for India, I was unable to confirm an appointment date by email with Qamar Qureshi, the institute's director of research, and he has been out for the Diwali holiday. Naha agreed to escort me onto the grounds when we got back to Dehradun, so I could pester the man in person.

It's ten past nine on Monday morning. I'm waiting for Naha at the front gate. The guard wheels an office chair out into the sun for me. He is dressed in a uniform with a fringed hip sash and a magnificent red plumed beret, as though it were royalty under his protection, not wildlife biologists. Just past the gate there's a small grounds building surrounded by chain-link fencing and topped with concertina wire. A macaque walks casually through the barbed loops.

Naha crosses the institute lawn to retrieve me. As we walk to the main building, he explains that Qureshi is in Monday morning meetings. He walks me to Qureshi's office and promises to let him know I'm waiting.

Qureshi's desk accessories are wildlife-themed—a zebra-stripe pencil cup, a tiger-stripe water bottle. To my left, a sliding glass door opens onto a terrace that, yes, macaques have twice used as an access point, ransacking desktops in a pandemonium of airborne papers and office supplies and then, finding nothing edible, dashing out the way they came. The incidents were related to me, mostly in pantomime, by a man who sits at a desk at the back of Qureshi's office. He speaks no English, and his function is unclear to me. He wears a striped short-sleeved dress shirt and a vest, also striped. It's all stripes all the time in here.

A secretary walks in and places two open file folders on Qureshi's desk. Forms are flagged with Post-it notes. "Otherwise he will sign anywhere!" She laughs. "Does Sir know you're here?"

"Oh, yes. He's in a meeting."

She makes an ominous clucking sound. "His meetings are indefinite. I wish you luck." Behind her, the striped man is tilting into sleep. A macaque minces along the edge of a rooftop at the far side of the courtyard.

Qureshi arrives around 11:00, with a few of his researchers. He is lean and tall and has a warm, sociable demeanor. In place of a perfunctory "How are you," he gives me "How are you finding India? How's your tummy?"

Before delving into the science, we talk more generally about India and its animal predicaments. "The entire country is almost a sanctuary," says Qureshi. He signs as he speaks, working his way through the Post-its. "In the sense that our laws are quite stringent." Since the 1972 passage of the Wildlife (Protection) Act, it has been illegal to kill or capture wild animals without a permit or a state declaration that a particular species is "vermin." Qureshi glances over the top of his reading glasses. "And people are for this."

Hindu deities often take the form of an animal, or they are part animal, or parts of several animals, or their spouse is an animal, or they ride around on one. I share the story of my first visit to Delhi, when a live rat dropped from someplace above the sidewalk and landed on my foot. "You are blessed!" proclaimed the man I was walking with. "The rat is the conveyance of Lord Ganesha."

Qureshi's researchers have been listening intently. "Everything is a deity!" hoots a project fellow named Uddalak Bindhani. "*Basil* is a deity! It is one of Vishnu's wives."

"If you think about it," says Divya Ramesh, a young behavioral ecologist with an easy smile and a pierced eyebrow, "it's a really nice thing, because people have this great association with nature."

But even Hindu tolerance has limits. Especially if the individual is a farmer. India's top agricultural pests also happen to be sacred animals. Elephants represent the deity Ganesh, and monkeys, Hanuman. The wild boar is an avatar of Vishnu. The nilgai, or blue bull, is actually an antelope, but *gai* means cow, and cows are sacred. When state officials wanted to begin culling the animal, they first pushed through a name change. Nilgais are now roj, or "forest antelope."

Despite the endlessly recurring "monkey menace" media

froth, the government neither of Agra nor of Delhi has declared macaques vermin. And if they did, they'd be hard-pressed to find exterminators. "You'll not find anyone to kill a monkey," says Nilanjana Bhowmick, a Delhi-based journalist I've recently met. The Veterinary Services department of Delhi's municipal government has a hard enough time finding people to catch macaques even just to relocate them, which is the current strategy. Even non-Hindus avoid the job, as monkey catchers are often harassed and threatened.

Deepening the problem: offerings. Tuesdays and Saturdays, devotees visit Hanuman temples to make a *puja*. To the icons inside, they present coconuts and garlands of marigolds; to the living representatives hanging around outside, samosas and Frooti mango pop. Feeding wild animals, as we know, is the quickest path to conflict. The promise of food motivates normally human-shy animals to take a risk. The risk-taking is rewarded, and the behavior escalates. Shyness becomes fearlessness, and fearlessness becomes aggression. If you don't hand over the food you're carrying, the monkey will grab it. If you try to hold onto it, or push the animal away, Qureshi says, it may slap you. Or bite you. The *Times of India* put the number of monkey bites reported by Delhi hospitals in 2018 at 950.

Qureshi recalls visiting a Hanuman temple in Himachal Pradesh, where he had gone for a wildlife meeting. His hosts warned him not to carry anything valuable, because the macaques would grab it to ransom for food. Qureshi left his phone and wallet locked in the glove compartment of his car. "This one guy came and—" Qureshi stands and pulls his pockets inside out. "Really! They put a hand in your pocket and properly search you!"

I have a macaque story too. Mine is set near Bundi, in Rajasthan, on a trail through the brush to the ruins of a fourteenth-century fort above the city. I knew there were monkeys up there, because their silhouettes are visible along the parapets at dusk. I went in the morning. I had bananas. I was asking for it. I wanted to know what it was like to be mugged by monkeys. My friend Steph followed behind to document the crime with a tightly clutched iPhone. The first shot is of me, looking down, preoccupied with my footing, orange plastic produce bag dangling from one hand. But look closer. A little tan head has popped up from behind a boulder farther down the trail. Out of the camera's view, another monkey lurks. Outlaws lying in wait for the stagecoach. As I approach the boulder, the first macaque steps into view. While we stand there sizing each other up, the other one shoots out from behind me and grabs the bananas. Slick! I wouldn't call it an attack. It was more of a purse-snatch, over too quickly to create any fear.

In 2008, the Delhi city government passed legislation prohibiting the feeding of wild monkeys, but according to one news story, no fines have been issued. Outside the Hanuman temple in Delhi's Connaught Place, I watched a man approach a group of macaques. There was a side-glancing furtiveness to him, as if he were approaching prostitutes, not monkeys. He quickly handed over a bag of tomatoes and watched as a portly female sat on her haunches, expertly freeing the pulp and sliming the pavement with the skins. A temple employee watched it happen but did nothing.

Because, Qureshi insists, the employee understands the importance of the gesture. "You want to go to heaven? Then you feed them. You want to book a berth, a nice house there? You feed."

"And these are the same people," adds Ramesh, "who are crying, 'Get rid of these monkeys!'"

Qureshi closes a folder and sets down his pen. "Many people, when you interview them, they say, 'Don't kill them!' They just want them to disappear." It's the same everywhere: wildlife NIMBYism. Squirrels in the park are adorable. Squirrels digging in your planters are deplorable.

Qureshi adds that the other problem with government-controlled culling—referring here to the shooting of wild boar and nilgais—is that while it is permissible to kill them, the law forbids eating the meat. "And here"—he means India—"you don't kill a species for the sake of killing. Only a psychopath does that."

The great hope is that science will come up with a way to impose birth control on problem animals. It's true that Qureshi's team has been working on an immunocontraceptive vaccine for macaques, but it's not true that it would render the animal sterile "within minutes," as the *Times of India* reported. Nor would it be orally administered, as other news outlets have reported. Qureshi rests his elbows on the desk. "A birth control pill for monkeys is a far-fetched dream." You'd need to make sure enough animals were eating enough of it on a regular basis and somehow prevent other species from doing the same.*

Oral contraceptives are most practical in a controlled, single-species condition. Like a sewer. To control Norway rats, some U.S. cities have begun using an oral contraceptive called Contra-Pest, which relies on two active ingredients. The first, VCD (4-vinylcyclohexene diepoxide), depletes the ovaries of eggs. VCD began its career as an industrial plasticizer, but when human health and safety tests revealed it to be an endocrine disrupter, it

---

* Researchers in the UK are gearing up to attempt it, though not, of course, with monkeys. After decades of mass extermination efforts, a National Anti-Grey Squirrel Campaign, and an order of Parliament failed to eliminate "this most undesirable alien"—to quote his lordship the Earl of Mansfield at the June 29, 1937, meeting of the House of Lords—science is stepping up. Researchers have been testing immunocontraceptive bait set out in tree boxes designed to exclude the nation's beloved red squirrels, whose turf the gray immigrants have steadily overrun. So far, as determined by biomarkers in the bait, most of the targeted squirrels in test forests seem to be taking the dose. Good luck, Britannia. Only 120,000 redcoats remain, while the American invaders number in the millions.

was repurposed for rodent birth control. Because VCD takes time to work, a second compound was added, this one for the boys, too. Triptolide impacts sperm and egg viability for as long as the animals consume it. It's not yet clear that the two together reliably confer permanent sterilization to a community of rats, but a few American cities are giving it a go. It would not, however, seem to be a solution for widely roaming monkeys with plentiful dining options.

The Wildlife Institute of India is running a trial of an injectable immunocontraceptive vaccine called PZP (porcine zona pellucida). The zona pellucida is a protein coat with sperm receptors that surrounds the egg. Give the female a vaccine of foreign (pig, say) zona pellucida and her immune system will be primed to make antibodies against her own zona pellucida. These antibodies glom onto the receptors, leaving sperm with no access to the egg. Fertilization stymied.

Logistical hurdles abound. Like many vaccines, PZP requires booster shots to keep the immune system on alert. This is, of course, a challenge with free-ranging animals. It's time-consuming and costly enough to round up and inject a population of animals the first time around. Administering boosters adds to the time and expense, and on top of that you need some sort of permanent marking—a tattoo, say—to enable booster-givers to know who's had the first shot and who has not.

In the United States, a synthesized zona pellucida has been tried—mainly on populations of geographically confined animals. The wild horses of Assateague Island were good candidates, because they move in herds and it's a small island, so it's relatively easy to give all the injections at one go. And then do it all over again in three to six weeks, after which a yearly booster seems to do the trick. For the tens of thousands of wild rhesus monkeys that roam Indian cities, it makes little sense even to try.

Here's the other problem with zona pellucida vaccines for

monkeys. Females that don't become pregnant will quickly cycle into heat again, and each time they do, males will respond with breeding-season behavior. Meaning they're more aggressive more of the time—not only toward other macaques but, it's believed, toward humans. This happened with white-tailed deer in some U.S. trials of PZP. The bucks weren't aggressive toward humans but they roamed around more, looking for sex, and their wanderings took them across roads and highways and that wasn't good for deer or for drivers. Partly for this reason, immunocontraceptive research in the United States has focused on a vaccine that instead blocks the effects of sex hormones. GonaCon stops females from cycling. After an initial injection and a single booster shot, 92 percent of mares in North Dakota's horse-dense Theodore Roosevelt National Park remain infertile seven years later. The study is ongoing and the hope is that infertility will prove to be permanent.

Is there any immunocontraceptive vaccine that might confer permanent sterility from a single shot? The National Wildlife Research Center and the U.S. Bureau of Land Management are currently testing one on a subset of a population of wild horses that has outgrown what its rangeland can support. The shot contains two active components (BMP-15 and GDF-9). Antibodies against these components hinder the egg's ability to communicate with the cells that surround and support it, so it never matures. Because this vaccine wouldn't require marking animals and tracking them down for a booster, it would seem to hold promise for treating urban macaques.

Qureshi sees a broader problem with immunocontraception, or any form of monkey contraception, for that matter. That is, people will expect the problem to start going away as soon as the treatment starts. "But you are not killing the animal," he says. "They will live their life." City macaques live twelve to fifteen years. Qureshi estimates it would take seven or eight years before the population drops enough for the effects to become noticeable to the average macaque-aggrieved Indian. "People will say, 'You spent all that money and the problem is not gone?'"

Qureshi politely excuses himself. He has another meeting to go to. Ramesh walks outside with me to hail an autorickshaw and hops in beside me for the short drive to my hotel. Along the way, we pass a dry streambed where people dump garbage. Each time I've passed this spot, pigs or monkeys are picking through the mess. I ask whether there are efforts, as in Colorado, to get the garbage under control.

Ramesh answers that a cleanup campaign is underway. Where before there were only community dump sites, now the city has garbage trucks that drive from building to building. In one locality, the truck plays an inspirational ditty.

I ask how it's going. Ramesh laughs. "The buildings are super multistory, so nobody bothers to go down and put it in the bins. They just drop the bag out the window."

Humans.

One mildly satisfying element of India's "monkey menace" is that the afflicted are often, for once, the upper classes. Urban monkeys prefer parks with trees and other landscaped spaces—the habitats of the well-to-do. From the boughs and greenery, they soon find their way onto roofs and terraces, through open windows, and into the buildings. They raid the mansions and offices of lawyers and judges. Monkeys have turned up in the prime minister's digs and—to the delight of newspaper headline writers—the halls of Parliament.

"They walk around in chambers!" exclaimed Meera Bhatia, a lawyer who has advocated on behalf of the residents of an upscale, monkey-plagued housing development. I met Bhatia for coffee one afternoon in Delhi. She told me she belongs to an exclusive health club, the same one to which Prime Minister Narendra

Modi belongs. "They opened a new pool, and the monkeys were going in the water!"

Bhatia filed public interest litigation on behalf of her housing development, and in 2007, the Delhi High Court decreed that the city's veterinary division must undertake a plan of action. At the moment, the burden weighs heaviest on one man: head veterinarian R. B. S. Tyagi, with whom I have an appointment in half an hour.

Tyagi's office is on the eighteenth floor of the South Delhi Municipal Corporation (SDMC), the bureaucratic heart of Delhi. As if to bestow the proper mindset for your visit, the elevators take ten minutes to arrive. Aggravated civil servants must have jabbed the Up button so often and so peevishly that at some point it gave out. A sign is taped to the wall—DO NOT PRESS REPEATEDLY.

I wait. A janitor pushes a dust mop across the lobby. He walks a perfect line, slowly, almost ceremonially, like a bride coming down the aisle. Another man dust-mops the black marble tile outside the front doors. Say what you will about the government of Delhi, but they've got that dusty floor thing licked.

Tyagi waves me in. He directs me to sit in one of two chairs arranged in front of his desk, even though the other chair is still occupied, the man's business with Tyagi apparently incomplete. Or perhaps he just feels like sitting there a while longer. On the wall behind Tyagi is a framed photograph of a koala. I have barely introduced myself when Tyagi begins talking. "We have been trapping monkeys as per directions of the High Court of Delhi. At present we have two monkey catchers. After trapping we relocate these monkeys to the Asola Bhatti mines. Details in this regard may be obtained from the chief wildlife warden of Delhi government, Dr. Ishwar Singh. Have you met him?"

I have tried. For weeks before I left for India, I would dial the number listed for Ishwar Singh on the forest department website. Day after day, no one picked up. I later came to know that only a sap tries to reach an Indian official through a phone number on a government website.

Tyagi and Singh—the SDMC and the Forest Department—have been trying to pass the buck for years. The SDMC maintains that because monkeys are wild animals, their control is the responsibility of the forest department. The forest department, in turn, contends that monkeys in cities living on handouts are no longer wild, and therefore do not fall within their purview.

I have heard about a plan to surgically sterilize Delhi macaques. I ask Tyagi for details.

"This is again the subject matter of the chief wildlife warden of Delhi government, Mr. Ishwar Singh."

Though I know the answer, I ask Tyagi why no fines have been issued for feeding monkeys at the temples.

"As you know, dealings of the monkey issue are religious. I would like to request you to discuss the issue with Mr. Ishwar Singh, chief wildlife warden . . ."

I finish it for him. ". . . of Delhi government."

"Yes. He will apprise you."

"Feeding is against the law, correct?"

"This is the subject matter of the chief wildlife warden." DO NOT PRESS REPEATEDLY.

While Tyagi and Singh dicker over whose duty it is, Delhi's better-to-do residents take matters into their own hands. Commercial buildings and affluent families hire monkey wallahs, men who patrol with leashed langurs. Indian langurs are the formidable black-faced species I saw in Pauri Garhwal and Rajasthan. They're larger than rhesus macaques, and the macaques keep their distance. A langur sitting on the side of the trail in Bundi was surely the reason I saw no macaques on the way up the hill to the fort. When I looked him in the eye, he raised his upper lip and flashed his canine teeth. Like pushing back a coat to reveal a sidearm, the gesture had the desired effect. I lowered my gaze and moved along.

"The use of langurs is banned." A contribution from the man in the other chair! He introduces himself. He is one of Tyagi's veterinarians. "It is illegal." Because of the Wildlife (Protection) Act. Meera Bhatia told me people in her circle quietly do it anyway. "I don't know whether we can find out *exactly* how many langurs the prime minister's house has, but . . ."* Then she veered off into a story about a macaque that got inside the All India Institute of Medical Sciences and took to pulling IV needles out of patients' arms and sucking the glucose like a child with a straw in a pop bottle.

After a crackdown on the monkey wallahs, SDMC hired ten men and trained them to impersonate the calls of the langurs. You may go on the internet to see and hear them at work. It is not true, as one newspaper reported, that they dressed in langur costumes. (It appears to be true, however, that the Sardar Vallabhbhai Patel International Airport hired a man to dress in a bear costume to scare langurs off the runways and prevent flight delays.)

"And the impersonators? How did that work out?" I've started addressing my questions directly to the veterinarian in the other chair.

"The problem is not sorted out. At best, the animals will simply move from one place to another. It is not a permanent solution." The same can be said for: Avi-Simian Shock Tape, slingshot brigades, plastic windowsill snakes, langur urine (one man told the *New York Times* that he "had 65 langurs urinating on prominent homes"), and life-size "langur dolls."

Tyagi removes his reading glasses and puts his full attention on

---

* When Donald and Melania Trump visited the Taj Mahal in 2020, the security detail included paramilitary forces, Armed Constabulary units, National Security Guard commandos, and five langurs.

me for the first time since I sat down. "Tell me. What is *your* solution to handle this situation in India?" I believe he would truly, desperately love to hear a good suggestion, a fresh idea, something, anything, that might appease his vexatious monkey-loving, monkey-hating public.

I tell him about the success they've had in some U. S. towns, working to lock down the garbage. Even as I'm saying this, I know that in a city as large and chaotic as Delhi it would be a wasted effort. Tyagi withdraws his gaze. "They are not generally going on garbage. That is dogs." Dogs, R. B. S. Tyagi will talk about. Stray dogs are rightly his job. Delhi has more dog bites than monkey bites, and the dogs, unlike the monkeys, carry a respectable threat of rabies. But dog attacks don't sell newspapers the way monkey attacks do. (Nor, for that matter, do the number 1 cause of animal-inflicted deaths in India: snakes. Around forty thousand people die from snakebite every year in India. My Google news alert on *snakes* and *Delhi* has delivered but one hit: a video of a macaque making off with a snake charmer's snake.)

Tyagi replaces his glasses and reaches for a printout on his desk. "This morning I was collecting information about stray dogs in the U.S." He reads aloud: "'Stray dogs have become one of the most serious public management problems in American cities.'" I think of the cities I know, and the many, many problems that plague them.

"*Stray dogs?*"

He continues reading. "'Packs of wild dogs roam American city streets.' Is it true?"

The paper references a story in the *Grand Forks* (North Dakota) *Herald* about a problem on a few Native American reservations. Kind of a leap, that. It makes me wonder: To what extent does the media in India overstate the problem, or even fabricate monkey assault stories? Take the *Times of India*'s "stoning" reference. In a different newspaper's account, a police investigator stated that the deceased had been sleeping beside a pile of bricks that collapsed on top of him when monkeys jumped onto it. Monkey-inflicted death by stone, yes, but hardly a stoning.

I suspect the press lathers up some of these cases, but macaques are certainly compromising quality of life at some of these housing complexes. A keyword search of "monkeys" on India's National Consumer Complaint Forum delivers eight-hundred-some pleas to authorities to, as it is frequently phrased, "kindly do the needful." Here is a typical missive, from Ravi Choudhary of Sector 46, "listing down some of the problems which we face on a daily basis":

1. Breaking of flower pots
2. Breaking of gate, lamp post
3. Children bite
4. Terror in the residents
5. Breaking of electric cables, [fouling of] water tanks
Temporarily any langur should be allowed to roam in the sector. Thanks.

I ask Tyagi if the statistic of 950 monkey bites per year is accurate.

"Monkey bites are recorded in the hospitals. That is another department."

Is there any question for which this man has an answer? "Dr. Tyagi, do you like monkeys?"

"Yes, of course. I am a veterinarian."

"I think you like koalas."

Tyagi swivels to gaze at his koala photograph. "Cute animals. *Very* cute. I went to Australia in 2010. I took that photograph." The memories seem to soothe him. He writes something on a pad, tears off the page, folds it in two, and hands it to me. It feels like a kindness, a small melting of the heart. I unfold the paper. It's the mobile number for Dr. Ishwar Singh, chief wildlife warden, Delhi government.

Over a span of twelve years, Tyagi's monkey catchers have delivered 21,000 macaques to an abandoned plaster of Paris mine in the southern reaches of Delhi. It was christened the Asola Bhatti Wildlife Sanctuary, and fiberglass walls were erected. There was a plan to plant fruit trees to feed the simian residents, but that didn't happen, so truckloads of food are driven in. I picture barren ground packed with dusty monkeys eating rotting produce.

The monkey area is closed to the public. If you ask, you won't get in, or not without a lot of paperwork or string-pulling. In the autorickshaw that takes me away from the SDMC, I text Nila Bhowmick, the journalist I met: "Feel like a road trip?"

The challenge appeals to her. We decide to just show up and see what we can see.

The staff plays hot potato with us. "You must go across the hall and speak to the forest officer." "You must see Mrs. Prasad in the natural history center." "The roads are closed due to mud." Nila cheerfully keeps at it, walking up to anyone we encounter. Back outside the administrative building, a woman in the parking lot points us in the direction of an elegant-looking man in white: white tunic and breeches, white turban, white mustache looped at the ends like ornamental cursive. He gives his name as Gurji, and he needs a ride out to the plant nursery, which is not far from the monkey area. Just like that, we are in. Nila translates as we drive along.

Gurji works in the nursery, but for three years he had been a macaque caretaker. What Tyagi would not say, Gurji is happy to provide. The city spends 40,000 rupees a day on food for the macaques: apples, corn, cucumbers, cabbage, sprouts, bananas. "Bananas are constant."

It's a six-mile drive to the monkey area. The road is deeply rutted and, yes, in fact, muddy. The sanctuary is enormous, beautiful, and wild, with low mesquite trees and bushy acacias. We pass nilgais and spotted deer. It seems like a fine place to be a monkey, though we have seen none yet. Gurji tells Nila he misses working with them.

"They were your friends," she says.

Gurji laughs. "Monkeys are no one's friends. They'll come and take the food. That's it."

We pull up beside a simple raised platform—a macaque feeding structure. A half dozen of them perch on the edges, eating corn and cabbage.

The sanctuary walls are poorly designed for keeping in monkeys. They climb them with ease and irritate the people who live nearby. With all the food and room to spare, why do they stray? It's the usual problem with translocation. An animal is not translocated into a vacuum. It's set loose in another animal's territory.

"They fight with the new arrivals," Gurji says. "They chase them away. The weak ones go."

We drive Gurji over to the nursery. It's a single-room structure guarded by a langur. It's feeding time, corn and cucumber. The langur ignores the cucumber and ravages the corn, kernels spraying from either side of its mouth. Gurji walks with us to a cluster of simple buildings across the road, once a residence colony for the mine workers and now a small village. A burly male macaque darts up the fiberglass wall meant to keep him out. He has a low brow and glowering mien that make me think of Joaquin Phoenix. He doesn't seem weak. Why would he want to leave the sanctuary?

"They all go in and out," says Gurji. "In my village, you can't even eat a roti without a monkey coming and taking it from you." He shrugs. "They're monkeys. What can you say?"

Dr. Ishwar Singh answers on the third ring. Asked about the forest department's plans for macaques in Delhi, he replies, "Laparoscopic sterilization!" He says it with grand flourish, as though introducing a special guest. And then he hangs up. It's the shortest interview I've ever done. I call back. I text. I email. No answer. I am the sperm, and he's the immunovaccinated egg. All receptors blocked.

I contact my acquaintances at the Wildlife Institute of India. Project scientist Sanath Muliya provides an interesting answer. He knows of no plans for surgical monkey sterilizations in Delhi, but laparoscopic vasectomies and, for the ladies, tubal ligations, have been underway on and off at eight Monkey Sterilization Centres (MSC) in the states of Himachal Pradesh and Uttarakhand. Since attaining the status of "pest (undeclared)," in 2006, a total of 150,000 macaques have been tied off, stitched up, and tattooed—with an ID number, I'm guessing, or a check mark, or something. Does 150,000 sound like a lot of monkeys? The forest department doesn't think so. In March 2013, the chief conservator for the Himachal Pradesh Forest Department sent a memo to MSC staff. "It has been noticed that the pace of monkey sterilization has not been satisfactory." The staff are directed to step up the pace "to approximately 90–100 monkeys per day in each MSC."

Photographs of the Himachal Pradesh Monkey Sterilization Centre posted online show two operating tables going at once. Assuming an eight-hour day, that's six surgeries and tattooings per hour that these surgeons need to be doing. Ten minutes per monkey!

But it wasn't the veterinarians who were hobbling the pace. The memo orders forestry officials to triple the ranks of monkey catchers, not medical staff. Even with the gussied-up title of Monkey Capturing Team, no one wants the job. Officials next tried recruiting the public at large, placing a 500-rupee bounty on macaques. Pushback ensued. As an activist pointed out to a BBC reporter, "Monkeys will be captured in a crude manner. . . . Many can get hurt."

Muliya says the public takes issue even with the vasectomies themselves. "They think it is inhumane." Plus, the monkeys scratch their incisions and the stitches come out. "That's what made us pursue PZP as an option," he wrote to me in an email.

Six months after I return home, the *Hindustan Times* would announce a new approach. "To control the monkey menace, the Delhi government's forest department is pinning all its hopes on an injectable contraceptive." But it wasn't PZP, or any other vaccine. It was RISUG (reversible inhibition of sperm under guidance), a gel injected into the vas deferens to block the sperm's path. The article quoted an affidavit "filed by Ishwar Singh, the Delhi forest department's chief wildlife warden . . . before the Delhi High Court . . . [stating] that the RISUG is a viable option after three failed attempts to get NGOs to conduct laparoscopic sterilization." *Laparoscopic sterilization!* The exclamation point was gone now.

The advantage of RISUG, said the *Hindustan Times* piece, is that it is injected. So there would be no stitches to scratch and pull out. But this wasn't true. Sanath sent me the official forest department description of the procedure, which ended with "and the site will be sutured." The advantage of RISUG over vasectomy is that it's reversible. This is of course no advantage with the macaques of India, but it is an advantage for human men (and the women not presently interested in bearing their children). The reason we know that RISUG works on rhesus monkeys is that the human version was recently tested on them.

The woman who'd carried out the tests is Catherine Vande-Voort, division director of reproductive endocrinology and infertility at the California National Primate Research Centers. We spoke by phone, and she confirmed that incisions were necessary. Though I had called to talk about macaques, I was curious about the future of (human) male birth control. A variation of RISUG is headed to clinical trials in the United States. Could people trust it?

VandeVoort thought things looked highly promising. "If you

can block a male monkey's fertility, you are golden," she said. Where tens of millions of sperm would constitute a respectable ejaculate for a man, a macaque launches hundreds of millions per wad. "And monkey sperm compared to human sperm look like they're jet-propelled. I've had people who are used to doing human sperm evaluations come into the lab and they look at the monkey stuff* and they're like, *My god, how can you even count them, they move so fast!*" In other words, if it works on a macaque, it ought to work on a human.

However. As a way to reduce a population of wild animals, any form of male contraception comes up short. One busy untreated male can take up a surprising amount of the slack created by his sterilized cohorts. For male sterilization to have a significant impact on a wild population, you'd need to treat 99 percent of the males. Whereas with females, treating around 70 percent should get you there.

Partly for that reason, the sex hormone–suppressor GonaCon is only registered for use on females. Additional reasons are set forth in "Observations on the Use of the GNRH Vaccine GonaCon™ in Male White-Tailed Deer." When the bucks' testosterone was suppressed, their scrotums shrank, their antlers came in all weird, and they failed to develop "the muscular appearance . . . of mature bucks in rut." Pictured next to a bull-necked control, they're pipsqueaks. Is humiliation an emotion felt by deer? I posed this question to Doug Eckery, who oversees fertility control research

---

*   To get the monkey stuff, VandeVoort's team developed a low-intensity penile electroejaculator. Why not just use a vibrator? "Oh, we tried. Oh my god, we tried. You could get a good erection but they wouldn't ejaculate." They also tried a dummy mount with an artificial vagina. Not happening. "A monkey is not smart enough to understand what we want from him, but too smart to have sex with a dummy." She emphasized that the device does not hurt or burn. Quite the opposite. She told me about an orangutan who'd come running at the sound of her voice. It was memorably uncomfortable: "Having an orang gaze longingly into your eyes as you try to set up your gear."

projects at the National Wildlife Research Center and serves as its assistant director. "I don't know," he said sensibly.

I asked Muliya how close the state of Himachal Pradesh has come to sterilizing 70 percent of the female macaques in conflict zones. Unknown, he said, because no one has followed up with a population count. I tried to imagine how it's possible to take a census without knocking on doors or mailing out forms. How do you know you're not counting the same individual more than once? Like the anxious San Franciscans whose sightings* led them to believe the city's Presidio woodlands held hundreds of coyotes—when in fact it was one wide-ranging pair and their pups. How do people do this?

---

* And, likely, what they heard. In a 2017 study by R. Kyle Brewster and colleagues, subjects were asked to listen to recordings of one to four coyotes howling and "yip-yapping" and then guess the number of animals making the sounds. Regardless of background (urban, rural, suburban), subjects overestimated the number of coyotes by a factor of two. Likely leading to "the misperception that coyotes are more abundant than they actually are," and lots of anxious yip-yapping to local and state authorities.

diam. ⅞ in.

diam. ¾ in.

a

<----- 4 in. ----->

5 in.

diam. ⅝ in.

b

diam. ⅞ in.

c

diam. 1¼ in.

d

<----- 9½ in. ----->

Fig. 53. Scats of the cat family

a. Bobcat.   b. Ocelot.   c. Jaguar.
d. Mountain lion (upper sample, Chisos Mts., Tex.; lower, Olympic
   Mts., Wash.).

6

MERCURIAL COUGARS

How Do You Count What You Can't See?

For fifty-seven years, California's mountain lions had a price on their heads. The rancher complained because the cats kill livestock, and the hunter, because they prey on deer. And the state listened. From 1906 to 1963, California paid a bounty for each hide or scalp or set of ears brought to a county courthouse or shipped to the Fish and Game Commission. The payments were logged in ballpoint in a series of ruled leatherbound logbooks that now reside in the state archives in Sacramento. On the inside cover of each logbook, someone has penciled the totals for the year, and a foldout map displays the number of payouts county by county. California excelled at counting dead mountain lions.

Counting the live ones has always been trickier. You cannot fly overhead and take photographs, as you might with herds of wildebeests or walruses hauled out on beaches. You could bring in volunteers to tromp through the woods, as you might on a drive count of deer or a line transect of sloths or an Audubon Christmas Bird Count, but it's unlikely any cats would be seen. Cougars are elusive loners. Their presence is knowable mostly by their "sign"—tracks and scat and other unique traces they leave on the land. The most prolific lion hunter in California's history, Jay Bruce, killed more than five hundred cougars in his years on the state payroll, but only once in all that time did he catch sight of one that wasn't under pursuit by his hounds. Up until the 1970s, bounty logs and county kill maps were the closest thing California had to a statewide cougar count. A county with few dead cats was a county with few cats, period.

Ironically, today, if a wildlife agency wants to know how many cougars live within its governance, it may make use of the same expertise that once brought them down. That is why the only state that prohibits the hunting of mountain lions—California—still has houndsmen on its payroll: to count, not to kill. The same skills serve: "cutting" (searching for) sign and, once fresh tracks are found, putting hounds on the scent to tree the cat. What's different is that the dogs will hang back and the lion will be darted and lowered from the tree. Because California seeks not only to estimate the lions' regional numbers but also to assess their genetic health and monitor their habitat use, a handful per region will be GPS-collared and a DNA sample taken.

The Statewide Mountain Lion Project is being undertaken by the California Department of Fish and Wildlife. More specifically, Justin Dellinger. Dellinger's job title sounds like something you'd hear if you asked an animal-besotted ten-year-old, What do you want to be when you grow up? Mountain Lion and Gray Wolf Researcher. That is what it says on his business card. Dellinger has a PhD in wildlife biology, but it was not just his academic standing that landed him the job. It was also his upbringing. His woodsmanship. He grew up in South Carolina, where he divided his time between the woods and his grandparents' horse barn. His hometown is small enough that any time Dellinger was interested in a girl, he had to "go ask Mamaw and Papaw" how he might be related to her. As the first in his family to go to college, he is a source of parental pride, and a little melancholy. He explained to them that he "had to disperse for genetic purposes."

I first met Dellinger at the CDFW Wildlife Investigations Laboratory, where he has a desk that he occasionally and not very joyfully sits at. He wasn't immediately located, so I sat in the waiting area, in the company of the taxidermied specimens I've come to expect at all government offices with *fish* or *wildlife* or *game* or *forest* in their title. A mountain lion crouched, snarling, on a faux rock ledge above the receptionist. A hawk was coming in for a landing beside a rack of hunting information, its talons extended as though reaching for a brochure. Eventually someone brought me back to Dellinger's office, cramped and narrow and rendered more so by a skein of antlers on the linoleum. He finds these and other "l'il wildlife trinkets" at mountain lion kill sites—meaning places where lions have devoured prey. Dellinger is a trophy scavenger, but not a trophy hunter. Killing bucks for their racks is, he said, "not necessarily something I understand." He hunts to be outdoors tracking and to "fill the freezer." As someone who lets slaughterhouse workers kill the birds and animals she eats, I respect that.

At this moment, the meat Dellinger is consuming is pepperoni, on a pizza at the only restaurant in Alturas, California, that is open at 4:30 in the afternoon. A 4:30 dinner is what happens when breakfast happened at 3:30 a.m. and lunch was a granola bar and a tangerine. There is soot on his clothes and face from this morning, when he was following a cougar through a stand of charred pines. It's not that he's had no time to wash up and change. I think it just didn't occur to him to do so.

Though Dellinger is comfortable living out of a tent—is probably happiest living out of a tent—he is by no means a hermit. He doesn't reject civilization; he's just uncharmed by it. His home is less than two hours from San Francisco, but he has never been.

While we wait to order, I ask Dellinger to explain how the

counting gets done. I have braced myself for math. The classic animal population estimation method, capture-mark-recapture, relies on proportions and a tolerance for word problems. Say a biologist wants to know the number of woodchucks in a forest. She sets out traps and she Marks, with leg bands, all the wood-chucks she catches (M). We'll say she captures 50. She returns them to the forest. A week later, she puts the traps out again and notes how many she Captures this time around (C). Let's say 41. She also notes how many of them have leg bands—that is, how many are Recaptures (R). We'll say 27. She lets them go. She gets out her calculator. Using the formula* M × C over R, and show-ing her work, she can estimate how many woodchucks live in the forest. In this case, 76. (In a higher-tech version of the technique, it's images of the animals that are captured and recaptured, using motion-triggered camera traps.)

Because the Statewide Mountain Lion Project is a more quali-tative survey, Dellinger's method is unique. He calls it "collar and foller," *foller* being *follow* in a South Carolina drawl. Between bites, he describes how the counting part goes.

"It's like, We found this animal, a male, right here." He puts his finger to an imaginary map. The male is given a tracking col-lar. "Now say we come across another tom track. We pull out the telemetry equipment and check on that first tom. We see that he's nowhere around here, so this must be a different male. Next day we find a female track. Now we've got a minimum count of three." And so on.

"You don't have to collar any of them, honestly, if you have a few good sign cutters and you use basic reasoning. Let's say you cut a male track over here on this ridgeline." That would be the

---

* The formula for calculating how much wood a woodchuck could chuck is $0 \times$ (the name is an English corruption of a regional Algonquian word, *wuchak*).

ridgeline between the oil-and-vinegar caddy and the table's edge. "Now about five ridge lines over"—the back of my chair—"my houndsman finds another tom track, and they're not even close to traveling in the same direction, and they're both from last night. Chances are, those are two different males."

To someone skilled at cutting sign, the absence of it is also telling. Very quickly upon arriving in an area, Dellinger knows whether or not there are lions. If not, he moves on. Still, it will take eight years to cover the whole state. I ask him why more people aren't hired to help. A waitress stops to clear the salad plates.

"Say that one more time? I was distracted." By the busy hand and the remaining pizza slice. "I didn't want to lose it."

One reason the state doesn't hire more trackers is that they don't exist. Aside from the houndsman currently on the job, Dellinger knows of only two others in the state whom he would work with. Both are in their eighties. As for wildlife biologists: "Maybe two percent of them can do this."

I have questions about the "this." How can you tell a mountain lion's gender from its tracks? How do you know where to start looking? How can you be sure a track is fresh from the night before, rather than from the day or the week before?

These things are more easily shown than told, and that is why I too will be eating breakfast at 3:30 a.m.

Tracking wild animals always struck me as something hushed and hidden. I envisioned someone walking with intense focus, head bent, through deep forest. Stopping to examine a broken branch, kneeling beside a watering hole. Moccasins might be worn.

So far it's been louder and less verdant. Dellinger looks for sign while standing at the handlebars of an ATV, driving slowly down a dirt logging road. Driving covers ground faster than walking, and tracks show up well in the fine-grained dirt. (Winter snow provides the other clean canvas for tracks.) We are deep in a forest, but the pines are charred and bare. Twenty thousand acres of Modoc National Forest burned last year, and that is where Dellinger is looking for lions this week.

Where I see blackened trunks in a lifeless moonscape, Dellinger sees the bright green shoots that have sprung up since the fire, thriving in the newly admitted sunlight. Tender new growth is what deer love to eat, and deer are what mountain lions love to eat. Dellinger uses the term "deer specialists." That's what it says on *their* business cards.

To find mountain lions, one heads to the places that would be, as Dellinger puts it, appealing to them. Mountain lions want to be where there is food and water—for them and for their prey—and unstrenuous routes through their territory. A male cougar may walk ten-plus miles in a night of hunting and checking up on his females. A saddle, or pass, between hills or a ridge along the top of them provides an easier, faster way to move through the landscape. Dellinger starts to say that cougars are lazy, then corrects himself. They're *efficient*. A "she" cougar, in particular, can't afford to squander calories. Almost always, a female is either pregnant or has young to feed. A male cub may be as big as or bigger than her by the time it heads off on its own. "Depending on the size of the litter, she may have to kill every day," Dellinger says. "That takes a toll." When he arrives in a new county, he takes out a topographical map and looks for saddles, ridgelines, and draws (small valleys with streambeds).

And national forest logging roads. "A road like this one," he says to me over his shoulder, "that's pretty straight, that gets them from point A to point B pretty quick—they're going to like. And this one is peripheral to the water down there, so it's good hunting for them." Why is there a logging road in a national forest? Because national forests began as—and to some extent remain—managed tree farms for the nation. They were set aside in part, to quote the Organic Administration Act of 1897, "to furnish a continuous supply of timber for the use and necessities of citizens of the United States." (And free grazing space for their cattle. The only animals I've seen in the woods so far this morning have been cows.)

Away to our right, the sun crests a ridge. Bye-bye, moon sliver and Nivea-blue sky. Cue the bird chatter and side-lit glory of a California wilderness dawn. Dellinger is missing it. His mind stays in the dirt. Today he's stopping more than he normally would, to highlight interesting tracks for his backseat city dweller. He just idled the engine to show me a set of badger tracks. Badgers are un-weaselly members of the weasel family who went their own way, specializing in bullying through ground squirrel burrows to eat the occupants. They evolved long, sturdy digging claws that give their tracks a spindly Edward Scissorhands appearance. To spend a morning cutting sign is to marvel at the surreal variety of feet and dance steps in the animal kingdom. Earlier we saw tracks of a mule deer stotting. The verb *stot* means to spring into the air and land with all four feet at once. (Various theories exist for why deer and antelope stot, as well as several names for the practice, of which *pronking* is the author's favorite.)

Dellinger stops again, this time for the fine-toothed imprints of a ground squirrel's claws. It's been scurrying all over the place, perhaps having caught wind of you-know-who with the Tim Burton manicure. Ground squirrels perform a cousin of stotting, wherein they leap and land with all four feet bunched together, leaving a track that looks like it was made by one large paw. Dellinger says he's watched people track a ground squirrel thinking they were on the trail of a lion. His boss periodically sends "aides" along; most are quietly sent off to cut sign in an area Dellinger knows to have no cougars, lest they drive over or trample tracks they've failed to see. When I first emailed Dellinger asking to tag along, he wrote back, "How are you at tracking?" Attached to the email were two photographs: examples of what I'd be expected to recognize as a mountain lion track in the dirt. A Leatherman multi-tool had been set alongside for scale. Because I could not, at a glance, see any tracks at all, I was puzzled. It looked like a photo shoot for Leatherman.

From what I have learned today, I can now identify deer and badger footprints. I could definitely track a cow. And I can tell coyote and fox from bobcat and cougar. Canid tracks are more elongate and may include claw marks, because unlike cats' claws, dogs' claws don't retract. The tell for a cougar or bobcat track is a pair of prominent notches ("cleats," in tracker lingo) at the rear of the large center pad. These are often visible even if it's a partial track, which, in more gravelly sections of the road, it often is. We brake for cleats.

Dellinger slows to a stop. He tells me to stay on the ATV, which suggests he's found mountain lion tracks and is worried I'll hop off and step on them. He gets onto one knee and lowers his face to the roadway. A tangerine in a cargo pocket bulges like a tumor.

It is in fact a mountain lion. Dellinger lays a small ruler alongside. Sexing a cougar track is a simple matter of measurement. If it's wider than 48 millimeters, it's a tom. Less indicates a she. And how does one know a she track from the track of a juvenile male? By who else is around. If it's a juvenile, the mother's tracks should be somewhere nearby. This one is a tom.

Dellinger presses the side of his fist into the dirt beside the track for a comparison. "See how this"—the extremely fresh fist-print—"is a lot crisper?" Moist soil holds together better. Taken to the end point, it's the difference between sand-castle sand and hourglass sand. In the heat of summer, by ten or eleven in the morning, the moisture from the previous night's dew will have mostly dried, and tracks from the night before will have lost their sharpness. This is another reason to get out of bed at 3:30 a.m. As is this: the side-arm sun of early morning casts shadows behind the ridges of a track, outlining its contours. If we came back at noon, even Dellinger would miss this track.

Dellinger believes this print to be that of the male he collared near here the week before. The goal today is to find and collar the female who eluded Dellinger yesterday, so we climb back on the four-wheeler and continue to stare at dirt.

Dellinger points to the road ahead of us. "Watch what happens to his tracks as we come up on this curve. See how he's saving steps? He's cutting the corner." If you can read sign, you can infer an animal's intent from the lay of its tracks. If a mountain lion is chasing prey, its tracks are messy and scuffed, and some will overlap. If it's in traveling mode, with a set destination in mind, the tracks are clean and widely spaced. The stride of this lion is shorter. "He's just amblin'," Dellinger says. Looking for someone to eat.

I will admit that, despite a statistically detailed knowledge of the extreme rarity of cougar attacks on humans, these tracks make me nervous—not now, on the back of the ATV, but a half hour later, when I leave the road and go off into the trees to pee. Dellinger never worries. As he put it, "We're not on the menu." Nor does he believe encounters with cougars are on the rise. "Californians are like, 'Lions are everywhere now!'" What's on the rise are home security cameras. Doorbell cameras are the mammograms of wildlife biology. Dellinger slows for a rut. "All it is, is a change in technology." Someone posts a doorbell shot of a cougar. It's reposted, goes viral. News crews show up. The whole neighborhood's talking about it. One sighting turns into five.

Dellinger upshifts for a straight stretch. "They've always been there. We just never saw them." He says he'd bet his paycheck that every day of the year, at least a dozen Californians come within pouncing distance of a mountain lion and never know it.

Almost 10:00 a.m. now. It's warm enough to unzip my jacket despite a steady breeze. Even if we found fresh tracks, the scent would be too dissipated by heat and wind for a hound to track. (The houndsman is off on another logging road; he and Dellinger are in touch by cellphone.) As the sun heats the air, scent molecules become more energetic, bouncing off one another and spreading out, becoming a diffuse cloud of diluted smell. The wind further scatters the scent. Even under ideal smelling circumstances, there will be a point, many points, where a hound loses the scent. This is called, yes, "a lose." To relocate the scent, good hounds will run a zigzag, sweeping wide left and right until they pick it up again. I tried this on a street near my office, after a young man passed by in a reek of Axe body spray. I let him turn the corner and disappear from view, then waited a few minutes. By zigzagging hound dog–style, I was able to track him to his destination, a cheesesteak place on the next block.

Dellinger is ready to pack it in. Tomorrow's another day. Except that it isn't. He neglected to mention that he has to drive to Redding for a "wolf meeting." I picture large canids in business casual. Sensing my disappointment, Dellinger offers to demonstrate how he uses tree-climbing gear to reach a darted mountain lion to lower it down. Without the lion, this is only lightly interesting. He also offers to drive me out to a "community scrape" that he found yesterday. At places where males' territories overlap, or where they make use of the same pass through the hills, say, lions will leave a calling card. Much as the dogs in my neighborhood spritz urine on the same sad shrub in our yard, cougars will kick back the "duff" with their rear paws, depositing scent onto fallen pine needles and other forest floor detritus.

Dellinger looks for scrapes under large trees. The bigger the tree, the deeper the duff.

We arrive at the spot and look at some scrapes. While visually unspectacular, a scrape is interesting for what it reveals to those who know the code. When mountain lions scrape, they typically face the direction they're traveling, which is helpful if you're looking for them. And, as with tracks, it's possible to tell how fresh the scrape is—here, by noting how many new needles have fallen on top. Dellinger is explaining how to "sex a scrape." A female kicks back with her rear legs together, leaving two parallel gashes in the duff, whereas a male scrapes with one foot at a time, angling out to the side, "because of its anatomy." By this he means "its balls." Dellinger was recently chastised for incautious language in a press interview. He had, among other things, likened maneuvering a woozy, darted cougar out of a tree to "tryin' to get your drunk friend into a taxi-cab and he's got his hands on either side of the door, resistin'."

I envy people able to read the natural world in this way. I move through the woods the way I flip through Chinese editions of my books, seeing shapes and patterns and having no clue what they might mean. Earlier, Dellinger showed me a line in the dirt from one side of the road to the other, like something made by a kid dragging a stick. It was in fact a "drag mark," but no child passed through here. The line was made by the dangling hoof of a dead fawn being carried in a mountain lion's jaws to a more secluded spot to be eaten. I of course missed the tracks on either side.

When I first met Dellinger, I told him that what I found interesting about the Mountain Lion Project is that it blends modern wildlife biology with its roots in natural history. The early naturalists spent weeks at a time out on the land, tracking and observing, deciphering behaviors, discovering new species. You could sense the excitement in the titles of their journal papers.

"Anecdote of a Combat Betwixt Two Hares." "A New Duiker from Zanzibar." I'm sure the authors of "Conservation Phylogenetics of the Asian Box Turtles (Geoemdidae, Cuora): Mitochondrial Introgression, Numts, and Inferences from Multiple Nuclear Loci" felt some excitement, too, but they didn't have those long, glorious stretches alone in the wild.

Dellinger knows the new stuff, but at heart he's a happy throwback. Standing over the skeleton of a deer earlier, Dellinger talked about how he's noticed that mountain lions in less arid parts of the state will pull out a deer's rumen and drag it away before feeding on the carcass. And lions in drier parts of the state don't seem to do this. An ungulate's rumen teems with bacteria that break down the plant matter it consumes; Dellinger's theory is that the behavior slows the rot of the carcass, which might boost the animal's odds of survival in humid climates where meat spoils more quickly.

Naturalists were the original biologists, and hunters and trappers were the original naturalists. No one knew more about a species—the wheres, whens, and whys of its movements through the land and the seasons, its relationships with prey and rivals and mates—than a person whose livelihood depended on that knowledge. The first natural history museums looked very much like the dioramas at a Cabela's. As natural history formalized itself and science became a paying career, rivalries and resentments grew. In 1941, the aforementioned houndsman Jay Bruce wrote a letter to his Division of Fish and Game superior, requesting that he not share Bruce's new report "Cougar in Relation to His Neighbors." "The naturalists have already stolen too much of my discoveries and never give me credit for it," the letter states. "Everthing they should have known, but didn't, they have implied was their own material."

Wildlife biology has always been a kind of snooping. Long before scientists were spying on animals with wildlife cameras or tailing

them with radio collars, they were poking around in their excre-
tions. As in human espionage,[*] it's done because you can't just
ask. You can't inquire of an animal what it eats or how healthy
or stressed it is, but you can sometimes learn the answers from
its scat.

"Droppings analysis" got rolling in the 1930s. The decade saw
a steady progression of learned men prying into the toilet of common woodland creatures: Hamilton on the diet of big brown bats,
Murie on coyotes, Dearborn on foxes, minks, and coyotes, Hamilton again on skunks, Errington on badgers and weasels. Before
that, if you wanted to know what a species was eating, you opened
a few hundred stomachs. You can imagine how amassing enough
organs to draw valid conclusions would be an unappealing prospect for most biologists, and certainly all the animals. Albert Ken-

---

[*]   The late Sayre Stevens, head of the CIA's Directorate of Science and Technology, had a trap placed in the plumbing beneath the toilet at Blair House,
the guest accommodations for visitors of the president, while Soviet premier
Nikita Khrushchev was in town. The captured stool was brought to physicians within the CIA's medical intelligence program to see what information
it might divulge. Specimens from Egyptian king Farouk and Indonesian president Sukarno (in this case urine from an airplane toilet) were likewise delivered
for testing. Given that the tactic predates clinical DNA analysis, how much
intelligence could a BM really provide? What could brown do for you? I posed
the question to Jonathan D. Clemente, a practicing physician and contributor
to the *International Journal of Intelligence and Counterintelligence,* currently at work
on a scholarly history of medical support for clandestine operations. Not much,
is the answer. "They were looking for blood in the stool, maybe parasites.
Whether they got any useful information, I kind of doubt it." More fruitfully,
Clemente said, CIA physicians have worked under assumed names at high-
profile medical centers where foreign heads of state sometimes travel for care.
Why bother with a turd when you have direct access to the man himself and his
medical records? Only once, to Clemente's knowledge, has critical information
come out of the enemy's toilet. He shared the story of a U.S. Military Liaison
Mission that had been spying on a field camp of Russian soldiers. At one point
the solders ran out of toilet paper and began using pages from their code pads.
MLM operatives went through the soldiers' trash and triumphantly delivered
the browned pages to the National Security Agency.

rich Fisher's 1900 "Summary of the Contents of 255 Stomachs of the Screech Owl" made me feel tired and sad, though also vaguely festive, owing to the author's "Twelve Days of Christmas"–style presentation: "91 stomachs contained mice . . . 100 stomachs contained insects . . . 9 stomachs contained crawfish . . . 2 stomachs contained scorpions . . ." Droppings provided a kinder, less taxing alternative.

They still do. Dellinger's master's thesis is on the diets of gray wolves. "I spent a lot of time," he recalled earlier, "walking around looking for poo." (*"Poo"*! There's the California Department of Fish and Wildlife talking again.) Old habits die hard. As we walked around a deer kill site, he bent to pick something up, saying, "Here's some bobcat scat." He held it out to me and then, quickly reassessing, let it drop.

Eventually someone hatched the idea of counting piles of scat to estimate species populations: shit as proxy for shitter. The technique became known as a pellet census: yet more biology-minded humans moving through the wilderness, staring at the ground. Provided your census takers can tell fresh feces from stale, and you know the average number of times a day the species in question defecates, it's possible to work out how many individuals a fecal tally of a set area represents. Possible, but not easy, and probably not all that accurate.

First, your census takers must know their shit. Raccoon scat, for instance, is most reliably distinguished from opossum scat by scent, the latter possessing a foul smell. Ungulates pose a quandary in that they often travel in groups and "dung" while walking. Making it hard to know: Are you looking at the droppings of two individuals or—quoting scatologist Ernest Thompson Seton—one "peripatetic defecator"?

Nor is it a straightforward matter to make the call between fresh and old. No one knew this better than a rat searcher. These were men, "specially trained in the habits of the rat,"* who, among other tasks, boarded ships at England's docks and counted up fresh pellets in order to estimate the vessel's "rattiness." Not as simple as it sounds. Fresh droppings in a hot engine room could be misleadingly shriveled and dry, and old droppings on a wet deck were misleadingly plump and fresh-looking. Mold was an unreliable indicator, as a 1930 study by Liverpool's assistant port medical officer proved. On certain diets—sunflower seeds, for one, and bran—the pellets grew moldy within twenty-four hours, "whilst on other diets, excreta kept under exactly similar conditions [showed] no signs of mould for several days." And depending on what sorts of edibles were in the cargo hold, it could also be tricky to tell a rat dropping from some other animal's. The small, hard black turds of a rice-fed rat were easily mistaken for those of a mouse. Despite all these complexities, the rat searchers' estimates, when checked against post-fumigation roundups, were impressively accurate, and prompted "a healthy rivalry among the men."

Figuring the daily defecation rate for a species posed its own challenges. Some researchers tried fashioning a "fecal harness" and outfitting a number of representative animals. (The harness held in place the "fecal bag"—a sort of feed bag in reverse.) The results strayed in unanticipated ways. One researcher's harness for shrub-grazing goats proved too restrictive; the creatures were unable to assume their preferred bipedal noshing stance, which allowed them to reach the higher leaves. A subsequent goat-diet researcher published plans for an improved harness that, despite its nineteen

---

*   And impressively well dressed for the job. In a 1930 photograph online, a West Indian rat searcher sports a double-breasted jacket with eight brass buttons and the kind of hat more often associated with commercial airline pilots. He carries a spiffy metal box that holds either a rat or a sandwich, I do not know which.

leather straps, allowed the goats to rear up on their hind legs. In a minor setback, several of the nonharnessed goats, being goats, ate the leather straps off their pals. Science is never simple.

An alternative would be to spend time spying on animals in the wild. Again, not as straightforward as you might think. "Dunging" frequency, wrote David Welch, in a 1982 study on the "dung-volume method of assessing occupance," differs by time of day and by season. Rabbits in Wales dropped an average of 446 pellets a day in April, when food was plentiful, but only 376 a day in January. The rate also varies by what foods the animal is eating. I'm not just pulling this out of my fecal bag. The Liverpool study details the "enormously" varied fecal output of rats on different diets. A rat fed rice will pass an average of 21 pellets a day; a bran-fed rat averages 128 ("very large, buff-colored cylindrical") pellets a day.

It was around this point that a JSTOR pop-up appeared on my laptop screen. "Want to connect with leading experts in pellets?" I kind of did. What would they be like? How many could there be? Was I one now?

The future of turd science is bright. Analyzing genetics from scat promises to be a faster, less costly version of capture-mark-recapture. Rather than counting recaptures of marked animals, one would count reappearances of genetic fingerprints in the collected scat. Dogs trained to whiff out mountain lion scat will soon be brought to these same areas Dellinger has been surveying. Their handlers will bag the scat found by the dogs and bring it back to the CDFW Wildlife Investigations Laboratory for genetic analyses. Scat can also deliver information about the health and genetic diversity of cougars in different regions.

If the genetics work turns up population data similar to what Dellinger is getting from his tracking and collaring work, it means scat-detection dogs and gene sequencing can be trusted to do the work on future surveys. If all goes as hoped, Justin Dellinger will be replaced by a heap of shit.

I think he'd miss being out here, though he says no. He says it'll free him up to spend more time researching hazing techniques and other ways to keep mountain lions apart from humans. Because whenever the two cross paths, at least in California, controversy boils. As Dellinger puts it, "For some people, ten is too many. For some, ten thousand is not enough." Interestingly, it is not commercial ranches requesting the majority of California's mountain lion depredation (kill) permits. Between 70 and 90 percent are granted to backyard famers—people with between two and ten animals. (The state has few large commercial ranches.) From a distance, the killing of a cougar is, for many in my state, an affront. *Build a safe nighttime enclosure for your animals! Keep your pet inside at night! How is the life of a beagle or goat worth more than the life of a wild mountain lion?* A bellows to the flames is the common impression, based on heavy media coverage of a pocket of cougars isolated by the Los Angeles freeway system, that the species is imperiled throughout the state. California's cougars are neither endangered nor threatened. But they are big and they are beautiful, and those are the animals people fight for hardest. It's the everlasting politics of charismatic megafauna.

And megaflora, even. The bigger the tree, the deeper the duff. Or something.

WHEN THE WOOD COMES DOWN

Beware the "Danger Tree"

What a Douglas fir does, it does very slowly, and that includes dying. Possibly the least attractive feature of a nine-hundred-year life span is the century or two spent dying. Decomposition drags on for another hundred years or so. A tree is the rare organism to which the comparative *deader* is often and accurately applied. A recently dead, or "dead hard," conifer progresses to "dead spongy," then "dead soft," limbs and top rotting and dropping off, until the last piece of standing trunk topples and the tree enters the final classification, "dead fallen." At some point in its protracted twilight, a tree that stands near a road or path or building may earn a new classification: "danger tree." Because if it falls, anyone it lands on will spend a very, very short time dying.

The victims of arboreal manslaughter may be, unlike the perpetrators, quite young. The *Australian Journal of Outdoor Education* published a summary of cases, since 1960, of children (and in two instances, their teachers) killed by falling branches or trees during school camping trips: six killed sleeping in their tents, one while swimming near a eucalyptus grove, and another six killed while hiking, including two teenagers crushed when the top of a mountain ash tree broke off and rolled down a hill onto the trail.

Wind is a common accomplice. The journal *Natural Hazards* reports that in the United States, between 1995 and 2007, trees toppled by strong winds caused the deaths of nearly four hundred people. My husband, Ed, and I were twenty feet away from a similar fate, awakened early one windy morning by the crack of a large branch breaking off an oak and landing near our tent.

Some trees kill in the normal course of life. The Coulter pine drops a cone as heavy as a bowling ball. According to "the largest review of coconut-palm related injuries," sixteen Solomon Islanders were struck by falling coconuts between 1994 and 1999. Balinese newspapers have three times in recent years reported on cases of bodies found beneath durian trees. The fruit of the durian tree makes an excellent murder weapon: big, heavy, and covered in hard spikes. The "suspect," being a tree, cannot hide the weapon; a bloodied fruit lay beside one victim's head. It is difficult for authorities to generate caution or concern. Confronted with the sign "Dropping Pine Cones, Proceed at your own risk," most will proceed.

The term "danger tree" is itself somewhat hilarious. It's like "danger mitten." The staff of Vancouver Island's MacMillian Provincial Park, home to a stand of "legacy" conifers, find no humor here. Because the most geriatric trees are also the tallest and most majestic. They are the trees people pay money to hike and drive amid and the ones the public very badly doesn't want cut down. This creates a conundrum and, very occasionally, a tragedy.

In 2003, an Alberta couple were passing through MacMillan's Cathedral Grove, a stand of massive, centuries-old conifers, when a fierce snowstorm hit. They pulled off the roadway to wait it out. One of the ancient firs, overburdened by snow and weakened from rot, fell onto their car and killed them.

Since then, MacMillan has maintained a relationship with

a certified danger-tree assessor. Twice a year and after any big storm, ongoing for fifteen years, Dean McGeough roams the woods looking for signs of dangerous decrepitude. Today is one of the semiannual inspections. Over the course of the day, Dean will flag trees he deems in need of mitigation: a limb lopped, or a top, or something more drastic. This is the part where, statistically, most of the manslaughter goes down. By a large margin, the people whom trees kill most often are the people bringing them—or pieces of them—down. An on-the-job fatality for a faller, as chainsaw wielders are known in these parts, is sixty-five times more likely than it is for workers in general. These are men with compression bandages stuffed in their pockets the way my grandmother had Kleenex. Men whose fabric of choice is cotton-Kevlar. Though it isn't usually the blade that kills. It's the tree. Sometimes it's the one they're cutting, but more often it's a bystander. The tree may bend the branch of a neighbor as it falls, causing it to slingshot back at deathly speed. Pieces of other trees lodged in the branches—a "swinging snag" or an "insecure hang-up"—may pull free and come down on the faller.

British Columbia has an active Forest Safety Council, and two members are here today. The title given on their business cards is Falling Safety Advisor. Earlier today I met a Falling Supervisor. The word *falling* has lost its slapstick for people in the logging industry. Someone will mention a guy from a long-ago job, and someone else will go, "Is he still falling?"

The most dangerous trees to fall are (duh) danger trees. A healthy tree with sound wood can be made to fall in any direction. Like this: Rather than cutting straight through the trunk, the faller stops partway and goes around to the opposite side and makes a sloping undercut. Now when he goes to finish the cut, the trunk will tilt down onto the slope made by the undercut and fall in that direction. A rotting tree is hard to control in this manner, its fall impossible to predict with certainty. If a conifer is rotting from the top down, the softened portion may break off as the tree starts to lean and come down on the faller. Or the whole rotted trunk may "telescope"—collapse straight down into itself. Or a rotted portion of trunk may suddenly crumble and change the direction of the fall. Think of those osteoporotic olds whose bone has grown so porous that one day a hip gives way when they shift their weight. (All these "overmature" trees may explain why the lumber company that once owned the grove donated it to the province in the first place: lotta punky lumber.)

Ideally, no one should be anywhere near a danger tree when the wood comes down. That is why very tall, very old, very dangerous trees are not cut down but, rather, blown up. Explosives aren't exactly crib toys, but they can be detonated from a safe distance. So regardless of what comes down and from what direction, no faller will be felled.

After Dean finishes his inspections, expert faller blaster Dave "Dazy" Weymer will start work. (The nickname dates to his twenties and has to do with weed, not flowers.) Dazy has been blasting trees for thirty-five of his sixty-eight years. Both his father and his grandfather were loggers. He grew up in logging camps. He was, he says, "a bit doomed to be a logger." I first saw Dazy in a YouTube montage, a sort of highlight reel of explosions and screaming chainsaws against a booming soundtrack of insistent strings and kettle drums. You need ear protection just to watch his videos.

The forest floor in Cathedral Grove is hardly like a floor. It's an obstacle course of decomposing branches and logs, their surfaces and outlines obscured by a damp, spongy pelt of mosses and ferns. It is difficult to predict when your foot will connect and what will happen when it does. It may come to rest on a log or it may push straight through what appears to be a log but is in fact crumbling, log-shaped mush. You will stumble and fall, but you won't be hurt, just moist. Moist fallen.

While Dean makes his rounds, he and Dazy bring me up to speed on basic tree anatomy. The tree, I am learning, is not entirely unlike the human. The older, harder wood that runs through the tree's core serves as the skeleton that supports it. Surrounding this spine of "heartwood" is the "sapwood," the flesh through which courses, slowly, so slowly as to possibly demand a different verb, the blood of the tree—the sap.

Bark is of course the tree's skin. It protects the flesh, and it is— again, like our skin—both an entry point for infection and a part of the immune system. Conifer bark secretes resin (aka pitch), a thick, sticky goo that seals wounds, traps bark beetles, kills pathogens. Also like us: a tree's crown thins as it ages, and the point at which its circumference is greatest is called the butt. And there my trees-as-people comparison sputters out.

"There's one I blasted." Dazy has a deep voice that projects well when he needs it to, which he often does because he's holding a conversation across a grove of trees or over the sound of an idling chainsaw. He's pointing to a Douglas fir. These trees stand out from the others here by their bark—thick, with deep vertical rifts.

From ten feet away, looking straight at it, the blasted fir looks no different from the live intact legacy trees all around it. Only the top third is gone, and to see the top third of this formerly 180-foot tall tree, you would need to crane your neck all the way back. Removing just the upper third makes the tree lighter and more stable—less dangerous—and at the same time preserves the grove's medieval Sherwood Forest vibe, what the tourism professionals call "visitor attractiveness." At eye level, the living, the dead, and the blasted look the same: enormous mossy tree trunks. As Dazy says, "You wouldn't know it wasn't just another pretty tree."

Elderly trees perform their own, more subtle version of what Dazy does. It's called retrenchment. The tree's trunk circumference and roots continue to grow, but it stops getting taller and the limbs of the crown die back and drop off. This makes it less top-heavy. More importantly, there is less "sail," meaning fewer surfaces for the wind to catch and less blowing about of the crown and risk of what forestry people call "windthrow"—a tree uprooted and blown over by powerful gusts.

I lean way back to try to see Dazy's handiwork on the fir. This causes me to lose my footing and topple backward off a log. Falling Author. Dazy extends a hand. It is noticeably unlined for his age, probably because when he's outdoors he's usually wearing gloves. If other fallers read this, he will no doubt get grief about his lovely hands, but I believe a man named Dazy will handle it.

Here is another reason not to cut a danger tree down to a stump. Dying and decomposing trees, far more so than young living trees, provide real estate for wildlife. Rot-hollowed trunks become dens for bears. Dead tree branches are hunting perches for raptors. Soft, rotting sapwood is easily excavated by woodpeckers and other cavity nesters. For this reason, a "danger tree" is often classed as a "wildlife tree," too. Blasting off the upper third of the tree facilitates the process. It speeds the decay of the remaining trunk by letting rainwater seep into its interior via the jagged, open terminus—the point where the blast took place. Dazy holds back a branch for me. "Biologists love blasted tops," he says as I step through the opening. Provided, that is, that the work is not done during anyone's nesting season.

Dean has marked a large Douglas fir for action. He breaks off a leathery-looking disk, one of a half dozen protruding from the bark. "This is a conk," he says, handing it to me, smiling vaguely. Dean has a kind of ongoing low-grade smile, though never really seems pleased. A conk is the tip of the iceberg, rot-wise. The symptoms of fungal infestations are often hidden until the disease is well-entrenched. By the time conks show up on the outside of a tree, the inside is far rotten.

Still, there's no real rush to take action. This tree has had conks the entire fifteen years Dean has been monitoring it. The bark comes off easily in vertical chunks, like wax drippings off the side of a candle. Dean breaks off a piece of bark and crumbles it between his fingers. Insects take advantage of the punkiness to work their way in and lay eggs. Contributing to yet more punkiness. "See this white powder?" Dean says. "This is frass." Frass is insect excreta, and my favorite new word of the day. It replaces *kerf*, which refers to the width of the space left by a saw-blade cut and is a useful word for Scrabble.

Dean walks out to the tree's drip line, the outermost reach of the branches overhead. This typically indicates the end point, underground, of a tree's roots. He shows me where the root mass is starting to lift on one side, because the tree is leaning. Danger tree! Dean adds it to the work list for tomorrow.

Lately, the trees of Cathedral Grove are succumbing to a root rot called *Armillaria* that spreads underground, an infected tree passing the fungus to its neighbors where the roots touch. Cedars prevail. They have chemicals that resist many of the fungal rotters (thus the wood's popularity for roofing shingles and outdoor furniture). The current situation in the grove is ideal for cedars. They need a fair amount of light, so as their less rot-resistant neighbors perish and fall, the cedars benefit from the newly admitted sun. It's all cyclical, Dean is saying. At some point drought will take the cedars, and another species will thrive where they've perished.

Dean does a lot of tapping and sounding to gauge the extent of a tree's inner rot. He's been dropping Latin names faster than I can misspell them. Dazy keeps it simple: heart rot, butt rot, root rot. Dean and Dazy used to teach a falling safety class together. Dazy would talk about technique, and Dean covered regulatory matters. Dazy would throw some *fuck*s into his first lecture to put the students at ease. Dean is not a cusser. His gear is immaculate, and he fills out paperwork promptly and legibly. He is exactly who you want keeping records on dozens of two-ton trees that might topple over onto people.

As different as these two men are, they are similar in the degree to which they don't fit my lumberjack stereotype. A few minutes ago, the group was comparing notes on their diets. Dean has two friends who each lost forty pounds on the keto diet, eating bacon "like it's going out of style."

"Man, I could do that," a falling safety advisor said dreamily.

Dazy volunteered that he is doing high-fat/low-carb, for his heart, but remains wary of bacon. "I try to sorta dwell on avocados. And fish."

"Fish," Dean agreed. "There you go."

Dean has tagged six trees for blasting tomorrow morning. We set a time to meet back here and call it a day. No one goes for beers. Carbs and all that.

The explosives are stored in an unmarked silver shed in the woods, five miles up a dirt logging road. *Shed* is the wrong word. Technically, an explosives storage structure is a "magazine."* This one has walls six inches thick and filled with gravel, so yahoos and hunters with poor aim can't shoot through them and blow the surrounding forest to mulch.

It's 5:00 a.m. The sky is still black, the Milky Way at maximum milk. A half dozen men from a road crew mill around in the headlights of trucks, carrying bags of Austin Powder Company explosives. I watch Dazy load five "sticks" of Red-D into the bed

---

* You can learn all about these structures on www.explosivestoragemagazine .com, which I at first took to be an online periodical about explosives storage. But it's just a redundancy. The industry does have its own periodicals, however. For instance, the *Journal of Explosive Engineers,* which I would subscribe to for the title alone.

of his truck. It comes in plastic tubes and looks more like cookie dough than dynamite. Like many products in Canada, Austin's are bilingually labeled. "*Explosifs*, Explosives"—a rare instance where the French is briefer. At a supermarket in town I saw a bag of "*nourriture pour oiseaux sauvages.*" Birdseed. Dazy wires a Day-Glo sign to the cab of his truck: TRANSPORTING DANGEROUS GOODS. Now if we're in a fiery crash on the way to the grove, the emergency responders will know to keep their distance. When we arrive, a morning meeting is convening around the hood of someone's truck. Dean is here, and the falling safety advisors, and some men to cut up, or "buck," downed treetops. Because the trees are near the highway, coners and flaggers are also here, to halt and direct lanes of traffic.

Dazy steps into his climbing harness and gets ready to ascend the first tree, a fir. He buckles climbing spurs onto his lower legs. By kicking the spurs into the sides of the tree—left, right, left— he ascends the trunk. Holding his upper body is a flip line looped around the trunk and into his harness. After every few kick-steps, he uses the flip line to pull his body in close to the trunk, then flips the now slack line a foot or so higher. And repeat, all the way up to the height at which he'll bore the hole for the explosive. Dazy has no fear of heights and has never fallen. "That seems to me a once-in-a-lifetime sorta maneuver," he said, when I'd asked.

It's cold and drizzling and barely light out. A safety advisor lends me a work coat. Wood chips in the pockets. I can hear the flaggers' chitchat over Dean's radio handset. They're at either end of the work zone, alternately stopping and waving on single lanes of traffic. "Hey," one radios to the other. "Here comes yer girlfriend."

Dazy lets down a rope, and the falling safety advisor ties the chainsaw to it. "There's probably a special knot, but we're not going to use it."

The chainsaw ascends and Dazy unties it and then lets us know the rope is coming back down. Sawdust and noise begin spewing from his perch. When the hole is bored, the chainsaw Rapunzels down, and a backpack with the explosives goes up.

Fifteen minutes later, Dazy's work is done. He climbs back down, trailing fuse. Dean spools it out to the detonation site, three hundred feet distant. We all follow. The flaggers radio that traffic is stopped in both directions, and Dean blows the air horn, twelve blasts. I'm the guest detonator. I get to stomp the "thumper." This sets off a chain of minute explosions that travel, in an instant, along the shock tube fuse. Now comes the boom, followed by two sharp cracks as the tree's top crashes through branches of an adjacent tree, then the thunderous *whump* as it hits the ground and a coda of exuberant whooping from everyone except Dean. If a tree falls in the forest and no one is there to hear it, that's a shame.

Dazy leads us back to the blast site. The accomplishments of "fragmentation and heave" lie all around us. A team of buckers slice the fallen top. The remaining tree, from our viewpoint down here among the mosses and ferns, looks just the same. Though of course it is different. It's safer.

The safety advisor is still grinning. I am, too. I'm not sure why big (controlled) explosions cause humans such glee. We seem to be drawn to extremes: huge, tall, loud. It's the pull of awe. It's one reason we care about whales and not sprat,* why people hug trees and step on clover.

No surprise, then, that Dazy's work in this grove has from time to time drawn complaint. He once tried to talk to a protester, to explain that these trees were dying, and that they'd be coming down soon(ish) anyway. To which the protester replied, "We think the trees know when it's their time to fall down." Of course,

---

* The Abrau sprat is among 455 critically endangered fish, none of which are featured in conservation fundraising campaigns. Who will save the eightgill hagfish? Who cares about the razorback sucker and the delta smelt?

it is not knowledge that prompts a tree's fall, but some fatal brew of wind and gravity and damage and rot.

I can't judge. We all have emotional connections to certain branches of the tree of life, and for some that branch is trees. We are irrational in our species-specific devotions. I know a man who won't eat octopus because of its intelligence. Yet he eats pork and buys glue traps for rats, though rats and pigs are highly intelligent, likely more intelligent—I'm guessing, for I have not seen the SAT scores—than octopuses. Why, for that matter, is intelligence the scale by which we decide whom to spare? Or size? Have the simple and the small less right to live?

Trees, the elders in particular, seem to evoke an urge to protect and defend. Perhaps that's because the trees can't do it themselves— or not in ways easily evident. A tree can't run away or fight back against anything larger than a beetle. Trees are vulnerable, peaceable, innocent. Plants in general have that vibe. Don't be fooled.

8

THE TERROR BEANS

The Legume as Accomplice to Murder

Like the FBI, the USDA has its lists of top criminals. Noodling around on the Federal Noxious Weeds List and other roundups of most wanted invasives, I came upon a plant called the rosary pea (or in India, the jequirity bean): *Abrus precatorius*. What caught my eye was a photograph of the plant's seed, a striking red and black bean instantly familiar to me, because I have two of them on my desk at home. They were given to me on a rain forest walk in Trinidad, by a guide who called them jumbie beads and said that locals wear them to ward off evil spirits. What he didn't say, and maybe didn't know, was that the pretty seeds of *Abrus precatorius* are the source of abrin, arguably the most lethal phytotoxin (plant toxin) on earth. Abrin is on the U.S. Health and Human Services list of Select Agents and Toxins, alongside the likes of ricin and Ebola virus. Possession of anything over a gram of abrin is a federal crime.

Possession of rosary peas, however, is legal. The internet has thousands of rosary pea necklaces and bracelets* on offer, as well

---

* But oddly, no rosary-pea rosaries. And only one 1931 JEQUIRITY BEAN PURSE *POISON IF EATEN*—as the seller listed it.

as crafting websites selling the beans in bulk to people who make these items. I looked at my beans and I thought of all the times the grandkids had been over. What would happen to a toddler who picked them up and swallowed them?

Likely nothing much. The seed's hard casing stands up to gastric juices and travels the gut intact. Fortunately, toddlers don't have molars during the "oral exploration" phase, when they try to put the things they encounter into their mouth. A parent whose child got into someone's rosary peas might only know about it because the toddler had begun shitting jewelry supplies.

Virginia Roxas-Duncan is a supervisory biologist with U.S. Army Medical Research Institute of Infectious Diseases, which researches biowarfare countermeasures. She wrote about abrin for the *Journal of Bioterrorism & Biodefense* and played with rosary peas as a child in the Philippines. "He had diarrhea," she said, of a playmate who once ate some. "But the following day we were playing again."

Even individuals who chew their rosary peas will likely pull through fine. Attempted suicide by *Abrus precatorius* is not uncommon in rural southern India, where the plant is easy to find and other agents of self-death are not. In 2017, the *Indian Journal of Critical Care Medicine* published a review of 112 suicide attempts. Six ended in death. In 14 percent of the cases, there were no symptoms at all.

It's much the same story with castor beans, the source of abrin's higher-profile cousin ricin. Like rosary peas, they are easy to come by legally, as both the plants and the seeds are sold by nurseries as ornamentals. (Though if you clean out their entire stock, as one questionably stable Washington State individual did, staff may alert the FBI.) *Clinical Toxicology* reviewed cases of castor seed ingestions logged with a Midwest poison control center over a ten-year span, 84 cases in total. Forty percent were suicide attempts—using a median of 10 castor beans. The other 60 percent were unintentional ingestions with a median of 1 bean—likely those intrepid diapered oral explorers. The seeds were crushed or chewed in 60 percent of the cases. No one died or was seriously ill as a result. Mostly there was vomiting and diarrhea.

Oddly, swallowing pure ricin, rather than castor seeds, appears (from mouse data) to be even less likely to be fatal. Montana State University biochemist Seth Pincus studies the toxin's therapeutic potential, and is developing treatments for people exposed to it. In his own tests, mice consumed the equivalent, for you or me, of about a Coke bottle's worth of concentrated ricin before succumbing. Pincus's theory is that oral bacteria may be absorbing the pure toxin, and stomach acid and enzymes degrading the remainder. Whereas if someone consumes ground castor seeds, the plant material acts as a sort of time-release mechanism, protecting the ricin in the mouth and stomach and delivering it fully operational to the intestine.

Just by the way, it is not ricin that makes castor oil an effective purgative. As the International Castor Oil Association takes pains to assure us on its website, the ricin is left behind when the oil is extracted from the seeds. Unless death by diarrhea* and dehydration is the intent, castor oil is a useless murder weapon. Casey Cutler had not visited the International Castor Oil Association website before he headed out to an Arizona Albertson's in the summer of 2005 to buy castor oil for the purpose of extracting ricin. George Smith, a senior fellow with GlobalSecurity. org, details the case on theregister.com. Cutler owed money to a

---

* A technique favored by Mussolini's squadristi thugs. Political foes were force-fed large quantities of castor oil—up to a quart, according to The Straight Dope. *Who does that?* Moreover, why? To kill by dehydration? To humiliate? I could find no satisfying answer, not even from the International Castor Oil Association, which, despite large quantities of emails, had no comment.

drug dealer and had hatched a plan to offer ricin masquerading as recreational drugs, should the man show up to collect. While Cutler was tinkering with his castor oil, his roommate began feeling ill. Fearing it might be ricin poisoning, the roommate went to the emergency room. It was just flu, but at the mention of ricin, medical personnel called in a potential terrorist situation and a Phoenix SWAT team descended upon the apartment. Cutler served three years for, essentially, possession of a laxative with criminal intent.

Cutler had one thing right: ricin via a needle in the arm would—if you actually had some in the needle—most certainly kill a man. By injection, the lethal dose (for a mouse) is in the neighborhood of a millionth of a gram. In 1978, Bulgarian dissident Georgi Markov was assassinated by a speck of ricin shot into his thigh with a pneumatic spy umbrella as he stood at a crowded London bus stop.

Assassination by abrin injection dates at least as far back as the nineteenth century, when a spate of cow killings was linked to a group of leather workers in the south of India. The technique is laid out in detail in *Pharmacographia Indica: A History of the Principal Drugs of Vegetable Origin, Met With in British India*. Ground rosary peas were made into a paste and shaped into a stout needle, called a *sutari*. This was then dried in the sun, honed, and affixed to a wand. As the cow was whacked,* the point would break off under the skin, leaving little trace of the crime.

---

* A few *sutari* crafters branched out into murder for hire. An 1890 issue of India's *Police Gazette* profiles the "great poisoner" Dooly Chamár, eventually caught and sentenced to "transportation for life." As a punishment, a lifetime of riding public transportation in India made a certain amount of sense; however, that's not what was meant. It meant he was exiled. Also initially confusing was the locale of exile—a white-sand idyll in the Andaman and Nicobar Islands. Before the dawn of tourism, the islands housed a penal settlement, where British colonials tested torture techniques far more heinous than anything dished out by Indian Railways.

And now I understand why intercepted terrorist communications have, on occasion, mentioned plans to explode suicide bombs that contain ricin or abrin. The shrapnel would act as tiny *sutaris*,* injecting the poison into otherwise survivable flesh wounds. "To give the conventional bomb a more lethal effect," explained a piece in the online edition of *The Diplomat*. Whereupon the paragraph broke for a subscription come-on: "Enjoying this article? . . ." I don't know, I had to say, does one *enjoy* an article about the mass slaughter of innocents via poison-tinged shrapnel?

You may be thinking, Did these terrorists perhaps intend for innocents to inhale the toxins? Perhaps so. Ricin administered in this manner is approximately as lethal as it is by injection. "Massive pulmonary edema," Pincus volunteered. "You drown in your body fluids." However, to kill a crowd of people in this manner, your terrorists would have to possess equipment and expertise sufficient to create a cloud of extremely fine ricin aerosols—ideally no larger than one or two microns. Otherwise the mist won't stay airborne long enough to pose a threat to large numbers of people. (Fine aerosols, compared to droplets, also penetrate more deeply—that is to say more dangerously—into the lungs, a discovery I was very much not, in the midst of a COVID-19 pandemic, *enjoying*.) Anyway, the terrorist cells in question—al Qaeda in the Arabian Peninsula (ricin) and Jamaah Ansharud Daulah (abrin)—did not have sophisticated aerosol dispersal systems. They had humans wearing bombs, bombs more likely to incinerate a toxin than to disperse it.

It's also possible that these groups planned to add ricin or abrin to their bombs for the simple purpose of terror. Being terrorists

---

*   The militaries of various Western nations have experimented with this from time to time, but they did not call them *sutaris*. They called them "flechettes"— tiny toxin-dipped arrows, as many as 35,000, packed into a bomb. Both Canada and Great Britain tested flechettes on animals. Despite impressive efficacy, neither nation, perhaps recognizing the rather dim threat of a projectile that sounds like a feminine hygiene product, added them to the arsenal.

and all. Because regardless of how many people a ricin incident harms, it seeds fear in a million more.

"Know the facts. Protect yourself." So go the ominous, ever-present phrasings of public awareness campaigns. You see it on websites about HIV, dengue fever, and Zika virus. Lead poisoning, identity theft, date rape, TOXIC BEANS.

The capital letters are not mine. The words appear that way on a Utah food-handler testing and certification site. The threat they refer to is not jequirity beans or castor beans. (The castor bean isn't actually a legume. It's a spurge.) The threats are kidney beans, red or white, broad beans, and lima beans. Fail to boil these common edibles for at least ten minutes, and you may find yourself in significant gastrointestinal distress. As did a thousand-plus viewers of a Japanese TV show that recommended grinding white kidney beans in a coffee mill, toasting for three minutes, and sprinkling on rice. According to the journal article "The 'White Kidney Bean Incident' in Japan," a hundred people were hospitalized.

For more on the evil a kidney bean can inflict upon a human, I steer you to "Foreign Body (Kidney Beans) in Urinary Bladder: An Unusual Case Report." In 2018, in Jaipur, India, a young man pushed four kidney beans up his urethra "for the purpose of sexual gratification." As often happens in these cases, the items made their way beyond the point of easy recovery, and once discomfort outweighed embarrassment, the man sought medical help. An ultrasound revealed the kidney beans "floating" inside the man's bladder. Because the beans had been soaking overnight, they had, as any dry bean would, expanded and softened, complicating their extraction. Figure 3, "kidney beans removed piecemeal" is a photograph: a helping of broken, slightly mashed beans in a stainless steel surgical basin—more appetizing than most things removed by forceps in a surgery suite, but probably not, owing to the urine presoak, more delicious.

Is there something uniquely dangerous about beans? I posed this question to plant scientist Ann Filmer, recently retired from the University of California, Davis. In her reply, she included a link for a website she had put together on poisonous garden plants. I was taken aback to note that nine of the 112 plants in Category 1 (Major Toxicity: "may cause serious illness or death") were currently, or had recently been, growing in our yard: oleander, lantana, night-blooming jasmine, lobelia, rhododendron, azalea, toyon, pittosporum, and hellebore. Another, the houseplant croton, was growing in an orange ceramic pot in my office.

In other words, it's not beans. It's plants, period. If you can't flee or maul or fire a gun, evolution may help you out with other, quieter ways to avoid being eaten. Over the millennia, natural selection favors eaters who turn up their proboscis at you, and eventually they all steer clear.

Given the surprisingly large number of deadly garden plants, why is it that ricin gets all the press? Why don't terrorists and assassins extract toxins from these other plants? The answer to the second question likely resides in the first: ricin gets all the press. The Markov murder made ricin a bright, blinking thing on the terrorist radar. It became the go-to poison for two-bit killers and survivalist crackpots. You don't have to go to the dark web to find instructions for extracting ricin from castor seeds. A quick Google search will take you there. But unless you're a criminal *and* a chemist—like Walter White, who extracts a toxin from lilies of the valley in the fourth season of *Breaking Bad*—you probably lack the equipment and know-how to turn any of these other plants into an accomplice for murder.

Ricin's notoriety has given it a sinister cachet other phytotoxins lack. If you're trying to build some cred in terrorist circles, it sounds better to say you're making ricin than to say you're trying to extract something from a rhododendron. This was pointed out to me by Andy Karam, a counterterrorism professional and the author of *Radiological and Nuclear Terrorism*.

And yet. None of this explains why ricin and abrin are the lone plant poisons on the HHS Select Agents and Toxins list. Seth Pincus had an answer for that. Ricin and abrin, he explained to me when we spoke, are "promiscuous" toxins. Ricin makes its trouble by binding to galactose, a carbohydrate on the surface of living cells of all types. (The outermost layer of skin cells is dead, so touching ricin powder would not pose a danger. Save a stamp, would-be postal assassins.) Most other deadly toxins—cholera toxin, say, or botulinum toxin—wreak their havoc exclusively in one site: cells of the colon, say, or on nerve cells.

I forwarded Pincus a web page from a Chinese chemical supply vendor. "RICIN in stock with best price," it said. (I'll say: $150 for a kilogram of ricin, 99% purity.) Abrin came up too, similarly priced. A half dozen such sites pop up when you search for either toxin via its Chemical Abstracts Service number. One vendor advertises free samples for many of its wares—including "horse spleen," though not ricin or abrin.

Pincus hadn't been aware of the website. He used to get his ricin from a researcher at the University of Texas. When ricin was added to the Select Agents and Toxins list, working with it became an extraordinary hassle, and the researcher decided to unload it. She offered her supply—maybe 10 or 20 grams—to the Biodefense and Emerging Infections Research Repository. "They said, 'Great, we'll send someone down to pick it up.' Sure enough," Pincus recalled, "this huge armored truck pulls up with police-type cars escorting it. And now you're telling me I can buy a hundred times that amount on the internet?"

It appears that way, I said. I'll inquire! A day later, two emails from China had arrived in my email inbox. The first one was typed in red letters. It came from someone at the chemical supply finding aid LookChem: "Any information violates the international or national laws . . . is not allowed to be published on Look-Chem. Once we find this kind of information, we are obliged to report it to the state organs." Its impact was diluted somewhat by the second email in my inbox.

"Nice to contact you," wrote Cathy, a sales manager at the Kaimosi chemical company, to whom the LookChem people must have forwarded my email. "I'd like to avail myself of this opportunity to establish business relations with you." Cathy apologized, saying that they did not have ricin in stock, but offered to make up a custom batch. "Our product is of mature technology, good quality and cheap price," she assured me, "and has been sold overseas and won wide praise." Cathy asked how much I required and when I would need it by.

"I don't know how far you should take this," said Pincus, when I showed him the email. FBI agents used to drop by his office from time to time (where he had a castor plant growing in the corner, unnoticed by his federal callers). Just to chat.

I asked George Smith whether the FBI monitors traffic on these websites. He said they did, but he imagined they'd long ago struck this one from the list. He highlighted the line on the Kaimosi ricin page that reads, "Store in a cool, dry place." A purified protein would have to be refrigerated, Smith said. "If anything, it's minimally treated castor mash." Most likely I was dealing with the kind of individuals who would "sell you any white powder for a hundred bucks a kilogram."

Given the challenges of aerosolizing ricin and the less-than-genius reputation of the typical "castor seeds putterer" (quoting Smith), I imagine the feds don't worry much about ricin as a weapon of mass destruction.

According to Pincus, they *should* worry—though not about ricin in the air or the food supply. They should worry about this: "You can take a ricin gene and dump it into a highly contagious virus, like influenza." Now you've got a bug that not only infects millions but also kills millions. (Of course, you'd first want to have a vaccine to protect the millions you *don't* wish to kill.) "I've heard people with homeland defense say, 'Oh, we keep a good control on whatever genes are synthesized commercially, blah blah blah.'" Pincus takes no comfort there. He related another story.

"For therapeutic reasons, we wanted to synthesize a gene that encoded the toxic part of ricin, that would be expressed—i.e., produced—in human cells. This should have set off a lot of red flags if anyone was looking for that kind of thing. But, man, we ordered it, and we had the gene two weeks later. So if you think the Select Agent list protects us from sophisticated terrorists . . ." I'll finish the sentence for him. It's a load of horse spleen.

Also: Terrorists? Try the military of some rogue nation. Try our own nation. Beginning in World War I and carrying on through World War II, the U.S. military experimented extravagantly with ricin. They mixed it in with shrapnel in grenades. They loaded it into four-pound aircraft bombs. They rigged up sprayers with dissolved ricin (or sometimes just castor seed powder). Nothing worked as they'd hoped.

Eventually they shipped some off to the nation's wildlife research labs, one in Colorado and one in Maryland, to try it out on rats.

Warfare and pest control long strolled hand in hand. Both, after all, seek to destroy grouped adversaries as efficiently as possible. Up until the nuclear era, any newfangled deadliness trained upon human enemies also tended to be tried out on enemies of the furred and feathered variety. A United Nations summary of control efforts used on the African quelea (or "locust bird"), for example, reads like a timeline of military weaponry: "guns, explosives, flame throwers, jellied gasoline, and contact poisons."

During World War II, the chemical warfare people and the agripest control people were united by a common foe: the brown rat. Aka the Norway rat, aka the sewer rat, aka—quoting a Denver Wildlife Research Laboratory press release—"Hitler's ace agent in this country." Supply routes for the raw materials for existing rat poisons had been cut off during the war, and rats were living high: ". . . sabotaging factories, destroying food needed for our allies, and spreading disease among our armed forces." This was not the first time rodents had been portrayed as enemy sympathizers. Posters for a California ground squirrel eradication campaign during World War I featured squirrels in tiny spiked German helmets. "Mrs. Squirrel" has been decorated with the Iron Cross, one of the highest military honors of the German Empire, which she wears as a necklace.

In June 1942, an unusual wartime alliance was forged. Division 9 (chemical weapons) of the National Defense Research Committee (NDRC) of the U.S. Office of Scientific Research and Development joined forces with the Denver Wildlife Research Laboratory (now the NWRC) in a quest for new rat poisons. The former suggested promising toxins from the arsenal, and the latter tried them out on traitorous vertebrates. Ricin, under the code name "compound W,"* was a candidate, as was sarin. The standout rodenticide, first tested in June 1944, was a phytotoxin the Division 9 people referred to as 1080. It was cheap and quickly lethal to rats.

Long before the ministries of war and agriculture got wind of it, 1080 had been used in rural Africa, in its natural plant form.

---

* Did the makers of the wart-removal product Compound W realize this when they named their product? I don't know, because Prestige Brands, which owns Compound W, doesn't return calls, their online media query form is a dead-end, and they're not on Twitter. But while we're on the topic of inappropriate names, let's consider "Prestige Brands." Because here are some more of their prestige brands: Fleet enemas, Nix for lice, Beano for flatulence, URISTAT, Nōstrilla decongestant, Summer's Eve douche, Boil-Ease, Efferdent denture cleaner, and Boudreaux's Butt Paste.

Then, too, both rodents and humans were potential targets. Because the toxin imparts almost no detectable taste, the aggressor would simply crush the plant and drop it down the enemy's well. While I might question the efficacy of the raw plant material, the lethality of the toxin eventually isolated—a fluoroacetate given the code name TWS—is well documented.

TWS was accidently discovered by a team of Polish chemists who later shared it with Allied intelligence. According to a declassified Division 9 memo dated April 20, 1945, a related fluoroacetate was under consideration as a "contaminant for water supplies" but never used. Personnel quoted in the memo had been shown footage of poisoned dogs and deemed it "an extremely revolting spectacle." (1080 is 17 to 35 times more deadly to dogs than it is to rats.) Both felt strongly that an agent causing death in this "horrible" manner "could not possibly be used by any civilized nation against an enemy even of the most highly depraved type."

So off it went to the Denver Wildlife Research Laboratory to test on rats. One of the laboratory's press releases describes a 1080 trial run at an infested New Orleans grain elevator company. The chemical was dissolved in water and left in half-ounce, rodent tea party–sized cups along known rat runs. Within twenty-four hours, the release stated, 3,690 rats* were dead. Less prodigious but still impressive body counts are given in the 1945 "Summary of Field Reports on 1080," compiled by the NDRC's Rodent Control Subcommittee. One thousand and thirty-two pounds of 1080 was cooked up by chemists at Monsanto and shipped off to rat-besieged army and navy installations and public health departments for field trials.

---

*   That is a serious fucking rat infestation. *"Huge,"* agreed a spokesperson for GEAPS, the Grain Elevator and Processing Society ("the knowledge resource for the world of grain handling and processing"). The GEAPS man said in an email that even in less developed nations with shoddy storage facilities, "that many rats seems extremely high and uncommon."

Along with the kill tallies, the summary details the edibles each location used as bait. These ranged from standard rat bait fare— barley, oats, sweet potatoes, coconut, chocolate, peanut butter—to more culinarily creative concoctions. Naval Base Guam mixed their 1080 with dried eggs, Mazola, and fresh bacon grease, while the Ninth Service Command fashioned a 1080 meatloaf using horse meat and bread crumbs. First Service Command blended 1080 into Meat Hash C-rations. The secret ingredients of the Texas State Board of Health? Popcorn and chicken feed. The summary's authors shared a few of their own favorite bait recipes, complete with cooking directions. ("Blend the 1080 and the flour. Dust the flour-poison mixture over the cubed vegetable, with continuous stirring.")

All was not copacetic, however. Dogs were taking the bait. Or eating the bodies of the dead and dying rodents. And dogs, recall, are exquisitely sensitive to 1080. One test reported fifty had died. The government agencies put their heads together. They made poison lemonade out of lemons. 1080 was registered as a predacide for ranchers to use on coyotes.

The ranchers had a new problem now: how to keep the pre-dacide from also being a man's-best-friend-icide. Especially since afflicted coyotes tended to run off and, in the words of Rodent Control Subcommittee chairman Justus Ward, "regurgitate a siz-able quantity of undigested poisoned food . . . over a considerable area." Which ranchers' dogs would find and eat.

Ward turned to Colonel C. P. Rhoads, chemical weapons whiz at the Defense Department's Edgewood Arsenal. For the pre-dacide version of 1080, Ward politely asked, "Perhaps you could suggest a drug which could be included with the 1080 to reduce the tendency of a coyote to vomit." And for the 1080 rat-killing formulation, might there be something that would do the oppo-site? An emetic—a vomit-inducer—"so that poison baits which were eaten by dogs would be thrown up much more rapidly than is done when emesis is caused by the 1080 itself." But would that not also make the rats throw up the poison—and thereby survive? It would not, because, to quote a then-secret memo from Birdsey Renshaw, of the National Defense Research Committee of the U.S. Office of Scientific Research and Development, "rats cannot vomit."

World War II ended, but the toxin screening program continued. Over a span of forty-five years, some fifteen thousand potential poisons and repellents were tested by the Denver Wildlife Research Laboratory. Under pressure from environmental and animal welfare advocates, the chemists grew choosier. They looked for poisons that were not just cheap and deadly but also more specific to the animals they sought to control. The poison DRC-1339 seemed to fit the bill, literally and colloquially. The big-flock crop-eaters—blackbirds, starlings, cowbirds, grackles— were all extremely sensitive to it. This was exciting news for the NSA: the National Sunflower Association.

For forty years, the NSA has supported the interests of sunflower farmers, most of whom are in North and South Dakota, smack in the migration path of tens of millions of blackbirds and lesser feasting flocks. You can appreciate the challenge. They are *attempting to keep birds from eating birdseed.* Surveys of blackbird damage to North Dakota sunflower fields between 2008 and 2010 showed average losses of around 2 percent of the crop.

The National Wildlife Research Center has a branch in Fargo, North Dakota, devoted to the sunflower problem. It has been an enduring challenge. They've developed repellents, but there are application problems. Because of the sunflower's unique asana, its "downward-facing head position," aerial crop sprayers end up dosing the undersides of the flowers and missing the seeds. They developed a hybrid with tightly packed seeds that the birds couldn't pull free, but the fat content of the seeds was low. That is a grievous drawback for the sunflower farmer, because the real money in sunflowers comes from oil, not birdseed. (Sunflower seeds make up a modest percentage of most birdfeeder mixes, which is good, because it would be hard to put aside the irony of annihilating birds for a product used exclusively by bird lovers.)

All along the way, the NSA kept up the push for poison. Things came to a head in 2006, when Frito-Lay announced plans to begin frying its major potato chip brands in oil from NuSun, a sunflower variety whose seeds contain no trans fat. To meet the demand, sunflower growers needed hundreds of thousands of acres of new plantings. The NSA sought permission for an increase in the annual "allowable take" of blackbirds. The South Dakota Game, Fish and Parks Department pushed back; ring-necked pheasants, a popular game bird, also turned out to be sensitive to DRC-1339. The National Wildlife Research Center, in 2003, had published detailed data on the sensitivities of dozens of nontarget bird species to DRC-1339, data that the original screening program didn't have or, in some cases, hadn't published. Not just pheasants, but also cardinals, jays, robins, bobwhites, meadowlarks, mockingbirds, tree sparrows, and barn owls showed high sensitivity to DRC-1339. The information has quelled somewhat my enjoyment of the salty, bagged snack foods. *Blood potato chips!*

Killing one or two million of the seventy million blackbirds that descend on the northern plains each year is like trying to solve global warming with an ice machine. The poisoning campaigns seemed almost more like spite than pest control, a practice undertaken out of frustration and anger, rather than documented results. Population modeling undertaken by a team of NWRC researchers in 2002 concluded that if the annual allowable take of migrating blackbirds was increased to two million, the benefits to sunflowers farmers "would likely be negligible." Yet the killing continues. In 2018, USDA Wildlife Services destroyed 516,000 red-wing blackbirds, 203,000 grackles, and 408,000 cowbirds.

The irony is that the sunflower growers have long known what works best. As early as the 1970s, contributors to NSA member magazine *The Sunflower* had been recommending effective nonlethal approaches: "primarily habitat and crop management." Give the birds something equally delicious to eat: plant an inexpensive, strategically placed "lure crop." Rather than plowing under the post-harvest plant stubble, leave it for the birds, to distract them from other growers' fields. Apply drying agents so the seeds are ready to harvest sooner, before the flocks arrive. Don't plant sunflowers near marshes with dense stands of cattails or other attractive blackbird habitats. In other words, take the advice of that endlessly quoted general whose name sounds like a new sunflower hybrid:[*] Know thy enemy.

Lately, habitat management has meant killing cattails. As it had with 1080 back in the day, Monsanto supplied the poison: the contentious herbicide glyphosate (the active ingredient in Roundup). Glyphosate, I just read in a 2012 paper by four NWRC researchers, has been one of the chemicals used to dry out sunflowers for early harvest. Long sigh and downward-facing head.

In war there is always one last option: surrender. "An obvious bird management strategy," wrote George Linz and Page Klug, in their chapter of *Ecology and Management of Blackbirds (Icteridae) in North America*, "is to abandon a bird-susceptible crop and substitute other crops . . . that are not damaged by blackbirds." Or quit farming altogether and go into politics, like North Dakota sunflower grower turned senator Terry Wanzek. "We've surrendered," he told Associated Press reporter Blake Nicholson. "The birds won."

---

[*]   Sun Tzu.

9

OKAY, BOOMER

Futile Military Actions Against Birds

As far as I can tell from newspapers of the era, the practice known as crow-bombing reached its zenith near the tiny town of Asa, Texas, on February 6, 1953. Joe Browder, "who hates crows," took 150 pounds of dynamite and parceled it out to make 300 bombs. Half-sticks of explosive were packed, along with metal shards from a local foundry, into cardboard tubes and strung together in the scrub oaks along the Brazos River where the birds returned each night to roost. By one giddy estimate, 50,000 crows lay dead in an instant.

Why didn't someone call the authorities? The authorities were already there. The local game warden was part of the team stringing up the bombs after the crows had flown off for the day to feed. And there was the birds' crime: feeding. A common concern expressed in the newspapers of the time was that crows—"black bandits of the air," "feathered gangsters," the "menacing sable flood"—were raiding the nests of waterfowl, devouring eggs and chicks to the extent that duck hunters would not have enough of them to shoot. Crow-bombing, in fact, was a government-sponsored conservation effort. Who oversaw the bombing deaths of 250,000 Illinois crows during the winter of 1935? The Illinois conservation commissioner. Who offed 3,000 crows at a roost near Coupland, Texas, flew to a wildlife conference in St. Louis, and returned the following Friday to bomb a roost in Creedmoor? The secretary of the Texas wildlife federation.

Such is the inside-out history of conservation in America. It wasn't until the 1980s that the word came to mean what it means to us now. Wildlife and wilderness weren't conserved for their intrinsic value. They were conserved for hunting and fishing. Mighty tracts of wilderness were protected from agricultural and other development to ensure there would always be places and things to hunt and fish. And the ducks were protected from the crows.

Farmers, too, raged against the eating habits of large flocks. In 1952, the agriculture arm of the government gave dynamite a try. The Denver Wildlife Research Laboratory ran a series of "experimental bombing tests" in the trees along a mile-long Arkansas marsh where redwing blackbirds, grackles, starlings, and cowbirds would come to roost after a long day of rice pilferage.* The study aimed to compare the cost-effectiveness of different bomb materials: dynamite versus Primacord, lead shot versus steel, pasteboard mailing tube versus ice-cream carton versus can. Because of this work, we now know, for instance, that the "average kill per

---

* Let us here lay to rest the myth that birds' innards will burst if they scavenge the rice thrown at weddings. Over the years, this enduring bit of misinformation migrated as far as Ann Landers's column and the Connecticut state legislature. In 1985, Representative Mae Schmidle proposed "An Act Prohibiting the Use of Uncooked Rice at Nuptial Affairs." The Audubon Society called hooey, pointing out that migrating birds feed on fields of rice. Some churches ban the practice anyway, not because it's perilous for birds but because it's perilous for guests, who could slip on the hard, round grains and fall and then fly off to a personal injury lawyer.

bomb" for dynamite and lead shot packed into a mailing tube is
1,820 birds, at a "cost per bird killed" of less than a cent.

What the experiment did not reveal was the extent to which
killing what amounted to 1 or 2 percent of an enormous flock
had a noticeable effect on farmers' losses. That assessment in fact
already existed, in the form of data collected over a decade of
bombing crows in Oklahoma. From 1934 to 1945, 127 crow roosts
were dynamited "to reduce predation on waterfowl eggs and damage to grain crops," writes longtime USDA ornithologist Richard Dolbeer in his chapter of *Ecology and Management of Blackbirds
(Icteridae) in North America*. An estimated 3.8 million crows were
killed, yet, Dolbeer writes, "no evidence was obtained to indicate the explosions influenced total population levels, agricultural
damage, or waterfowl production."

Here's the thing with killing as a wildlife damage control tool.
It isn't just mean. It doesn't—barring wholesale eradication—work.
On a visit to the National Wildlife Research Center, I spent a
morning in the archives, reading oral history transcripts of some of
the long-timers. One in particular stayed with me. Weldon Robinson began his career as a government bounty hunter, killing coyotes
for $3 a head. He was soon hired by the Denver Wildlife Research
Laboratory, where he rose through the ranks. By 1963, Robinson
oversaw four units, including Bird Control, Predator Control, and
Agricultural Rodents. He was the czar of nuisance wildlife offing.
Somewhere along the way, Robinson's oral history reveals, he had
an epiphany regarding the animals whose numbers the agency had
for so long fought to reduce: "Mother Nature adjusts."

Robinson had landed on the phenomenon of compensatory
reproduction. Destroy a chunk of a population, and now there's
more food for the ones who remain. Through a variety of physio-
logical responses—shorter gestation periods, larger broods, delayed
implantation—a well-fed individual produces more offspring than

one that's struggling or just getting by. With ample food, both the well-fed parents and their well-stuffed young are more likely to survive and reproduce. Coyotes, for instance, may have three pups when food is scarce, and eight when it's plentiful. That figure comes from *Resolving Human-Wildlife Conflicts*, a technical press book by Michael Conover. Conover is a former director of the Jack H. Berryman Institute, which sponsors research into methods of human-wildlife conflict resolution (plenty of them lethal). Conover adds that killing coyotes opens up territories for males that would otherwise not breed. Bottom line: in order to end up with a net reduction in a coyote population, humans would have to destroy at least 60 percent of them annually.

Robinson shared his catchphrase for it. "Natality is more effective than mortality," he said to the interviewer. Catchy, but a bit abstruse. She changed the subject. He went right back to it. Here I pictured him leaning forward in his chair. "Natality is more effective than mortality," he repeated. Whereupon the interviewer said she had no more questions, and they looked at some old photos and marveled at how a colleague named Johnson Neff used to have hair.

Another thing I ran across in the oral histories was a mention of the practice of overestimating flock sizes in order to secure federal funds for bird control. The center's head of bird depredation control was explaining to the interviewer how he'd estimate the size of a flock as it flew between two points overhead. The technique wasn't perfect, but it was, he said, better than just making up a big number, as some states were doing. "'There's twenty million birds! We need more money.' [*Chuckles.*] Yeah, the farmers would do the same thing. 'I got two hundred thousand birds in my field!' I need more! [*Chuckles.*]"

It wasn't even clear, in the case of blackbirds eating rice, that the take represented a significant loss. By the estimates of a 1971 Fish and Wildlife Service Resource publication, the average loss per acre was between half a bushel of rice—by volume, about four gallons—and just over one bushel. That is less, the document goes on to say, than the amount that would typically fall to the ground as a combine harvester moved through the crop (a loss of 1.6 to 8 bushels per acre).

And then there's this: birds provide significant pest and weed control services to the farmer. Based on an examination of nearly five thousand blackbird stomachs, Foster Ellenborough Lascelles Beal of the Bureau of Biological Survey (a forerunner of the NWRC) had this to say: "Judged by the contents of its stomach alone, the redwing is most decidedly a useful bird. The service rendered by the destruction of noxious insects and weed seeds far outweighs the damage due to its consumption of grain." The report was signed by none other than the secretary of agriculture. Today you have to go to organic farming organizations* to find this sort of information. (The Wild Farm Alliance, for instance: Based on stomach contents, "each Tufted Titmouse is worth about $2,900 to the pecan industry.")

Bird-bombing was soon abandoned, Dolbeer wrote in the blackbird book, "for various obvious reasons: labor, expense, hazards involved, large rate of crippling . . . and lack of effectiveness in solving problems." It was odd, then, that the Arkansas bombing experiment report concluded with a thumbs-up. The results, stated researchers Johnson Neff (Neff! With hair!) and Mortimer Brooke Meanley Jr., indicated "the effectiveness and economy of roost bombing."

Were Meanley and Neff, like Joe Browder, simply men with a hatred of crop-raiding birds? And/or a love of setting off bombs? Bird-bombing seemed to peak after World War II. I began to

---

* Or Mormon history books. In 1848, a flock of California gulls flew over from the Great Salt Lake and descended like the hand of God upon swarms of insects devouring the settlers' crops. Which explains why to this day, the state bird of Utah is the California gull.

wonder if it was the product of some lingering zest for combat, some wartime hangover of misplaced patriotic zeal. An obituary for Meanley put that notion to rest. His World War II service took the form of rehabilitating soldiers by taking them out into nature. "He would comment on the unbelievable good fortune of meeting his military commitment by taking soldiers on bird walks." Both Meanley and Neff were respected ornithologists.

How, then, did this gentle bird lover, author of *Natural History of the Swainson's Warbler*, end up dynamiting blackbirds in an Arkansas bayou? The same way, I suppose, that entomologists end up developing insecticides and wildlife biologists find themselves having to put down bears. The jobs are few, and that's where they lie. For someone whose know-how is birds, their control was one of few ways to make a living.

I can't say why Meanley and Neff endorsed bombing. Perhaps they never read Mr. Beal's blackbird report or the paper on the lackluster outcome of the Oklahoma crow-bombing campaigns. It would seem no one read Neff and Meanley's paper, either, because blackbird control quickly shifted away from bombs.

And over to chemical warfare. Five years later, Neff and Meanley were out in the fields again, scattering strychnine-treated grains around the roosts of blackbirds and cowbirds. Both species, wrote Dolbeer, "generally avoided the baits." Or maybe just avoided Johnson Neff and Mortimer Brooke Meanley.

The war on thieving birds has occasionally strayed beyond the realm of metaphor and taken the form of an actual military operation. In October 1932, the Australian minister for defense agreed to dispatch a pair of machine-gunners, led by a Major G. P. W. Meredith, to rout mobs of emus trampling and helping themselves to farmers' wheat in the state of Western Australia. (The defense minister had denied the farmers' original request, which was to borrow some machine guns.) In return, the military asked only for the feathers of the fallen, to be used as a flourish on the hats of the nation's light-horsemen.

The emu proved a tougher adversary than Major Meredith and his gunners had anticipated. Though flightless, the bird is a fast sprinter, reaching 30 mph with proper motivation. Emus blend with their surroundings and travel with a lookout that, in this case, would sound a warning long before the birds were within range of the guns, whereupon the flock would scatter in a rising billow of dust. On day three, with a confirmed body count of just twenty-six birds, Major Meredith shifted tactics. He set up an ambush, concealing his gunners in the scrub above a dam where emus would come to drink. Around 4:00 p.m., a sizable flock was spotted in the distance.

"The constant craning of necks and the cautious manner of their approach showed they had not forgotten the happenings of the last few days, but relentless thirst drove them on," wrote a correspondent for the *West Australian*. When the birds were a few hundred yards out, Major Meredith gave the order to fire. As the dust settled, the men got up to count the bodies. An underwhelming fifty birds lay dead. Excuses were made. The machine gun jammed, someone told a reporter. Someone else conjectured that the majority of the bullets were passing harmlessly through the birds' plumage, because the emu has "more feathers than flesh." Major Meredith believed hundreds more had been hit but survived. He credited the emu with an almost supernatural ability to "face machine guns with the invulnerability of tanks." He sounded wistful. "If we had a military division with the bullet-carrying capacity of these birds it would face any army in the world."

On day six, Major Meredith withdrew in defeat. "Emus appeared in huge flocks along the road," observed the Perth *Daily News*, "as if to give a mocking farewell." And that was that, almost. Twelve years later, as World War II raged, the upstart wheat farmers of Western Australia would again ask for military intervention. This time they wanted "light bombs dropped from low-flying planes." The request was denied.

Meanwhile, out on the Pacific Ocean, the albatrosses of Midway Atoll were shaping up to be an equally insuperable foe.

The islands of Midway sit halfway between North America and Asia, in the Pacific Ocean. As such, Midway Atoll was strategically important to the United States, and in 1941, the nation built a naval air station there. Midway was (and is) also important to a dozen or so species of seabird, including the tens of thousands of Laysan and black-footed albatrosses that return to the islands each year to lay eggs and raise their chicks. Because the birds had had no predators on the islands, they greeted newcomers—both human and mechanical—not with fear but with a mix of nonchalance and curiosity. They soared over the navy's runways with blithe disregard for the large, loud metal birds with which they shared airspace. Collisions—bird strikes—became an issue.

"We got a bird in the carburetor air scoop." An aviator machinist named Jerry is speaking into the microphone of a navy public affairs man in a 1959 government filmstrip. "Caused a total loss of power in our number three engine."

The microphone tilts back to the interviewer's mouth. His moustache is trimmed to an inverted V, a soaring upside-down albatross of a mustache. "What would happen if you took a gooney bird in the air intake of a Super Constellation radar plane?" ("Gooney bird" was the de facto military handle for albatrosses.) "Do you think you would crash on takeoff? Total loss of plane and personnel?"

"Yes, sir, more than likely."

The interviewer turns to face the camera. "And there you have it. From people who know what they're talking about. It's hard for them to understand the continued existence of these gooney birds here on Midway."

The scene shifts to an office at Naval Air Station Barbers Point on Oahu. We are introduced to Rear Admiral Benjamin E. Moore and his pointer. Admiral Moore stands beside an easel on which is propped a posterboard highlighting key gooney bird cost statistics. "We had five hundred thirty-eight strikes last year." The pointer taps the text BIRD STRIKES 538. He then breaks down the cost of these strikes, beginning with the wages of the men who repair the damage. "Two thousand five hundred twenty man-hours times two dollars an hour equals five thousand forty dollars." Admiral Moore steps to a second easel. The pointer taps FLIGHTS ABORTED 33. Each aborted flight requires the pilot to dump three thousand gallons of fuel to reach a safe landing weight. The pointer directs our gazes to JETTISONED FUEL 99,200 GALLONS and below that, $17,500 COST. The admiral returns to the first easel, where someone off-camera has snuck in and changed the posterboard. A final, decisive tap of the pointer: $156,000 TOTAL.

Admiral Moore steps to his desk. To his left, an American flag hangs on a pole in the sad, limp manner of all indoor flags. The admiral sits. The mood darkens. "It's either remove the gooney birds from Midway and keep the men, or keep the birds and eventually bury a flight crew of twenty-two men. I hope I'll never have the task of explaining to some mother or wife that her son or husband was killed by . . ." Admiral Moore pauses to raise a photograph that the easel sprite has slipped onto his desk. He glowers over the top of an enlarged photograph of an albatross standing calmly on a lawn. ". . . *this*." A dramatic orchestral swell accompanies a fadeout and the words THE END.

The filmstrip is titled *The Second Battle of Midway*. And the battle was a lengthy one, longer than the first, longer than World War II itself. In the beginning, the strategy was straightforward massacre. Because the birds were many, the cost of ammunition high, and the admiral budget-minded, the initial assault was done without guns. It took place in 1941 and is detailed in a Fish and Wildlife Service Special Scientific Report, under the heading "Large Scale Elimination Experiment." Two hundred men carrying lengths of pipe or wooden clubs "spent 6 or 7 hours a day" hitting albatrosses on the back of the head. An estimated eighty thousand birds were killed. "For a brief time the hazard to aircraft was reduced," the paragraph concludes. "The following season there appeared to be as many albatrosses as before."

The strategy shifted to harassment. To persuade the albatross to nest elsewhere, personnel fired off everything they had. Under the heading "Disturbance," a 1963 government review of albatross control efforts at Midway lists "rifles, pistols, bazookas, mortars." "Some birds registered discomfort" but "no appreciable number" left their nests. Ten carbide exploders were set up at intervals of 135 feet along the edge of the runways where the nests were densest. "No decrease in number of flying birds." Personnel burned rubber tires and lit flares to roust the birds with noxious smoke. They tried blasting an "ultrasonic siren," thinking, mistakenly, that the birds could hear ultrasound. At one point a Lockheed WV-2 Warning Star radar surveillance aircraft taxied to within a hundred feet of the nest area and switched on a high-intensity radar beam. With no observed effect.

Unable to harass the birds off the island, the navy entertained the possibility of physically moving them. In another test program, eighteen albatrosses were plucked from the nesting site near the runways, banded, and put on outgoing military transport flights: to Japan, the Philippines, Guam, Kwajalein, and—*aieeee!*—Naval Air Station Barbers Point on Oahu, nesting site of the military's highest-ranking albatross hater, Admiral Benjamin E. Moore. Fourteen of the eighteen flew (sans airplane) back to Midway in time for nesting season. The navy had not realized that although an albatross wanders thousands of miles, it always returns to the same spot to nest.

And no, you cannot just move the nest. The navy tried this, too. The birds would consult their internal GPS, see that the nest had been moved, go back to their actual birthplace, and proceed to construct a new nest. Next the navy tried moving the eggs. Ten thousand albatross eggs were carried to the adjacent island, where the sailors chased the local birds off their nests and quickly installed the kidnapped eggs. It was like those Folgers ads from the 1970s. *We've secretly replaced these birds' eggs with eggs from the neighboring island.* The birds, unlike the instant-coffee dupes, were not fooled.

Exasperated, the navy turned to science for help. On October 2, 1957, Hubert Frings, a Pennsylvania State University professor of zoology, got a call from Washington. Would he be willing to come to Midway Atoll for the December/January nesting season? In other words, would Frings make the supreme sacrifice of spending his between-semester break on a tropical island in lieu of hanging around Altoona, Pennsylvania? You bet. Accompanying him would be his wife, Mable, a librarian and bioacoustician with a special interest in cricket and grasshopper "chirp sequences."*

On the Fringes' very first day at Midway, another "large-scale elimination" was taking place. Despite the documented failure of earlier mass killings, there was talk of clubbing the entire Midway population. Mable went out into the fray with her reel-to-reel, thinking to record the albatrosses' alarm and distress cries, so that these might later be played in a more ornithologically informed

---

* It was Hubert and Mable Frings who debunked the folksy almanac claim that you can determine the temperature outside—or, if you are the Frings, inside your own apartment—by counting cricket chirps. (Thusly: Take the number of chirps in 25 seconds, divide that by 3, add 4, try to recall if that's the Fahrenheit or the Celsius formula, and add "download weather app" to your to-do list.)

effort to scare them away. There were no cries to record. "The birds," Hubert wrote in his memoir, "mostly sat quietly in place until clubbed." Mable's tapes captured only "the crack of ash on skulls" and, at one point, the distress cry of a young sailor: "I didn't join the Navy to crush the heads of innocent birds."

Hubert quietly raised the issue of morale and, if word got out, "the reaction of people stateside." His concerns were ignored. As before, the slaughter failed to meet its objective. Though 21,000 birds had been killed, "there were," observed Hubert, "almost as many still wandering around near the runways, and counts of bird strikes by planes remained the same." Besides, he continued, "even if total annihilation on Midway were achieved, it would probably be only a short time until the land available sign would bring in new settlers."

Hubert and Mable did what they could to provide kinder alternatives. They suggested clearing sea grape from the beaches on nearby islands to make them more appealing to house-hunting albatrosses. The plan has a cameo in the navy publicity film. We see Admiral Moore's pointer tapping a photographic blowup of Kure Atoll. The Navy Bureau of Aeronautics, he states in the film, planned to raze the shrubbery, but I could learn nothing more about it.

The navy did try its hand at "terrain modification," though this was on Midway itself. Someone had suggested that a row of dunes near the runways was creating updrafts and that the albatrosses needed these to fly. And that leveling the dunes would therefore solve the problem. Hubert didn't think so. The wind off the ocean, combined with the sizable surface area of albatrosses' wings, provided all the lift required. He pointed this out, but the dunes were bulldozed anyway. If anything, there were now *more* albatrosses soaring over the runway, as the demolition opened up easier access to the area.

The weeks passed, and Hubert had to return to his teaching.

He and Mable agreed to do some experimenting back in Pennsylvania. The pair were keen to develop an albatross repellent. The navy, by now experienced shippers of live albatrosses, promptly dispatched two to the Fringes. A photograph in Hubert's memoir shows Mable in a short-sleeved dress and two-tone flats, lifting the lid on a plywood crate. Poking up over the rim are the heads of two Laysan albatrosses fresh from the final leg of their journey, the express train to Altoona. The birds appear, as ever, utterly unfazed by the strange things humans do to them.

The albatross-repellent trials were a disappointment. Mothballs did not bother them. Live snakes were presented and ignored. Recordings of distress and alarm calls likewise caused no reaction. This was more or less expected, because in order to elicit and record these calls, a Midway albatross was swung in a circle, and while this went on, nesting birds a few feet away "did not even look over to see what was causing the commotion." The albatross is a largely unflappable bird.

The Fringes returned to Midway the following January. They were running out of ideas. One day Hubert noticed a group of navy wives effortlessly shooing a flock of loafing albatrosses by walking onto their lawns holding tablecloths in front of themselves. "We made tests with various surfaces held in front of us, walking toward the birds," he wrote. ". . . A sizable flat surface was . . . quite repellent." Excited by his findings, Hubert set a meeting with Midway command. With an "integrated program" of men holding large colored squares, he believed, they "might very well be able to clear the island of nesting birds." He estimated that twenty to thirty personnel would be needed, daily, for the duration of the albatrosses' return to their nests.

The proposal was not embraced. It had followed on the heels of a proposal to harass the albatrosses by stringing low wires across the nesting area—by tripping them. Hubert's final suggestion was inspired by an observation that no albatross was ever seen flying under the metal aprons suspended from some of the hangar roofs. He wondered whether suspending long panels of fabric might stymie any albatross seeking to nest where it oughtn't. He envisioned colored panels 10 feet long by 3 feet wide, suspended between 20-foot poles in the beach along the runway. It sounded marvelous, if you were a wedding planner, or Christo. Less so to a navy pilot. Hubert broached the idea at the Officers' Club one evening. "Opinion," he wrote in his memoir, "was not unanimous on this."

It was around this time that attitudes toward the Fringses began to deteriorate. "Our work," he recalled, "was regarded . . . as useless and troublesome." And so Hubert Frings returned to his teaching, and Mable to her grasshoppers and crickets and to a new project, "reviewing the life styles of spiders." I end with a quote from Hubert's journal: "Of all the animals we have ever kept, these Albatross are the ones to which I have become most attached. I really love them and respect their independence and jaunty ways. This marks the end of a period of acquaintance with real aristocrats of life." I kind of feel that way about Hubert and Mable Frings.

The navy closed Naval Air Station Midway in 1993. No plane ever crashed. No airmen died because of the gooney bird. Bird strikes had continued throughout the navy's tenure, despite everything it tried. There were, one report stated, twice as many collisions at the end of one four-year albatross control program as there had been at the time the effort started.

In September 1958, *Flying* magazine ran a piece about Midway's albatross conundrum. An airman was quoted saying, "It's my bet that no matter what we throw at them, when it's all over, Midway will still be the Gooney territory and ourselves only transients subject to the whims of a bird that refuses to be conquered."

I hope the airman put cash on that bet. Because it is all over now, and the islands are very much gooney territory. Naval Air Station Midway is now Midway Atoll National Wildlife Refuge. Nothing goes on there but the happy hatching and rearing of seabirds and the quietly atoning efforts of Fish and Wildlife personnel, now working to restore the birds' habitat.

All around the world, wildlife continues cluelessly, tragically, to cross paths with large vehicles. And science continues, sometimes entertainingly, always earnestly, to search for solutions.

10

ON THE ROAD AGAIN

Jaywalking with the Animals

On July 26, 2005, the space shuttle *Discovery* hit a turkey vulture. The encounter, which happened during liftoff, was caught on video. You see the great bird soaring, perhaps delighting in the awesome thermals of a rocket liftoff, and then abruptly bouncing off the external fuel tank and plummeting, Icarus-like, into the exhaust plume. Because the craft had just begun to accelerate, the damage was small, and largely confined to the vulture. Nonetheless, recalls Travis DeVault, a wildlife biologist with the Sandusky, Ohio, branch of the National Wildlife Research Center, "it created a lot of heartburn."

The NWRC branch offices share the grounds of NASA's Plum Brook Station. Plum Brook is where NASA tests rocket engines and Mars landers and the like to be sure they'll perform under the stresses and unusual circumstances of space travel. Plum Brook engineers can create wind that blows six times faster than the speed of sound, and vibrations as ungentle as those of a rocket launch. A wayward buzzard seems laughable by comparison, but no one was laughing. Travis reminds me that it was a ding to space shuttle *Columbia*'s exterior on the way up that led to the tragic explosion on the way back down.

Plum Brook Station is headquarters to the National Wildlife Strike Database, co-managed by the FAA (Federal Aviation Administration) and the USDA. Plum Brook is federal chop suey. The FBI is here too, behind a row of closed doors with nameless nameplates, across from Travis's office. He doesn't know what they do in there but is impressed by their paper-shredding capabilities. "Comes out like *dust*."

In 2015 the National Wildlife Strike Database summarized data from twenty-five years of collisions between civil aircraft and wildlife. The report goes species by species:* number of strikes, the total cost of damage to aircraft, and the number of people killed and injured. There are two ways to be a bird of interest to the FAA (or, who knows, the FBI): weigh a lot or travel in groups.

By dint of their size, turkey vultures† rank high on the worry list: eighteen people injured, one killed, and costly damage to the aircraft 51 percent of the time. By contrast, of twenty-seven documented chickadee strikes, none caused damage of note.

Jet aircraft engines are given a "bird ingestion" test, but the birds used in the testing weigh two pounds. Turkey vultures aver-

---

* How is it possible to identify the species of a bird that has slammed into a nose cone or passed through the whirling blades of a jet engine? Forensic ornithology! Feathers, down, beaks, talons, and/or "snarge" (tissue scraped from the plane) are collected and shipped to the Smithsonian Institution's Feather Identification Lab. The United States Postal Service happily delivers snarge. They also, I learned on their website, will deliver live animals that are intact—including leeches, goldfish, scorpions (double container required), birds lighter than twenty-five pounds, and "small harmless cold-blooded animals," whom they have also hired, at least once, to man the counter at my local post office.

† To clarify, a turkey vulture is a vulture, not a turkey. Though turkeys, too, have crashed into planes. But only wild ones. Supermarket turkeys have never hit planes, but supermarket chickens have, because they are fired at jet parts to test their ability to hold up to birds strikes. The device that fires them is called, yes it is, the chicken gun.

age three. Travis sent me a link to footage of one of these tests, slowed down to reveal a progression of fan-jet engine blades slicing a bird like a meatloaf. With a bird the size of a vulture or a pelican, the blades may end up in pieces, too. Shards of fan blade slamming into delicately calibrated engine parts can have catastrophic consequences.

Travis also emailed me control tower footage of a Boeing 757 strike that he uses at trainings for airport biologists. You see an unidentified blackish bird—really just a speck on the screen—disappear into the mouth of an engine during takeoff and produce a near-instantaneous fart of fire out the back. It's the soundtrack that stays with you. You hear the pilot's call, "Mayday, Mayday, Mayday," and in the background the normally cheerful but now darkly sinister twitterings of a bird.

Odds are decent it was a starling—among the six most frequently struck bird species in the United States. To confuse winged predators, starlings sometimes fly in enormous shape-shifting flocks called murmurations that swerve and split and coalesce again, all without warning or seeming logic. The open maw of a jet engine passing through is bound to swallow a few. They're like avian krill.

Worst-case scenario: big birds in groups. It's no surprise that the species that brought Captain Chesley B. "Sully" Sullenberger down to the Hudson River was Canada geese, one or two per engine.

Travis DeVault's wildlife strike research has focused, at various times, on turkey vultures, blackbirds, and Canada geese, but tonight he'll be gathering data on a yet more dangerous creature, the species the FAA considers the "most hazardous wildlife to U.S. civil aircraft": the white-tailed deer.

From 1990 to 2009, the National Wildlife Strike Database logged 879 collisions between white-tailed deer and aircraft. The impacts injured 26 people and caused, on average, six times as much aircraft damage as other wildlife strikes. Only Canada geese, red-tailed hawks, and pelicans caused more aviation deaths. Deer strikes have happened on landing, during takeoff, and while taxiing. Obviously no deer have struck planes at cruising altitude. Less obviously, two animals struck planes that were parked.

White-tailed deer are thirty times as heavy as turkey vultures, and they travel in groups. Double whammy, both for aircraft and for vehicles on the road. Travis's focus, of late, has been this: Creatures on roads and tarmacs often don't get out of the way in time, even when there's time to get out of the way. He's looking for answers to the questions we can't just ask. Why do you stand there in the headlights? How can we help you? How do you *not* notice a space shuttle?

Travis is driving me around Plum Brook's six thousand woodsy acres. It's a wildlife biologist's tour of a space center: *See the bald eagle nest up there? And this is a great place to hunt for mushrooms. That building has some big thermal thing. There's another nest!* You can tell which buildings are rarely used, from the deer. A half dozen of them mill about on the lawn of the Hypersonic Tunnel Facility, nibbling and strolling like guests at a wedding reception. Clusters of them appear along the roads with the regularity of mile markers. For the first couple minutes, I'd lean forward against my seat belt and exclaim, "There's one!" At one point Travis turned to me: "You don't see many deer where you live, do you, Mary?"

Out where he lives, Travis sees a lot of deer, but he did not see the one he hit. This is often the case. "The deer you hit are the followers," he says. You brake, and your eye follows the deer you just avoided hitting. "You speed up, and here comes the next one."

As soon as it's dark, we'll be meeting up with Travis's research partner, Tom Seamans, to take some data for what I'm calling the deer in the headlights project.

It's approaching dusk now, the time of day with the highest

hourly incidence of deer-vehicle collisions—three times higher than the dark of night, according to a paper by Travis and four colleagues. Deer are crepuscular, a word born for dermatology but in fact meaning "active at dawn or dusk." November is the other standout risky drive time, because it's the rut—mating season. Hell-bent on reproduction, the deer fail to note the most blatant obstacle to the successful onward advancement of their genes: traffic.

We're seeing deer all along the roads partly because it's dusk, and partly because Plum Brook has a lot of them: about ten deer for every one person who works here. Also, a road that runs through woods is an attractive place to a deer. Food grows close by, and it's a clearing of sorts, so predators can't easily sneak up unseen. The open space of a road also appeals to birds that hunt insects on the wing, because it's easier to see and maneuver here. The ones that get hit draw the scavengers. Roadkill begets roadkill. (In hopes of preventing another turkey vulture mishap, the Kennedy Space Center grounds crews formed a "roadkill posse" to clean up carcasses with unusual alacrity in the days leading up to a launch.) On the simplest level, animals take to the road for the same reason people do: the going is easier.

The low speed limit at Plum Brook has helped keep wildlife populations booming. As long as cars aren't moving faster than a natural predator would, a prey animal will usually get out of the way in time, even if the driver doesn't brake. The hunted maintain what's called a spatial margin of safety. They're able to visually intuit the distance between themselves and a predator, and they have an uncanny sense of exactly how close they can let that predator come before they need to take off. Flight initiation distance, or FID, as that closest point is called, shrinks and lengthens according to circumstances. If animals or birds are feeding on something nutrient-rich and wonderful, they may wait until the last possible moment—the shortest FID—to abandon the feast. If the predator is coming at them at a run, its speed is factored in and they'll take flight when their pursuer is farther away. They almost always judge the safe getaway distance correctly. Unless the thing coming at them has an engine.

Mammals and birds, sensibly, perceive onrushing cars as predators. Their escape algorithms work well on a congested city street—though you may try, you will almost never hit a pigeon—but their judgment fails them on a freeway or a rural thoroughfare. Because what predator comes at you at 60 mph? Evolutionarily speaking, this is something new. "Fast cars have only been around for a hundred years," Travis says. He flips a visor down against the setting sun. "In terms of evolution, that's nothing."

Travis has speculated that this explains the baffling inability, among wildlife, to avoid what should be easy to avoid: "a large, noisy vehicle traveling along a predictable path." Evolution hasn't had time to upgrade the processors. Judging speed requires an ability to perceive and interpret "looming"—how quickly an object appears to be growing in size as it comes at you. Looming is harder to detect and visually process when the object is traveling quickly. The "looming-sensitive neurons," as some pigeon researcher went ahead and named them, are overwhelmed.

Travis and Tom have devoted considerable time to studying this. The pair's original study protocol was straightforward: "We drove a vehicle directly towards turkey vultures . . ." The vultures had been lured by a raccoon carcass anchored to a heavy metal plate, to keep the birds from dragging it off the road to a more relaxing setting. The vehicle, a Ford F-250 pickup, traveled without braking, at three different, constant speeds: 19, 37, and 56 mph. Flight initiation distance was measured by dropping a beanbag out the window onto the asphalt to mark the moment a vulture moved to retreat, and then measuring the distance from beanbag to carcass. The FID for 37 mph was not significantly different from the FID at 56 mph, suggesting that, as Travis and Tom had predicted, an unnaturally fast "predator" overtaxes the prey's sensory and cognitive gifts.

No vultures were hit, though there were close calls, all of them at the fastest speed. To see what would happen at even faster speeds, Travis and Tom devised a video truck. Cowbirds—because they're common around here and hardy—were installed in a roomy cage (and, fret not, released afterward). One wall of the enclosure took the form of a video screen, on which the researchers played footage of a truck driving straight at a video camera they'd placed in the middle of a road. Cowbirds, they found, will take to the air when a vehicle gets to about 100 feet away, regardless of its speed. Up to about 75 mph, they have ample time to get out of the way.

Through the magic of variable speed video playback, Tom and Travis were able to accelerate the video truck to 224 mph—roughly the speed of a plane taking off. Because that's really what this research is about: flight safety and airplane damage prevention. It would be great to figure out how to prevent the deaths of hundreds of millions[*] of small creatures on U.S. roads each year, but that's not the ultimate goal of the work.

Faced with a truck going as fast as a plane, every one of the cowbirds would have wound up a statistic in the National Wildlife Strike Database.

Research on jaywalking pedestrians tells a similar story. Most of our decision-making is based on how far off a car is. We're not so good at factoring in the speed. Experimental evidence suggests that full looming sensitivity doesn't develop until adulthood. A young child on the side of the road and a car traveling faster than

---

[*] Or probably more. Figures for roadkill prevalence tend to be underestimated owing to the speed with which scavengers arrive to eat the evidence. Within twenty-one hours, one Mojave Desert survey found, all that remained of a crushed tortoise on the centerline of the road were scattered pieces and "two withered limbs eight meters from the point of impact." At ninety-two hours post-impact, the tortoise was but a "faded stain."

20 mph combine to encourage, quoting a team of European psychologists, "injudicious road crossing." Hence the need for injudiciously punctuated SLOW CHILDREN signs. It's not just that kids aren't looking when they cross; they're also not seeing.

For animals facing down a predator, fleeing is but one option. Mammals rely on a diversity of features and behaviors that, over the millennia, have increased their odds of staying alive long enough to pass on their genes. The skunk sprays a vile smell, the porcupine wears darts. When the "predator" is a speeding automobile, these tactics range from ineffective to tragically maladaptive. The turtle stops in its (and your) tracks and pulls its head into its shell. A deer may freeze to avoid being seen among the trees. Squirrels and rabbits zigzag and spazz halfway across a street. When your killer is a hawk that has calculated the likely intersection of its path and yours, changing course abruptly may save your life. When the killer is a land-based commuter, it foils her efforts to avoid hitting you.

When self-driving cars take over the roads, the lives of squirrels and skunks (and cats and small dogs) may no longer be spared by kindly drivers who swerve and brake. By the cold calculus of (human) survival, drivers are safer doing neither. The Centers for Disease Control and Prevention estimates that ten thousand people per year are injured when they take evasive action to avoid hitting an animal. That's only two thousand fewer than the number injured when the vehicle actually hits the animal. In 2005, the Insurance Institute for Highway Safety (IIHS) analyzed 147 fatal (to the human) vehicle-animal crashes, 77 percent of which involved deer. The initial impact rarely killed or even hurt people; almost always, they died because the driver tried to avoid the animal. The driver or motorcyclist braked, and the vehicle skidded and went off the road or collided with something less yielding than venison.

The eight exceptions were instances in which large deer—and in one case, a horse—crashed through the windshield. Taller is a killer. Because now the car's front end strikes the animal's legs rather than its torso. And when the legs are knocked out from under, the torso and head pinwheel over the hood and crash down onto the windshield and, if the animal is tall enough, the roof. Thus Volvo has a LADS—large animal detection system—but no SADS. "The camera looks for a specific signature," a Volvo communications manager said in an email. "A large body mass with four very thin long legs." The example he gave is a moose.

For a 1986 master's thesis, a team of Swedish bioengineering students staged a moose strike and filmed it at high speed, so the impact and its aftermath could be studied in slow motion. The aim was to provide a more nuanced understanding of the biomechanics of these often devastating collisions, and then use this understanding to develop a moose crash test dummy. An "ill and weak" bull moose "was put to death and quickly afterwards it was hit by a Volvo 240 at a speed of [50 mph]." The phrasing intrigued me. Apparently the Volvo 240 is a car that goes from 0 to 50 quickly enough to reach a moose in the fleeting moments between its death and the crumpling of its legs. For you could not suspend a carcass in a frame, as this would keep the moose from doing the very thing the authors aimed to study.

Anyway. What the film revealed. If the roof collapses as the passenger is thrown forward—and now I borrow the gently evocative phrasing of Swedish moose crash test dummy designer Magnus Gens—"crumbled steel interferes with the head's path." Borrowing the less gentle phrasing of "Moose and Other Large Animal Wildlife Vehicle Collisions," "axial compression . . . causes bony fragments to be pushed into the spinal canal." Moose falling on driver's head crushes neck vertebra, sharp pieces of which slice spinal nerves, causing full or partial paralysis. Also disturbingly common: broken face bones and lacerations from hitting the windshield in mid-shatter. The wounds become infected by "debris, hair, entrails, and feces." And finally, should the two of you manage to survive the impact, one of you now has a flailing moose in his lap.

Making matters worse: the long legs of a moose may boost the animal's eyes up above headlight range, eliminating the reflective shine that helps drivers see it in the dark. (The tapetum lucidum, which does the reflecting, is actually there to aid *their* vision, not ours. It boosts mammals' vision in low-light conditions by bouncing the light back into the retina a second time.)

If you plan to be driving in far northern regions where tall ungulates are likely to dart into the road, you might want to consider a Saab or a Volvo, as their roof pillars and windshields are designed and reinforced with input from Magnus Gens's awesome moose crash test dummy. Magnus received funding from the Swedish National Road and Transport Research Institute, which shortly thereafter received a plea to help design a camel crash test dummy.

A camel is taller and heavier, and therefore even deadlier, than a moose. More of the roof is likely to collapse directly onto drivers' heads. If they lean or duck sideways to dodge the hurtling ungulate, now their back is likely to be broken instead of their neck. Of sixteen Saudis whose cars struck camels, one study relates, nine wound up with complete quadriplegia. Along certain stretches of highway, camel density has been as high as nineteen per mile. These animals are not wild, but their owners often allow them to roam. Sometimes even, in days past, encouraged it. Because until recently, Saudi law required the driver to pay the camel's owner for the loss. "Therefore," reports a team of neuroscientists at Riyadh Armed Forces Hospital, "some camel owners have been known to push their animals onto the highways after sunset to claim compensation after the accidents." A pox on them. Debris, hair, entrails, and feces on their heads.

Summing up: Do not brake excessively or swerve wildly for a small creature, no matter how cute. Do swerve and brake and run off the road for a camel on an empty desert highway, because there's nothing to run into but sand. Never speed in moose country. About deer, I don't know what to tell you. The IIHS study suggests you should brake or swerve only when there's the space to do so safely, never to the point of skidding or losing control, because deer impacts don't reliably injure parties other than deer. The alternative is what, plowing into them? Who does this? People brake. And if they brake hard, the nose of their car drops down and the impact happens lower, perhaps at the level of the deer's legs, causing more of the torso to pinwheel toward the windshield. And the cars behind to rear-end you. What's a rational person do?

Let's ask the most rational driver of all: the autonomous car. If it slams its brakes, does it only do so when no one's tailgating? If it swerves, does it do so only if the path is clear? If either criterion is missing, will it go ahead and run straight over a beagle or a skunk? I posed these questions to Google/Waymo's self-driving car media relations person, but she refused to have relations with me. I got no answers and no one to interview, and soon she stopped replying altogether. Somewhere in the middle of our standoff, one of Uber's autonomous vehicles, traveling 43 mph, plowed into an Arizona pedestrian without braking or swerving. As if she were a squirrel. Seems like they don't have the answers, either.

In 2012, a North Dakota woman named Donna called in to a morning talk-radio program hoping to draw attention to a situation that had been bothering her. She'd been in three car crashes involving deer, and each time, it had happened near a DEER XING sign on a busy road. "Why," she lamented in a recorded encounter that would eventually top one million internet hits, "are we encouraging deer to cross the road in such high-traffic areas?" A short silence followed. "You seem to think," one of the hosts began tentatively, "that deer-crossing signs are telling deer where to cross?" As nicely as possible, he explained that the signs are meant for us, the drivers, to tell us to slow down.

They might as well be talking to deer. Drivers don't slow down when they see a DEER XING sign. This is true of Uniform Traffic Control Device W11-3, the standard black-on-yellow, diamond-shaped sign, and it's true of fancy blinking neon deer warning signs—including the model with three neon deer "activated in sequence to give the impression" of a bounding buck. I am familiar with this technology from a neon sign in front of a strip joint across the street from a San Francisco pizzeria I used to eat at. A bosomy nude danced a sequence of three moves, over and over. I bet you drivers slowed for her.

Warning signs work slightly better when drivers can see some evidence of real danger. This was demonstrated by a researcher who, every Tuesday at dusk, for an unnamed number of weeks, hauled a deer carcass into place a few feet away from a flashing deer-crossing warning. Cars slowed by an average of 7 mph. Likewise, a "realistic taxidermist deer mount" set up in the brush just back from the road, near a blinking neon sign proclaiming "Deer on Road When Lights Are Flashing," caused a 12-mph drop in the average speed.

Did drivers slow down to heed the danger posed by what they took to be a real deer, or did they slow down to gawk at a carcass or, more arrestingly, the odd spectacle of a taxidermy museum specimen in the weeds? I'd guess the latter. That's what drivers do: they ease off the gas and rubberneck. Who wouldn't slow to check out a stuffed deer? Maybe throw it in the back of the truck? *Pull over, Jeb, somebody left a full-body deer mount by the side of the road.* Now you need JEB CROSSING signs.

Nonetheless, a valid concept: pair the warning device with the danger, at least intermittently. Don't cry deer unless deer are around. The best results have come from systems in which the warning message is activated by a signal from solar-powered animal sensors. For instance, a sign that only lights up when something more or less the height of a deer breaks a radar or laser or microwave beam near the road.

So why are we not seeing these systems everywhere? The answer perhaps lies in the woe-laden tale of the Western Transportation Institute at Montana State University. In 2005, they tried out one such system in Yellowstone National Park. Alas, the contractors were unaware of certain laws and park regulations pertaining to what could and couldn't be put up on the side of the road. The system called for erecting eighteen poles per mile, which was far too many, and, PS, they had to be wood, not metal, and they couldn't be that garish color. There were software problems, faulty hardware, budget troubles, topsoil issues, signal failures, false negatives, false positives. Lots of false positives. Drivers were warned a deer was lurking near the road when really it was snow falling. Or a car pulling onto the shoulder. Or moving plants. *Caution! Milkweed swaying in wind!*

When the danger is strictly to the animal, not the driver or the car, forget about it. In 2009, the National Park Service decided to see whether they could save endangered tortoises by installing flashing lights and WATCH FOR TORTOISE signs along two stretches of highway in Mojave National Preserve. A short distance along, they placed a model of a desert tortoise on the side of the road. Then they hid nearby to see if drivers slowed or braked or even craned their necks in a way that suggested they were, in fact, watching for tortoises. Nope.

If only it were possible, as Donna believed, to communicate with the animals rather than the humans. With deer, this has been attempted in various disappointing ways: bumper-mounted deer whistles that don't work, roadside deer repellents that don't work. What has sometimes seemed to help is to mount angled reflectors along the highway to redirect the beam from approaching headlights and enhance "ungulate awareness" of an approaching car. A study in Wyoming recently delivered some promising results, though not with the reflectors they were testing. It was the study's control—reflectors wrapped in white canvas to keep them from reflecting—that worked. The researchers speculate that the deer might have been responding to the white bags as they would to the white butt and underside of a fellow deer's tail. Tail flagging—raising the tail and flashing the white—is thought to be what a startled white-tailed deer does to communicate danger to its companions.

I am skeptical, only because I have read the 1978 paper by researchers at Pennsylvania State University who tried to warn away white-tailed deer by erecting roadside plywood cutouts of deer rear ends with tails a-flagging. On some, the raised tail was painted white; on others, an actual deer tail had been nailed in place. Sadly, because who wouldn't want to see our nation's highways lined with plywood deer asses with decomposing tails, none of it worked.

The Wyoming group also theorized—and I'm with them here—that the reason the death rates were lower with the canvas bags is that drivers were slowing down to look at them. *Hold up, Jeb. White thing on a stick.*

Getting back to Donna for a moment. You could in fact direct deer to where you want them to cross the road—not with a DEER XING sign but with a wildlife-specific overpass. If the overpass is combined with fencing along the side of the roadway, the animals can essentially be funneled to the safe crossing point. The challenge is that crossing structures and fencing are costly, and it may be impossible to install them in the places they're needed. And with white-tailed deer in some parts of the country, at some times of the year, they're sort of needed everywhere. (Wildlife overpasses and underpasses tend to be used for species that migrate en masse to breed or find food or for populations whose genetic health is threatened by a highway that fragments their population.)

Here's the problem with deer and headlights. When it's dark, deer don't necessarily make the connection between the little lights and the big car behind. Even if they do, a pair of automotive headlights communicates practically no useful information as it hurtles closer. Pinpoints of light don't "loom" perceptibly, so it's not obvious that they're approaching, let alone how quickly. Travis and Tom have been testing something that could help: a rear-facing light bar that illuminates the vehicle's grill. The hope is that deer, now able to make out that there is in fact a large object barreling toward them, will get out of the way more quickly and dependably. Now that it's dark, we're headed out onto the Plum Brook back roads to collect some data. The idea is to compare flight initiation distance—FID—and the likelihood of freezing behavior (deer in the headlights) with and without the helpful lighting.

The main Plum Brook parking lot faces a field of corn stubble, that dry, dead snowdusted brown that is, to me, the color of an Ohio winter. Tom is squatting on his heels, attaching the rig to the front of Travis's truck. He looks up. "Do you hear the woodcock?" The buzzing I assumed was an insect is in fact part of the courtship dance of the male woodcock. In keeping with that rural paradox I will never quite grasp, Tom has a fondness both for wildlife and for hunting it.

While the men ready the equipment, I take their range finder and walk off toward a clique of deer in the field. A range finder is a sort of laser tape measure. Aim the beam onto an object, and the readout tells precisely how far away it is. The FID of these deer, by my estimation, is about three hundred feet. They're wary,

with reason. Range finders around here are used mostly by hunters, to calculate bullet-drop compensation when taking aim at
distant targets.

Travis yells that we're heading out. I leave off harassing the deer
and head back to the truck.

The back roads have few streetlights, making it hard to see
a deer in the distance. We're feeling them instead, with a heat-
sensing system called FLIR (forward-looking infrared). FLIR displays the world by relative temperature, on a dashboard-mounted
display. The image is rendered in a grainy gray scale, like a char-
coal drawing. Snowbanks are black. Skunks and racoons in the
roadside brambles glow an eerie white. They look like they're
made of that strange filament in old camping lanterns. On hot
summer nights, a mammal's body heat may match the heat of the
asphalt, and Travis and Tom have lost sight of deer standing in the
middle of the road twenty feet off.

"There . . ." Tom points into the murk. On the FLIR screen,
our first test subject stands in the rough off the righthand side of
the road, a couple hundred feet ahead of us. Travis accelerates
to a steady 37 mph. Tom's arm hangs out the passenger window
and his eyes are on the FLIR screen. He's holding a small bag of
gravel covered in reflective tape. As the deer turns back toward
the woods, Tom drops the bag. A few seconds later, the truck
arrives at the place where the deer took flight and we stop. Tom
gets out with the range finder and waits for Travis and me to back
up to the gravel bag Tom dropped a moment before. Then he
aims the range finder at us. The number it gives him is the FID
for this deer.

I'm going to fast-forward here, because I want to share the
results of Travis and Tom's project. Their light rig worked, a patent is pending, and the NWRC is seeking a partner to build and
market the technology. Although the deer's flight initiation distance wasn't significantly different, adding the rear-facing illu-

mination caused a dramatic decrease in the incidence of "deer immobility." With the rig in place, only one deer froze in place, as compared with the ten deer that froze in the headlights of the same truck without it.

Back inside the cab, Tom fills out a data sheet. I watch the FLIR screen. A ghost coyote moves through the trees. It stops, looks over at us, continues on its way. A security gate in the distance appears to be manned by a yeti. To see by heat is to understand something key to this work: There are other ways to perceive the world. If you want to communicate something to an animal, you may need to translate your message.

For instance: the light that Travis and Tom are shining on the front of the truck has a strong blue and ultraviolet emission, because that's the part of the spectrum across which deer see best—far better than we can. Deer vision was explained to me in some detail by a man who has studied it: Bradley Cohen, an assistant professor of wildlife biology at Tennessee Tech University. "At dusk, when deer are most active, UV light is the most available thing," he said. While we struggle to see detail in a purply dark dusk, a deer sees it clearly in a bright blue. Our twilight is their noon.

Laundry detergents that claim to give you "whiter whites and brighter brights" provide a service to deer. The manufacturers of these soaps have added an optical brightening agent that emits in the ultraviolet range. To us, it adds no detectable color. To a deer, the clothes are, Cohen says, "luminous." Joke's on you, hunter who washed his camo pants in Cheer.

Whatever advantages a deer has on the ultraviolet end of the spectrum, it lacks at the other end. To a deer, red and orange are perceived as an absence of color. Orange is the new black. So while a safety orange/red hunting jacket makes the hunter stand out to other hunters, to deer it may be more camo than store-bought camo.

Here's something else unusual about deer vision. Deer see best in an elongated strip across their field of vision—a "visual streak"—rather than in a central core of visual acuity, as we do. It would be like reading a book with your peripheral vision. Deer can't read, obviously, but the setup helps them detect a predator trying to sneak up on them.

Some birds have a visual streak, too. It's useful for hunting prey and for traveling. A migrating bird with a visual streak can study the whole horizon without moving its eyes or head.

Tom Seamans used to do a lot of bird research. He didn't study their field of vision so much as put strange things into it.

11

TO SCARE A THIEF

The Esoteric Art of the Frightening Device

Tom Seamans has worked at Plum Brook Station for thirty-one years. He has hair the color of a white-tailed's rump and a soft-spoken, folksy manner. I didn't take a photograph of him while I was here, and my brain, in coming months, will go ahead and substitute Orville Redenbacher in a quilted tan hunting jacket. Tom is a Cornell-educated wildlife biologist and a born tinkerer. He and Travis share a workshop where, you can tell, they pass many contented hours. I've been poking around while they put away the light bar and close up shop. In many ways, it's a typical workshop, and in a couple, it isn't.

The raccoon urine is one way it isn't. Tom and a colleague have been testing potential starling repellents. The urine will go into empty prescription drug vials with perforated caps that are then attached to the floor of starling nest boxes, like giant nightmare

Air Wick dispensers.* Starlings are cavity nesters, and a jet engine cowling is, up until the tragedy of ignition, a cozy postmodern bungalow. Starlings can build a nest in under two hours, potentially setting up housekeeping in the interlude between pre-flight inspection and takeoff. Bad news for the tenant, bad news for the landlord.

Bird deterrence research has a lengthy history here at the Plum Brook branch of the National Wildlife Research Center. Up through 1990, the focus was not on keeping wildlife away from aircraft and cars but on keeping it away from farmers' crops. When Seamans was hired, in 1987, the work was all about blackbirds and corn.† Tom has deep knowledge in the ancient art of bird-scaring. He's tested dozens of what are known in the human–wildlife conflict field as "frightening devices." Most are only temporarily frightening. It is easy to scare birds away, but much tougher to keep them away. Creatures habituate. They get used to the sound or sight that once alarmed them. They start to call your bluff.

Least—or most fleetingly—effective is the stationary predator decoy. The internet abounds with photographs of pigeons roosting on great horned owl replicas and Canada geese relaxing in the shade of fiberglass coyotes. The classic cornfield scarecrow may actually *attract* birds, because they start to associate it with food.

---

* To be sure it was the predator-suggestiveness of the piss smell that was potentially repelling the birds, rather than its foreignness, the team also tested vials containing something just foreign: Febreze Extra Strength Fabric Refresher. Science is here to tell you that starlings feel the same way about Febreze Extra Strength Fabric Refresher as they do about raccoon piss. Neither smell made a bit of difference.

† The NWRC's Sandusky branch was formed through the lobbying efforts of a group of corn farmers who incorporated themselves in 1965 as the Bye Bye Blackbird Association. The name did not play as well on Capitol Hill as it had in rural Ohio, and in 1967 the Bye Bye Blackbird Association shelved the snickering wit and became the Ohio Coordinating Committee for the Control of Depredating Birds.

To a flock of migrating blackbirds, it's the golden arches on the side of the highway, the Bob's Big Boy sign, a reason to pull off for a large, fattening meal.

Back in 1981, the human-wildlife conflict researcher and author Michael Conover tested a series of ultrarealistic raptor decoys. Taxidermied museum mounts of sharp-shinned hawks and goshawks were set up at feeding stations to see how long they'd keep away ten species of smaller birds that the hawks are known to prey on. A limp five to eight hours, was the answer.

For long-lasting fear, birds need to see or hear some consequences. In a follow-up study, Conover extended the useful scariness of an owl decoy by incorporating the sights and/or sounds of an actual starling grab. Some involved taped distress calls, some involved a live performance. Here is one of the takeaways: "Tethering a dead starling to the model was less effective than attaching a live starling." Unless you wish to be tethered to the talons of PETA, this is not a viable approach. Besides, in some species, distress cries serve to attract, not repel, the victim's flock mates—to help or sometimes just to gawk.

Depending on what species one is looking to disperse, a dead bird on its own can be strangely effective. Provided it is properly installed. I quote from the National Wildlife Research Center "Guidelines for Using Effigies to Disperse Nuisance Vulture Roosts." "The posture of the prepared bird should resemble that of a dead bird hung by its feet with one or both wings hanging down in a[n] outstretched manner." I spoke to two of the researchers who wrote the effigy how-to manual, Michael Avery (since retired) and John Humphrey, of the NWRC's Florida Field Station. In a 2002 study, they and a colleague documented the effects of rigging up a vulture carcass or a taxidermied vulture on six different communications towers. Vultures like to roost on these and other open-frame towers, and their slippery, pungent droppings make climbing the structures dangerous and gross for repair crews. Within nine days of hanging an effigy, the ranks of roosting vultures were lower by 93 to 100 percent. The birds stayed away as long as five months after the effigy was taken down (or rotted).

Avery was a convert. "Works like a charm."

As with a charm, there is no rational, non-voodooey explanation for why it works. "The answer I give people who ask," Humphrey told me, "is, 'We don't know, but if I went into a neighborhood and saw a person hanging upside down from a tree, I'd leave, too.'"

Avery agrees that, as he and his colleagues wrote in 2002, "It is tempting to speculate that the vultures recognize the taxidermic effigy as a dead roostmate and, not wanting to befall a similar fate, leave the area." He resists that temptation. "This is a fanciful, anthropomorphic notion."

True, admitted Humphrey, "it's not the best answer. But it's the only one I have."

Tom Seamans has a vulture effigy over in a dry-storage room in the workshop, and we are going over there now to visit it. It's one of the models used by USDA Wildlife Services in states with lots of vulture complaints. The body is Styrofoam, because Styrofoam lasts longer than bird, but the wings and tail fan are real, because feathers seem to be the key element.

Given that vulture roost dispersal strategies include shooting at them, the vulture effigy is a lovely if macabre development, and we have Tom Seamans to thank for it.

Like so many discoveries, this one was accidental. Tom straightens a shelf of birdseed bags. "There used to be a big rocket tower here, and the vultures hung out on it." Back in 1999, Tom needed to calculate the average body density of twelve problematic birdstrike species. (He was helping design a standardized bird dummy for testing jet parts.) He headed out to the rocket gantry with his gun. As the dead vulture fell, one of its legs got hung up on the structure, about two hundred feet up. "And I wasn't climbing out on that gantry to get a dead bird." So it was left. No one ever saw another vulture on the tower.

Tom wondered whether the effect could be replicated and put to use. First he tried just laying a carcass out. Nothing doing. It had to be hanging and spinning.

He doesn't know why it works, either. "My only guess is, it's just so unnatural. They're thinking, *Something is wrong here.*"

We can never know whether vultures think such thoughts, but humans assuredly do. The staff at Royal Palm Visitor Center in Everglades National Park tried using effigies to discourage the black vultures that fly over from a nearby roost and vandalize cars in the parking lot. Visitors would return from a day of fishing to find the rubber blades torn off their windshield wipers or the seals around their sunroofs peeled off. Effigies were strung in the trees around the parking lot, which discouraged vultures, but now the rangers spent large portions of their day explaining the effigies to weirded-out park visitors. Now, instead, there's a box in the parking lot with a sign: "Use Tarps to Protect Your Vehicle from Vultures."

Why do vultures do this? Do rubber, caulk, and vinyl off-gas a chemical that's also present in decomposing carrion? Is there one attractive compound common to all these items? Researchers with the NWRC's Florida Field Station tried to figure it out. Because if they knew what that compound—or family of compounds—was, they could use it to lure the vultures elsewhere, in the same way setting up a scratching post lures a cat away from the furniture.

Thus began a project that, from the description in the journal paper, made it seem as though the staid and dedicated scientists of the National Wildlife Research Center had taken to cooking crack in the lab: twenty-one vulture-damaged objects were "finely chopped using a razor blade" and heated to 131 degrees Fahrenheit. The vapors were trapped and then identified by gas chromatography. The idea was to find compounds common to all the materials, and then soak sponges with them and present them to vultures to see how they respond. Alas, the chemicals evaporated too quickly, and then the funding did too. The mystery—and the vandalism—endures.

It's also possible the vulture vandalism has nothing to do with the smell of the materials. My favorite explanation for the behavior comes from retired raptor conservation biologist Keith Bildstein. Bildstein observed the same kind of pulling and ripping behavior in striated caracaras in the Falkland Islands, and he had heard that keas, New Zealand mountain parrots that eat, among other things, dead sheep, also vandalize parked cars in this way. Bildstein noted that the movements and neck strength needed to tear hunks of meat off a carcass were similar to what a bird would need to pull away rubber and caulking. And the stretch and density of carrion muscle and tendon are similar to that of rubber and seals. He speculated that vultures who strengthen their necks by pulling on whatever muscly-tendony items are at hand (or beak) will have a competitive advantage during the avian scrum of a group carcass feed. In other words, it's a fitness regimen.

Tom lays the effigy back down. He does this almost tenderly. He likes turkey vultures almost as much as he likes woodcocks. "Sometimes," he told me earlier, "I'll just lay out in a lawn chair, tip back, and watch them circling around. If it's still, you can hear them sweep by."

Tom shuts off the workshop lights and locks the door. "Effigies," he says, "were one of the things we used at the 9/11 recovery site."

When the Twin Towers collapsed, close to a billion pieces of debris had to be searched for human remains. It was the largest forensic investigation in United States history. A thousand people from twenty-four agencies ultimately recovered twenty thousand pieces of human remains. "It was a process involving garden rakes and people literally on hands and knees," wrote Seamans and his colleagues in a 2004 paper describing the process. A large space, remote but not inconvenient, was needed for the operation. A recently closed Staten Island landfill with the awkwardly appropriate name Fresh Kills was selected.

On the third day, the birds came. "We knew what piles were rich in body parts by the way [they] descended on it," said the NYPD inspector who served as incident commander for the operation. "And you would have to fight [them] for the human remains." Seamans was part of the Wildlife Services team brought on to keep the birds away. I've asked him to tell me the story. We're sitting on a beige couch in an old FEMA trailer he and Travis use for watching and analyzing research video footage. The furnishings are low-budget generic. A human nest box.

"We would get there before sunrise," Tom recalls. "We'd try to keep them away before they landed. We'd walk the perimeter, scanning the sky for approaching birds, and then shoot off pyrotechnics to keep them from landing." Pyrotechnics are harmless explosives that sound like shotgun fire. When the birds habituated to the shotgun sounds, the team got out the actual guns and added some "lethal reinforcement."* One bird is shot, and the

---

* My favorite lethal reinforcement story involves Scarey Man, a scarecrow version of those floppy fan-inflated tubular attention-getters you see at strip malls. Unlike those things, Scarey Man also screams and is inflated only intermittently, bursting up like a jack-in-the-box. Birds start to habituate to Scarey Man in about a week, a 1991 USDA study found. Two of the researchers, Allen Stickley and Junior King, then tested whether lethal reinforcement could make Scarey Man scary longer. They'd dress up in a vinyl poncho and sit unmoving on the shore of a cormorant-plagued fish farm. When Scarey Man went off, the man would leap up alongside, "emit a high-pitched wail and bob up and down," and then fire a shot at the cormorants. The log of hours spent "impersonating Scarey Man" resides in the NWRC archives, along with Stickley's field notes. The birds seemed more entertained than scared. "March 1, 1992. 1456 hrs: Three birds came in and sat and watched me do my thing." Junior got bored and began wandering away from his post and shooting random birds. "I reminded him that the object is to make birds think Scarey Man is shooting," Stickley wrote, disgustedly. What is it with guys called Junior?

others take note. It is unpretty but effective. "The only time we would shoot was when they'd gotten to the point where they were completely ignoring the pyrotechnics." Of the thousands of birds run off during the ten-month operation, just twenty-three were killed.

Those twenty-three were used to fashion effigies. These worked pretty well at loafing sites—places where birds hang out together, resting and digesting—but not so much on the recovery site itself. "Because . . ." Tom thinks about how to put this. "The motivation was too high."

I say something about vultures and their ghoulish relish for things dead.

"Vultures?" Tom says. "We only saw one vulture the whole time."

I have made a species-ist assumption. The birds that came to eat the dead were herring gulls. Of course! Gulls and landfills. An estimated 100,000 gulls scavenged at Fresh Kills when it was an active landfill.

By chance, my next research jaunt is to a landfill outside San Jose, California, where the inventor of a robotic falcon will be demonstrating his product.

Falcons routinely prey on pigeons and other birds that size, but a gull is about as big as a peregrine and not lightly messed with. It's hard to imagine a falcon taking one on. Or a gull being afraid of a falcon, or a falcon robot. All will shortly be made clear.

For now, RoBird rests in an aluminum case in the trunk of Nico Nijenhuis's rental car. Nijenhuis is the creator of RoBird. He has traveled from the Netherlands, where his company, Clear Flight Solutions, is based, to give demonstrations to potential customers. But first, a slide presentation. We're sitting around a conference table in the Guadalupe Landfill administrative building. Between Nico and some other Clear Flight people is a man from Aerium, a company that provides drone-based bird-scaring services, including RoBird. I'm down at the other end of the table with the people in the Day-Glo vests: the landfill operations manager, its director of engineering, and a man whose card I did not get but who will later tell me, "I've been in waste for twelve years."

Listening to Nico, it doesn't take long to see why any bird, of just about any size, would be inclined to steer clear of a falcon. Peregrines dive at their prey going 200 mph. They, not cheetahs, are the fastest animals on earth. When a falcon arrives at its target, it does what it has come for swiftly and efficiently. "Direct stomp, clean kill," it says in my notes. And yes, they will take on a full-grown gull.

The downside of all that pursuing prowess is that falcons do not excel at the leisurely glide. Hawks and eagles and other "long wings," as Nico calls them, have the surface area needed to coast and ride thermals, hunting as they go.

Nico says something that kind of floors me. "Birds don't like to fly." He means the hectic up-tempo flapping variety of flight. Because it's tiring. Falcons hunt airborne for only five or six minutes at a time, and then they rest. This explains—or is an excuse for—the twelve-minute battery life of RoBird.

"Twelve minutes?" someone says. It might be me.

"Longer would be unnatural," Nico says.

A colleague jumps in. "Let's go see what RoBird can do."

We walk out to the parking lot, where landfill staff are handing out hard hats for us to wear when we get out to the "tipping face." This is where truckloads of garbage are tipped into the "filling canyon" and then compacted and, at the close of the day, covered with construction debris to discourage feral pigs and other nighttime scavengers. The freshly tipped loads are what the gulls are watching for. Even up here by the administrative building, a few are always drifting overhead. Their shadows slide across the pavement.

I introduce myself to the operations manager, Neil. Gulls rank low among Neil's problems. Far higher on the list are the realtors who work the fancy neighborhoods that surround the landfill and tell prospective buyers it will be closing in two years. The home-owners soon realize this was a lie, and they're angry. They've been trying to shut the place down. They complain about the smell, and they complain about the pigs, which wander onto their lawns at night to dig up grubs. The pigs are a holdover from the days when dumps would employ them for garbage disposal.

In the middle of our conversation, Neil turns without explana-tion and walks off.

"You said 'dump,'" someone offers.

"This is a sanitary landfill," someone else adds. "It isn't a dump."

"*Dump* is a four-letter word."

I guess I'll be driving over with the RoBird people.

A few minutes later, we stand on the lip of the man-made trash canyon. Below, a worker drives a compactor inside a bliz-zard of western gulls. Like driving anything in a blizzard, there's an element of heightened risk. The gulls make it harder for the drivers to see what they're doing. They also drop things, some-times surprisingly large things. The hard hats are not just for bird splat.

RoBird is removed from her case. Her exterior is realistically airbrush-painted and textured to approximate the aerodynamic contributions of feathers. Nico opens her cranium to show us where a small compass sits. Then he closes it back up and hands her over carefully, as one would an infant, to a young man whose shirt says "Pilot." The pilot draws RoBird back over his shoulder like a paper airplane and launches her while a second pilot, Ekbert, works the wing controls: a console with two simultaneously operated joysticks, throttle on the left and altitude on the right.

Nico has created something quite astounding here. This is an unmanned aircraft system—a drone—that has no rotors and makes no sound. It soars and dives as a falcon does, on the power of flap, lift, and gravity. The pilot doesn't just steer it around in circles; he mimics the moves of a hunting falcon. All RoBird pilots train with falconers.

If your company buys a RoBird, who will do this for you? No one will, because you can't buy a RoBird. You are buying the services of an Ekbert. A pilot will come out with a RoBird, as often as you need or can afford.

Alternately, you could hire an actual falcon and falconer. Bird-abatement falconers are a thing. They train for two to five years and are licensed by states or, in the case of airport falconers, the Federal Aviation Administration. The San Francisco Giants baseball team looked into hiring a falconer to deter the hot dog–crazed flock of gulls that circles the stadium in the ninth inning, defecating on fans and every now and then dropping down to interrupt play. Nico recommends that even RoBird clients bring in a falconer from time to time for some lethal reinforcement.

Out on the tipping face, gulls are moving off. The usual noisy swagger seems muted. I've seen birds take off when someone fires a pyrotechnics pistol, as Neil did here a few minutes ago, and this is different. "Pyro" causes a sudden mass liftoff, but hang on for a couple minutes, because the birds will be back. It's more of a startle reaction than the kind of lasting low-grade, let's-get-out-of-here nerves brought on by the presence of a predator (or a convincing robotic imitation). Firing off pyro amounts to, as someone here just put it, "exercising the birds."

And irking the neighbors. People don't want to hear fireworks exploding all day. And they don't want to clean droppings off their cars because twenty times a day a flock of gulls circles over their street from the local landfill. They complain, and then Neil has to hand out car wash coupons.

The landfill's recent approach to the gulls has been to ignore them. "It's just less of a hassle to keep them here," said Neil, back when he was still talking to me. Neil is done thinking about gulls, but I'm not, not yet.

12

THE GULLS OF ST. PETER'S

The Vatican Tries a Laser

f you make a gull sufficiently nervous, it will vomit.[*] While unpleasant for biologists who need to handle them, the habit affords easy insight into what the birds have been eating. Here is a partial list of what a herring gull considers edible—that is, things that have been vomited in the general direction of Julie C. Ellis, senior research investigator at the University of Pennsylvania: bologna, ants, strawberry shortcake, a large mackerel, a whole hot dog, intact mice, squid, a used sanitary napkin, discarded lobster bait, Vienna sausages, an eider duckling, beetles, rotten chicken drumsticks, a rat, a paper muffin wrapper, a loaded diaper, and a plate's worth of spaghetti marinara with mussels.

No gull has upchucked flowers at Julie Ellis. Gulls would eat the world but leave the plants. So when Dutch master florist Paul

---

[*] When I first heard about "defensive vomiting," I figured it was a way to become lighter and better able to take wing and flee. Nope. Nor is it done to repulse the predator. *Au contraire*, it's more likely, said gull expert Julie Ellis, "a way to distract a potential predator with some alternative food." Animals are different from us.

Deckers arranged three truckloads of blooms around the altar out-side St. Peter's Basilica on the eve of the pope's 2017 Easter Mass, he did not worry about gulls.

And yet. "What we saw was not to believe." When he'd left St. Peter's Square the evening before, six thousand daffodils lined the wide, shallow steps to the outdoor altar. He returned at 6:00 a.m., mere hours before the crowd would be let into the square, to floral mayhem. Potted daffodils lying on their sides in the cen-ter aisle and helter-skelter on the steps. Potting soil on the chan-cel floor. Long-stemmed roses had been plucked from vases and strewn about, as though a diva ballerina had come through for a farewell performance sometime in the night.

Yet the blossoms weren't eaten. It appeared to be an act of sense-less vandalism, as much a mystery as the vinyl-pulling vultures of the Everglades or Dipanjan Naha's cooker-smashing macaque. Why would a gull do this? Was there a biological motive? Are some species just dicks?

For answers I turned, via online meeting software, to Julie Ellis at her desk in the laundry room of her home. "The only thing I could imagine," she said, "is that they were looking for worms in the soil." Could be. But most of the ravaged daffodils I could see in the pictures Deckers sent were still in their pots. And the roses were cut flowers.

Fellow gull researcher Sarah Courchesne had joined the con-versation, from her car in a parking lot in Maine. Sarah studies herring and great black-backed gulls at Shoals Marine Laboratory on Maine's Appledore Island. "What do you think, Sarah?" said Julie. "Sarah? Looks like she's trying to get the sound to work."

Sarah's mouth moved silently and then her face froze and then at last words came through. "Well, that was a tense moment. Was it a form of grass-pulling, Julie?"

"I thought of that," Julie said. "Maybe." In breeding colonies, gulls use their beaks to pull up tufts of grass as a territorial display.

"Displaced aggression" is the term Sarah used. Like punching the wall instead of someone's face? Yes, Julie said, though gulls do the latter sort of thing, too. At breeding colonies, herring gulls will sometimes peck to death another gull's chick should it blunder into their territory. And then they, or another gull, may eat it. I read all about this in "Cannibalism in Herring Gulls," an article by Jasper Parsons, who watched a lot of it go down on Scotland's Isle of May. I showed off my new vocabulary word: *kronism*, the eating of one's offspring. They also do that.

Based on Sarah's observations, reports of herring gull cannibalism are overblown. "There are sizable numbers of times when they kill the neighbor's baby but don't eat it. Or they don't even kill it. Just peck at it."

"You know," Julie chimed in, "maybe blind the neighbor's baby, leave it lying there as it slowly dies. It's lovely, really."

Still, neither Julie nor Sarah would say that gulls, as a species, are dickish.* While 20 to 30 percent of herring gull chicks that wander away from their nest are attacked, a similar number, one study showed, are adopted by a neighbor who feeds and protects them. As with humans, as with bears, a few individuals are responsible for the bulk of the species' churlish (to us) behavior. Of the 329 herring gull chicks cannibalized during the 1968 Isle of May breeding season, 167 were eaten by just four cannibals. According to Parsons, one out of 250 herring gull pairs practices cannibalism. It has nothing to do with food shortages, he found. It just seemed to be what they liked to eat best.

Gulls have evolved a generalist's bill and a thick gizzard that withstands and regurgitates shell fragments (and baby gull beaks and diaper linings), so they can pretty much eat what they want.

---

* Especially as they don't have dicks. Like most birds, gulls mate by aligning their cloacal openings. The ornithological term for this is "the cloacal kiss." Which makes bird sex sound sweet and demure, until you remember that they also excrete through their cloaca.

Individual gulls are as different as individual humans. Some keep to the shore and fish for their living. Some hit the landfills. Still others commute into the cities, where they eat pavement fries and hot dog leavings (and, one study suggests, get coronary artery disease). A few eat their neighbors' children, and a few are hunters, nabbing flickers and songbirds on the wing. Sarah recently saw a great crested flycatcher in someone's vomit. "Although," she added, "having been through the esophageal chute both forward and in reverse, it was more of an *okay* crested flycatcher."

There is, or there was, a hunter gull that hung around St. Peter's Square, site of the aforementioned floral vandalism. We know this because the bird was caught on camera in 2014. You can watch it in slow-motion as it swoops in, beak first and irony ablaze, to nail the white "peace dove" that Pope Francis had just released. Every January the pope appears on a balcony with children from a Catholic youth group to read a message of peace and release a dove. The dove survived, but the tradition did not. In later years, a helium-filled balloon in the shape of a dove was released.

Julie and Sarah wish you to know that gulls have a "more endearing side," which we would see if we spent time in their breeding colonies. Gulls are devoted parents, and that includes the males, who stick around to help raise the chicks. This is not typical for birds. More typical is the starling, whose nest behaviors were observed a century ago by naturalist F. H. Herrick and described in his *Home Life of Wild Birds:* "In the space of four hours . . . [t]he female brooded her young over an hour, fed them twenty-nine times, and cleaned the nest thirteen times. The male made eleven visits"—whether he fed his chicks or just sat around and got in the way is not made clear—"attending to sanitary matters but twice."

Gulls are community-oriented, provided the community is their own. When a gull sights what it perceives to be a predator, it warns the rest of the colony. The frequency of its alarm calls communicates the intruder's approximate location, and others rush to harass it.* This of course contributes to their dickish reputations, because the intruder, in coastal resort towns, is often a tourist who has unknowingly wandered too close to a nest. "Pensioner Left with Blood Pouring Down Her Face After Seagull Attack," says the British ITV website, sensationalism pouring from its headline writer. (The accompanying photograph shows only a small patch of dried blood on the crown of the woman's head.)

Where India has monkeys, it seems, we have gulls: bloodying old ladies, stealing snacks from tourists' hands, boosting newspaper sales, vexing politicians. ("Seagull Attacks: David Cameron Calls for 'Big Conversation' About Issue" was the *Guardian* headline after a gull stormed a pet tortoise in Cornwall.) Gull researcher and College of the Atlantic professor John Anderson sees gull "attack" media hype all the time in the coastal town where he lives. "It's absurd," he said to me in an email. "Dogs barking and lunging at people are so NOT news, whereas a gull diving at you gets press." Anderson throws some blame on *The Birds*. "Alfred Hitchcock has a lot to answer for."

So let's focus on gulls' endearing qualities. I recently read a book about gulls by a former *Audubon* magazine editor. The book mentioned, and I then mentioned to Julie and Sarah, that gulls have a food-sharing call. How lovely is that?

"Hmm," said Sarah. "The call I hear them make when they find food is the long call, which is territorial. I would bet they are staking a claim to the food they find, rather than inviting friends to dinner. So I must plant a flag in the gulls-are-dicks hypothesis on this one."

---

* And good luck if it's you. In addition to beaking your head, gulls can be, as Ellis phrased it, "very adept at aiming their feces." She shared the story of a student on Appledore Island who, hoping to protect herself while traversing a nest-dense canyon, put on a raincoat and pulled the hood up tight. "A gull managed to shit directly into her mouth."

But that's a survival strategy. All of this is. It's about keeping fed, protecting one's progeny, escorting the genes to the next generation. It's gulls being gulls and, unfortunately in some cases, trying to do it too close to people being people.

But messing with the pope's roses is . . . who can say what that is? I better go see. Easter weekend is coming up, and this time the Vatican is prepared. Along with Paul Deckers and his team, they've flown in a "laser-operated scarecrow" and its creator, André Frijters. I keep hearing about lasers as an effective, seemingly benign bird-scaring method. Tom Seamans used one in the evenings at Fresh Kills. Staff at the Royal Palm Visitor Center tried one with the car-vandalizing vultures. Could keeping birds away be as cheap and easy as pointing to graphs on PowerPoint slides that no one understands?

Easter weekend at the Vatican is a joyful thrum of Catholic professionals. Nuns and priests jet in from all over, giving St. Peter's Square the look of a college campus on commencement day: the flowing robes, the specialty headwear, the selfie sticks. A dozen or so gulls are here, too—cooling down in the fountain, watching the crowds from St. Peter's marble haircut—but for now there's no need for deterrence. Six assistant florists bustle around setting up the flowers.

André Frijters is meeting me at the security gate in an hour, around 5:00 p.m. He can't begin his work until all the flowers are in place, because the laser beams must be set to reach every plant. I wander the souvenir stalls and pontifical tailor shops, the low and the high of Vatican shopping. I try, and fail, to talk my way past one of the Vatican Swiss Guards, partly because I'm bored, and partly because a man in ballooning striped knickers doesn't seem all that imposing.

From behind a barricade, I watch Paul Deckers orchestrating the final adjustments. For most of my time here, he is a figure glimpsed at a distance, always in a hurry, striding through my field of vision in leather hiking books and a fanny pack (also leather). The night I arrived, he alit, briefly, to chat as I sat with André at the cafe in their hotel. He shared the story of the 2017 gull debacle and how he'd gone on Dutch television when he got back to the Netherlands to crowd-source a solution for future Vatican gigs.

Of the 250 or so people who responded, most had not thought through the singular exigencies of the scenario. From where I stand, I can practically see inside the pope's bedroom. The residents of Vatican City don't want to be kept up all night by "sirens with yelling sounds," as one person suggested, or "bomb sounds." They don't want to attend Mass the next morning in a fog of "smelly odors." And although an effigy could be seen to complement the Crucifixion iconography already on display at St. Peter's, a dead gull suspended by pinioned feet was likewise not up for consideration. André wasn't even sure an effigy would work at night. Gulls evolved as diurnal birds; though some have begun taking night shifts to raid Roman garbage bins, gull night vision may not be sufficiently acute to process the creepy particulars of an effigy. I wondered about RoBird, but apparently Nico Nijenhuis hadn't been watching TV that day.

André telephoned Deckers the day after the televised plea and said to him: "I have been scaring birds for twenty-five years." He owns the company Vogelverschrikker (Dutch for *scarecrow*). André suggested bringing in a LaserOp Automatic 200, a sort of monochrome laser light show. Lasers are silent, seemingly humane, and they can usually be counted on to unnerve gulls for at least a week. They're mainly used in darkness or low light, when the beams are most visible. Wildlife Services, in the States, has used them to dissuade cormorants, gulls, and vultures from roosting on structures where someone doesn't want them—and their droppings—to gather.

The laser beam is green, a color some birds are thought to see

better than we do. There's a theory, set forth on some manufacturers' websites, that the lasers work because these birds perceive
the beam as a solid green rod slicing through the air and coming
at them. (I wonder whose theory this is and how many *Star Wars*
movies this person has seen. I checked: no "lightsaber" references
in the product copy. If anything, they call it "the stick effect.")
Some species—pigeons, for instance, and starlings—either don't
see it this way or aren't fazed.

At five sharp, André shows up and walks with me to the altar.
Tomorrow eighty thousand Catholics and tourists will come to
hear the pope say Mass and take communion from a roving pla-
toon of priests, but for now we look out on an audience of gray
plastic chairs. André has brought two LaserOp Automatic 200s.
One would suffice, but this is, you know, the Vatican. He wants
to be sure. The lasers are housed in a boxy white structure that,
out in a farm field, might be mistaken for stacked beehives. Here,
at the foot of the altar at St. Peter's Square, they look like, I don't
know, Ikea baptismal fonts?

André and I attempt conversation over the noise of a cleanup
crew with a leaf blower. "How did you get into bird-scaring!"
I'm yelling.

He leans in toward the side of my head. "Well, I was a farmer
myself!"

The possibility hadn't occurred to me. André looks like a
farmer the way Paul Deckers looks like a florist. He, too, is wearing leather—in his case, a lived-in black jacket. His hair and his
stubble are buzzed to the same length. His jeans sit low on his hips
and while they are not tight, they are for sure not farmer jeans.

André grew lettuce. "Iceberg and little gems!" The leaf blower
shuts off in the middle of this. "Little gems!" comes out like a
sports cheer.

A salad-loving menagerie went after his lettuces. Hares, crows,
wood pigeons. "The wood pigeons would eat the hearts. I couldn't

get them away with the gas bangers." (A gas banger, or propane cannon, is a pyrotechnics device left out in a field and set to fire at intervals.) "You can put netting on top of the crops, but it's expensive and a lot of work to keep it up. The crows would pull up the plants because they'd think there's a worm underneath." I ask André whether that might be what the gulls were doing with the daffodils. He says he thinks the damage was done out of curiosity: "The gull thinks, *There is something different, maybe I can eat it. So I try it first.*"

André likes birds. He mentions their helpful consumption of insect pests. "The birds were here before the farmers," he adds. "The farmer comes along and opens a restaurant and the birds are coming to eat, and that's generally how it goes."

Like the senator from chapter 8, André undertook the most effective strategy of all. He changed careers. One day, while he was still a farmer, André was on the phone with the manufacturer of Scarey Man, the noise-making, intermittently inflating pop-up bird deterrent. "He said to me, 'You know, I don't have an importer in Holland. Would you like that?' 'Okay,' I said. 'I like that.'" André became the Scarey Man man for the Netherlands. Customers would often ask for a cheaper product, and soon he began stocking pyrotechnics, hawk kites, effigies. "Everything you can buy in the world about bird-scaring, we had it in the warehouse." For a while he kept up the farm. "Between the scaring and the lettuce, it was too much." He went with the scaring. The hours are better, and the money. As with a gold rush, so with farming: the people who make the most reliable living are the ones who sell the supplies to everyone else.

André gets up to walk the perimeter of the floral display, squatting every few feet to sight the laser and note any points it might miss. I tag along with my notepad, interrupting his focus and bumbling into his sight lines.

"I'm just going to follow you around."

"I see that."

From somewhere beyond the square, a clamor of whoops and cheers coalesces and crescendos. It's a red-carpet sound—the A-list star pulling up and then stepping from the limo. "It's him," André says. "Francis. He's a rock star."

Hundreds of nuns are in town this week. I've seen nuns with green habits, pink habits. Bad habits! Nuns vaping, nuns cutting in line. Right now, nuns are running, laughing, elbows pumping and veils flying. The security gates on the north side of St. Peter's Square just opened, and a group from the front of the line are rushing for front-row seats at whatever services Pope Francis is about to hold inside the basilica. As they race behind the outdoor altar, a gull on a pole-mounted speaker takes flight. Two on the cobblestones follow.

The nuns have just demonstrated the world's oldest technique for scaring birds away from plants: have someone run at them making noise. The best means of keeping away crows and other "enemies of the corn," wrote Gervase Markham in 1631, is "to have ever some young boy . . . to follow the seed-man . . . making a great noise and acclamation." The practice was documented in detail for an 1869 British parliamentary investigation of child labor practices which was the subject of a recent museum exhibit in Nunney, England. The children were usually between six and nine—boys not yet strong enough for more laborious farmwork. They were paid a pittance to roam the fields "hallooing" and clacking wooden bird scarers. They worked a month in spring when the seeds were sown and came back in fall for another month when the crops were ripening, effectively obliterating any continuity in their education. The drawback from the farmers' perspective was that the boys were lazy, to the extent that, by one farmer's words, "each boy required a man to look after him."

Well, halloo, they should have just had that man do the scaring. Adult bird scarers are uncommon, but they do exist, even today. Hiring them can be cost-effective. I say this with some confidence because science has looked into it. The UK Ministry of Agriculture, Fisheries and Food maintains an experimental farm in north Norfolk, England, for testing, among other things, "scaring regimes." The site, a series of coastal wheat and rapeseed fields, was selected for the noteworthy annual depredations of a flock of three thousand resident brent geese. Researchers Juliet Vickery and Ronald Summers used the site to compare the cost-effectiveness of commonly used techniques—propane cannons and the like—with that of a "full-time human bird scarer" tooling around on an ATV six days a week. ("The farmer scared geese on the seventh day," the researchers wrote biblically.) The human bird scarer achieved a significantly greater reduction in the amount of time the birds spent grazing and the intensity—as quantified by "dropping density"—with which they went at it.

Even factoring in the initial outlay for the ATVs, the human scarer was more cost-effective. A few farmers would seem to have taken note. Dutch berry growers sometimes hire college students to work as bird scarers in the summer as the crop ripens.

A human bird scarer makes sense for a relatively small acreage. The flowers outside St. Peter's cover less than half an acre and require just a couple nights of protection. If ever a scenario called out for simple, cost-effective human bird-scaring, here it was. If I had seen Deckers's original call for ideas, I'd have sent an email saying, "Did you think about hiring a human bird scarer? Just some person to sit there at the altar and keep an eye out and run off any gulls?" Instead I sent the email after I got home from Rome. "No, we did not think about that," came Deckers's reply. "It's better for André that we did not!" And it was. The Vatican City State is gearing up to be a nation with one grocery store, one pharmacy, no movie theater, and two LaserOp Automatic 200 bird-scaring units.

André opens the lid of one of his laser units. Inside there's a digital display, like something Daniel Craig would be hunched over, trying to defuse the bomb that's set to take out MI5 and the whole London waterfront, or program his sprinkler system. "Only five buttons," André says. Buttons for setting the start times and the intervals between activations, and buttons for setting the boundaries of what needs to be covered. "A farmer can do this."

For a large farm, using human scarers would either be personnel-heavy or endlessly Whac-A-Mole. An automated system like André's holds great appeal: a series of solar-powered units that can be custom-programmed to cover the whole field. Is this the golden future of bird deterrence?

André drags a planter out of the beam's path. "In five or six years," he begins, but he doesn't go where I'm expecting, "no one will be using lasers. It's dangerous." Even the handheld kind sold for classroom lectures can damage a retina. When laser light is absorbed by pigments in the eye, it deposits energy and heats up the tissue. Because the light arrives in a tight beam—and is further focused by the eye's lens—the energy density is high. In terms of the damage caused, think of the difference between someone in stilettos stepping on your foot and someone wearing loafers.

André isn't wearing laser safety goggles. When asked, he points out that the beams are aimed downward, and that you would need to be looking into the source of the beam to harm your eyes.

Who stares into a laser beam? Adolescent boys, in 80 percent of the cases. That was the finding of a team of ophthalmologists who reviewed 77 cases and emailed surveys to hundreds of their peers. Some of the kids, they found, were taking part in laser staring contests, reluctantly admitting to emergency room personnel that they'd looked into the laser for 10, 20, in one case 60 seconds. (The boys often had behavioral or genetic disorders linked to self-injurious behavior.)

Before I left, I spoke by phone to a Purdue University biology professor and former National Wildlife Research Center researcher, Esteban Fernandez-Juricic. He was involved in a study examining the safety—to birds—of brief exposures to a bird deterrence laser. It was the first study of its kind, previous safety claims having been based on human focal lengths, spectral sensitivities, and eye configuration, which have, as he put it, "zero to do with the bird eye." Some species, Esteban said, don't react to a laser, even when it crosses their visual field. "Maybe some species cannot see the lasers. We don't know. So with these species that don't respond, we should take more care." But the opposite may be more likely to happen. Esteban voiced the thoughts of an imaginary frustrated farmer faced with a flock of one of those species: "*Come on, species, there you go! More to you!*" He has a wonderful, excitable Roberto-Benigni-wins-an-Oscar way of speaking.

Laser companies know about Esteban's project, and it makes them uneasy. They've been scrambling to manufacture large units for agriculture, and a lot of money is at stake. "I have stories which I cannot tell you," Esteban said. "It's . . . how can I characterize it . . ."

"Someone tried to influence your results?" I offered. "A bribe?"

"What you are saying might not be very far from what happened once." Esteban had to call the university lawyer. "I said, 'Oh my gosh. You need to get involved, because this is . . . ! *Whoa.*'"

André Frijters had not heard of Esteban, which was not surprising, because the results of the study had not yet been published. Three months after I got home, Esteban shared preliminary findings with me—no details, and no laser names. The results, he said, were unsettling enough for one bird-scaring outfit to travel to Purdue for a meet-up.

Esteban tried to calm his visitors. "It's not like, *You bad companies that are trying to do this!* It's an opportunity to work with the scientific community to modify the way the laser is operated, or the wavelength or the intensity."

One year later, the study remained unpublished and Esteban was not replying to my emails. I hope he's okay.

Around 5:00 a.m., about the time of the Easter 2017 flower assault, I walk over to St. Peter's Square to see what's happening. The lasers make firefly flashes as the beams touch the plants in their whizzing circuit of the altar area. They appear to be doing their job. From what I can see from back behind the security fencing, Deckers's flowers are unmolested. Thirty gulls are asleep at the base of a fountain in the center of the square, drawn to the heat of the cobblestones.

Along the colonnade, a dozen of Rome's unhoused are waking and quietly rolling up their sleeping bags. Police cars are parked thirty feet away, but the *polizia* have left these men and women to sleep in peace. Is this the kind and gentle hand of Pope Francis? I wonder to what degree his papacy is influenced by his namesake, St. Francis, friend to the poor and friend to the animals. Might the current papacy espouse more progressive approaches to nuisance wildlife? A ludicrous inquiry, I know. But I'm here. I might as well ask.

13

THE JESUIT AND THE RAT

Wildlife Management Tips from the Pontifical Academy for Life

The Vatican City State is a sovereign nation the size of Disney's Magic Kingdom. Like Disney World, it has places where tourists may go and places where only staff are allowed. Because I'm neither, and because I don't fit the Vatican's definition of press, I was told to write to the secretary-general of Vatican City State. It was like trying to enter the United States by writing a personal note to Donald Trump. Though likely went better. My note was forwarded, and promptly generated a gracious reply, which Google Translate presented as follows. "I gladly take this opportunity to offer you, my dear lady, the expression of my distinguished homage." Signed, somewhat incongruously, "the Vatican Director of Gardens and Garbage."

The very same director, in a blue Ford Focus,* has just pulled
up along the curb at one of the guarded entrances to the Vatican.
Rafael Tornini gets out to shake my hand. In person, he is a formal but not showy man. He is dressed in a dark blue business suit,
clean if slightly worn. We head out for his office. The streets are
narrow and largely empty. It appears to be a city without traffic, a
country without children.†

"And there is the border with Italy!" I follow Tornini's gaze to
the massive wall that surrounds the Holy See. A gull glides over.
There's your symbol of peace, I think to myself. A bird, any bird,
soaring over walls, ignoring borders! Peace, freedom, unity! It's
possible I've had too many espressos.

Tornini's office is modest. His view is of the leaves of some
thick vine. An interpreter joins us. Tornini says gulls cause no
trouble inside the Vatican gardens. "Here it's the green parrots."
They eat the seeds that the garden staff plants.

Nothing is done to get rid of them. "They're part of the system." According to Carol Glatz, the Catholic News Service
reporter who covered the peace dove fiasco, the Holy Father is a
bird guy. He used to own a parrot (and he taught it dirty words).

---

*   Pope Francis rides around in a Focus, too. What's up with the Vatican
and the Ford Focus? Nothing, says Erich Merkle, of Ford's marketing department. Coincidence. Merkle drives a Mustang, but he defended the lowly Focus.
"Visually, the lines are actually pretty handsome." He added that in Europe,
Ford sells a turbo-charged 252-horsepower Focus called the ST. "That thing's
a monster. Once you option it up? Put a spoiler on it, some Recaro bucket seats
in the front? It's really hot." Was the pope's ride in fact a Focus ST? It was not.
The pope opted for the "more rudimentary" model.

†   A few high-ranking members of the Swiss Guard live with their families inside the Vatican City State, but no baby has ever been born inside the
nation. When journalist Carol Glatz was nine months pregnant and working in
the Vatican press hall, a nun working alongside her hoped the baby would be
born right there. Because it would have been the first birth ever registered in
the Holy See. What splendorous benefits might befall a Vatican anchor baby?
Zippo. Vatican citizenship is granted by administrative decision, not birth.

Pope Francis has indeed steered the nation according to the worldview of St. Francis of Assisi, the OG animal activist. Shortly before Tornini took his current post, Pope Francis decreed that biological pest control be used in place of chemical pesticides. Insects that prey on problematic bugs have been introduced, and nest boxes have been mounted on tree trunks in the gardens to encourage bats, because the bats eat the mosquitoes.

We get back in Tornini's car to go see the Vatican bat boxes. They're very nice, they're wooden, they're tasteful. Soon we arrive at a low, sprawling mound of grass clippings. The Vatican compost heap! Far in the back, a week old by now, is some distinctly papal organic waste: elaborately woven palm fronds left over from Palm Sunday. Tornini pulls one out for me.

Open compost piles can attract animals. Here it is—that rare conversational segue to the topic of Vatican pest control. I ask the interpreter to inquire about rodent problems.

Ratti, I hear him say to Tornini. He turns to me: *Sì, sì*. The Vatican has rats.*

Sì, answers Tornini when I ask whether they set traps. He says something to the interpreter, who then adds, "They have to do an action against them, because it's a big population. And they really make damage. To the machinery, the wires. They try to keep everything as clean as possible, but—"

---

* After I got home, I went on the web to see if I could find out which pest control firm the Vatican uses. A search brought me to a computer-generated web page for what had to be a nonexistent company, Derattizzazione Roma. Under the heading "Our greatest services for Vatican deratting" was a crazed two-page list that, were it real, suggested a fabulously pestilent Vatican where deratization was underway day and night and even on Sunday, in every room of every building, where even the mice had rats. Urgent Vatican Deratization, Night Vatican Deratization, Deratization of Vatican Canteens, Sunday Vatican Deratization, Deratization of the Vatican Shopping Centers, Deratization of Vatican Technical Rooms, Deratization of Vatican Elevator Lifts, Deratization of Vatican Mice. The exuberant translation software had even created a drinking establishment for exhausted deratizers to repair to when their shift was done: Rodent Pub Vatican.

"So Pope Francis is okay with killing rats?"

Tornini has never even met Pope Francis, and now he is being asked to speak for him. The interpreter listens, then turns to face me. "He says you should ask him."

Of course I can't do that. Instead, I will do the closest thing available to a lapsed Catholic with no high-level connections. I have an interview with Father Carlo Casalone, a staff bioethicist at the Pontifical Academy for Life. The PAL is a Catholic think tank of sorts. Its members are appointed by the pope but are not necessarily clergy or even Catholic. The PAL guides Church doctrine on matters ranging from the perennial (abortion, euthanasia) to the more cutting-edge (gene therapy, artificial intelligence). As the sexual abuse scandal widened, the PAL played a role in drafting the Church's response. I told the PAL media manager I was interested in the academy's opinions on the designation of certain wildlife as pests. That is, under what circumstances should a species be exempt from moral protections against extermination or cruelty? I quoted St. Francis of Assisi. I did not mention rats. He replied right away. It was perhaps a welcome diversion from thornier inquiries lately clogging his inbox.

Walk straight down Via della Conciliazione from St. Peter's Square. Three blocks along, across the street from the souvenir store selling Pope Francis bobbleheads, with their unfortunate suggestion of tremor, you will see a boxy three-story building of caramel-colored stucco. A plaque beside the doorway announces that you have arrived at the Pontificia Academia pro Vita, the Pontifical Academy for Life. Although situated outside the physical borders of the Vatican City State, the PAL is officially a part of it, which means that while visiting, you undergo a sort of geopolitical transubstantiation. You are in Italy, and yet you are inside the Vatican.

Father Carlo's office walls are white and undecorated, unless a crucifix counts as decor. It was the same at Tornini's office. The extrav-

agance of the Vatican seems concentrated, like a laser beam, inside
the museum and the churches. Father Carlo is himself unadorned:
black pants, black shoes, black button-down shirt with the white tab
collar. His voice is low and quiet and he doesn't talk with his hands,
or not in the stereotypical manner of the Italian male. Though the
floors are marble, I imagine him making no sound as he walks.

Some of this anti-ostentatiousness may be the influence of Pope
Francis, who in turn is influenced by St. Francis of Assisi, the humble, nature-besotted Capuchin. When he became pope, Francis
took a regular clergy apartment. Like Tornini, he gets around in
a Ford Focus. This week I dropped by a Holy Thursday Mass that
included the ritual washing of a few congregants' feet. It was more
a gesture than a real cleaning—a splash of water over the instep.
"Francis gets right in there with a scrub brush," laughed my Catholic News Service acquaintance Carol Glatz. (And yes, he eschews
the red loafers.)*

So I'm curious. How far does the pope think we should go in

---

* I almost wrote "Prada," and then I read Dieter Philippi's exhaustive treatise
on *campagi,* papal shoes, replete with more than 100 photographs of custom-made
red papal footwear. Pope Benedict's official shoemaker was Adriano Stefanelli.
He made the red loafers that earned Benedict the *Esquire* honorific "accessorizer
of the year," as well as "the special slippers which the pope wears around his
apartment." Gammarelli, a papal and clerical tailor shop near the Vatican, has a
shoemaker who by tradition makes the ceremonial red loafers worn by newly
elected popes when they first appear on the St. Peter's balcony. And anyone else,
sniffs Philippi throughout his 119-page commentary, is lying. Silvano Lattanzi's
claim that Benedict wore a pair of his velvet slippers? "I am quite certain this
is a false assertion." The mules made for him by Raymond Massaro? "I do not
think the Pope has ever worn this kind of shoe." Ferragamo's wine red loafers in "the papal style"? "The Pope has never worn these shoes." The slip-ons
imprinted with the Vatican coat of arms? "The Pope would certainly never wear
a design of this kind." Those Prada red loafers with the ornamental seams? "A
fallacy. The Pope renounces ornamental stitching." According to Philippi, only
one commercial shoe company can rightly claim to have shod Pope Benedict.
While hiking during his summer 2009 holiday, Benedict wore Camper Pelotas
leather sneakers.

the direction of respecting and protecting the natural world and its wild inhabitants? Before I arrived, the PAL media manager sent me a copy of Francis's rather beautiful encyclical *On Care for Our Common Home.* "Each creature has its own purpose," he writes. "None is superfluous." He describes how St. Francis would burst into song when he gazed at the sun, the moon, or the smallest of animals. I read these passages to Father Carlo.

He listens, nodding. "Saint Francis began a new relationship between nature and humanity. If you read his poems, you find the expressions Sister Water, Brother Sun, Sister Moon."

"Would Saint Francis include Brother Rat?" Sister Boll Weevil, Uncle Blackbird Who Devours 2 Percent of the North Dakota Sunflower Crop?

Father Carlo says yes, yes he would. "He includes even death."*

"Did Saint Francis say anything specifically about rodents?" I hear myself say.

"No, he didn't. But the point is, brotherhood is not a simple relationship. With your brothers and sisters, normally you fight. You cannot think that there is an idyllic way of being in a relationship with someone. Every relationship among humans and the earth is not only connotated with positive aspects. At the same time you also have negative aspects. The point is, how do you deal with those aspects." He's good, this guy.

"Yes, and how *should* we deal?" It's well and good to say these things, but how do we act in a way that serves both human and animal fairly? Let's take the example of Canada geese on golf

---

*   Sister Death—it's a woman. Also the title of the seventh album by Alec K. Redfearn and the Eyesores. *Signal to Noise* magazine describes Sister Death as "a gorgeous amalgam of 20th Century Americana, cabaret . . . Eastern European folk, noise rock and minimalism." You never know where a footnote will take you.

courses. What is their crime? Befouling the turf.* Littering. For this, should we be allowed to call someone in to round them up and gas them? Do they deserve to die because a few well-heeled humans want to hit a ball into a hole and they need an obsessively tidy playing surface the size of the Holy See? Think of all the Sister Water that gets wasted watering the greens. Maybe it's time to eliminate golf, not geese!

Father Carlo collects his thoughts. Among them, surely: *Who let her in?* "We have to put the action in the context of where we are. What does it mean, the golf field, for the people that are working in it? If this is the only way that people can find employment in the region, you have to keep in mind also this aspect of the action you are performing.

"Secondly, maybe it's not necessary to kill the birds. You can

---

*   But not as heavily as the internet would have you believe. Goose Busters has them extruding 3 to 4 pounds a day. The Geese Police superintendent at the National Mall in Washington, DC, claims 2 to 3 pounds a day. A Boston city councillor: "as much as 3 pounds a day." The Canada goose fecal smear campaign appears to have hit its zenith in New Jersey's *Montclair Local:* "An adult Canada goose can weigh up to 20 pounds and defecates more than twice its weight daily." That would be *40 pounds a day,* coming out of a single goose. That's how much a *horse* makes. The reporter cited the USDA; a contact at their NWRC public affairs office steered me to the USDA "Geese, Ducks and Coots" fact sheet, which gives a daily total of 1.5 pounds. The author of the USDA fact sheet got his information from a Virginia Tech University Cooperative Extension goose fact sheet. *That* fact sheet says "studies have shown" but does not cite any studies. A Google Scholar search brings up just one researcher, B. A. Manny, who actually went out and weighed some turds. Manny's findings: the average total (wet) weight of a Canada goose's daily droppings was just a third of a pound. Where did Virginia Tech get the 1.5 pounds per day figure? The author did not reply to multiple emails, so it remains a mystery.

Poundage aside, the Canada goose is a frequent crapper. Twenty-eight times a day, on average, Manny found. In related research, a Canadian team reports that "sleeping geese sometimes produce small piles of droppings."

act in another way to deviate the trajectory. You have to move, to think, in a progressive way of intervention."

Like egg addling! I almost blurt. Rather than culling flocks of Canada geese, some municipalities seek out nests and either shake the eggs or coat them in oil and then return them to the nest. With the result that the parents are incubating blanks. To figure out the cutoff for humane termination of a Canada goose fetus, a team from the Michigan Department of Natural Resources examined tens of thousands of eggs. They then established a method to assess the age of an egg by seeing if it floats—an indication that there's more air inside than goose. The technique has been recommended by the Humane Society of the United States and by PETA, and while I wonder what the Catholic Church might have to say about goose abortion, I don't wish to addle Father Carlo, so I move on.

What about a predator, say, a coyote, that kills a person's pet? Is it ethical for the person to kill the predator? When the predator acts by instinct, to survive?

Father Carlo aligns a stapler on his desktop. "You have to keep in mind the emotional impact for the people."

"But how do you weigh these things, Father Carlo? The feelings of the person versus the life of the predator?"

There is a knock on the office door, and a man steps in carrying a tray of traditional Italian Easter cake and a carafe of water. "Ah, Sandro, *grazie!*" Father Carlo seems delighted by the refreshments or maybe just the break in the conversation. Sandro sets down the water glasses. Mine has a bit of brown gumpy on the rim. Father Carlo quietly swaps his for mine.

While we take a break to have our cake, I mention that I watched Pope Francis zipping around after Easter Mass on an electric scooter, shaking people's hands, utterly exposed to the crowd.

"Yes, and it drives the security people crazy." Father Carlo shares a story about the time Pope Francis needed to fill a new glasses prescription. Without telling security, he paid a visit to the optical shop. The optician was delighted but ultimately a bit disappointed. "Francis said to him, 'I want only the lens, no frames. Because I already have the frames.' He really said, 'I have the frames. I just need the lens.'" Father Carlo shakes his head, amused by the memory. "Can you believe?" His smile reveals a gap between his two front teeth. Because of the plain clothing and the regulation haircut—the absence of any indicators of personal style—the gap has the effect of an accidental intimacy, like bitten fingernails, or a bra strap that has slipped into view.

Presently Sandro returns with his tray to collect the plates and glasses. Father Carlo watches him retreat, then turns back to his guest. "Animals act according to their instinct, as you say about the coyote." He pronounces it *co-dee-oh-tay*, a most lyrical word. "Human beings, on the other side, have free will. They are responsible for the stewardship of creation. Their role is to help nature. Because we can study the system, and the animals cannot." He cites an example of reintroducing wolves into regions of Italy where deer and wild pigs were overpopulated, rather than culling the pigs and deer. "They asked the wolves to balance the ecosystem." He smiles. The gap again! I live for it.

I share the story of the Indian mongooses that were brought to Hawaii to kill rats in the sugarcane fields. What someone overlooked is that the rats are nocturnal, and the mongoose is diurnal. The mongooses ate a few rats and lots and lots of sea turtle eggs.

"Yes. Well." Father Carlo reaches for his briefcase. He has a train to catch. "In the complex systems of the world, we cannot predict the effect of all our actions. So? We have to act according to the principal of prudence."

Amen to that. Regarding the Italian wolves, I read later that they did indeed help lower the population of deer and wild pigs.

They were fruitful and multiplied and began turning to ranchers' livestock for food. And then the ranchers began agitating for a cull. It is ever thus. In the words of National Wildlife Research Center public affairs specialist Gail Keirn, "When it comes to wildlife issues, seems like we've created a lot of our own problems."

Perhaps no place better knows the challenges and grand *doh!* moments of animal stewardship than my next stop, the lovely island nation of New Zealand.

(No Model.)

J. A. WILLIAMS.
ANIMAL TRAP.

No. 269,766. Patented Dec. 26, 1882.

Witnesses:
J. E. Clark.
W. H. Kern.

Inventor;
Jas. A. Williams
per
F. A. Lehmann
Attorney.

14

KILLING WITH KINDNESS

Who Cares About a Pest?

To live in a penguin colony is to know no modesty. Anything you do—mate, preen, throw up fish for your young to eat— you do in plain view of the neighbors. The yellow-eyed penguin will have none of this. It builds a nest in the coastal tussock, out of sight from other yellow-eyed pairs. As with human suburbanites, the quest for space and privacy brings with it a longer commute. Every evening, yellow-eyed penguins of New Zealand's Otago Peninsula return from their labors at sea, crossing the beach, picking their way through the scrub, and trekking partway up a steep bluff to get home.

You can watch yellow-eyed penguin rush hour from a hide in a private reserve held by Elm Wildlife Tours. Shaun Templeton, Elm's dry-witted operations manager, leads today's outing. Shaun is youngish, with a shaved and well-tanned head. He has large brown eyes that remind me of a seal's, though maybe only in the context of this afternoon. Some beaches here have pinnipeds the way Coney Island once had humans.

Just past 5:00 p.m. the first penguin appears in the backwash of broken waves, bodysurfing. It lets the water carry it as far as possible and then, beached, it stands and crosses the strand in the deliberate plod that is a yellow-eyed penguin in a hurry. At the base of the bluff that rings the cove, it begins a series of considered uphill jumps, bending forward, pausing to cock the knees, then springing upward, its whole being applied to the task of moving one inch higher up the slope.

This is why, or partly why, the yellow-eyed is the world's most endangered penguin. The isolated nests and the long, exposed trek to reach them provide shopping and dining opportunities for predators: seals and sea lions, as always, but also newcomers— stoats (known in the United States as short-tailed weasels), rats, and feral cats. Only around four thousand yellow-eyed penguins remain in the world, forty-three of them currently at this bay. Stoats are known to have killed three chicks here already this season. Shaun keeps track, because he's here most weeks and because Elm contributes to the conservation efforts of the Yellow-eyed Penguin Trust.

It's dusk, the commute mostly over. We're watching the fur seals now.

"Have a look." Shaun indicates a loose scatter of bones: heads and spines of two barracuda, and some kind of octopus scaffolding. "A most impressive sea lion vomit." Sea lions eat their prey bones-and-all and then regurgitate what's undigestible, as owls do but less tidily. We happily note the absence of penguin solids.

Shaun makes his living from tourism but is by avocation closer to a naturalist. I don't know the precise parameters of that calling, but I think that if you find yourself applying the word *impressive* to any variety of animal vomit, you may be one. It has been a disastrous two decades for the yellow-eyed penguin. Shaun explains: On top of losing habitat to humans and getting tangled in their fishing gear, the birds have lately been killed off by disease—

avian malaria and avian diphtheria—and starvation. As the sea has warmed, the bottom-dwelling fish that these birds eat have begun to move farther out into colder water that is also deeper. Yellow-eyed penguins can dive deep, but not as deep as some of these fish are now living. Perhaps more than anything else, though, it's the unnatural predators: stoat and company.

If the present rate of decline continues, the yellow-eyed penguin will likely be gone from the planet in ten or twenty years. It is difficult to be here watching them and not feel somewhat slammed by this information. What a thing to lose! Go look them up. The candy red beak, the pink go-go boots, the yellow mask angling back from the eyes. They're the Flash, they're 1970s Bowie! I don't mean to imply that adorable, showy species are of more value or somehow deserving of more concern. It's just . . . *damn*.

Up on the bluff, a pair is greeting each other. You can tell by the ruckus. The Maori name for the yellow-eyed penguin, *hoiho*, means "noise shouter." *Shhhhh*, you want to say. *The stoats will hear you! They'll wait until you leave and come for your children.*

It is not just the yellow-eyed penguin that is in trouble here. It's any flightless New Zealand species (and many that still have wings). For tens of millions of years, these islands had no native land predators, so birds who arrived here no longer had need for swift escape. Evolution gradually discontinued some species' wings, applying the energy to something more contributive to survival. Then the predators showed up—stowaways and introductions from other land masses. Each year, invasive species kill around 25 million native New Zealand birds: kiwis, most famously, but also kakapos, blue ducks, blue penguins, and keas (the carrion-eating mountain parrots we met in chapter 11). Stoats are efficient killers and ready tree climbers. They like eggs very much, but they will also kill and eat young birds. Each year 40 percent of all North Island brown kiwi chicks, for instance, are killed by stoats.

Stoats. Who invited them?

It began with rabbits. In 1863, homesick European settlers formed the Otago Acclimitisation Society, one of several such groups in New Zealand at the time, and released six rabbits into the Otago countryside. It was hoped, they explained, that "sportsmen and naturalists would be able to enjoy the activities that made the remembrance of their former home so dear."

What followed is nicely if perhaps hyperbolically summed up by one landholder's "rabbit arithmetic": $2 \times 3 = 9,000,000$. Two rabbits in three years equals 9 million rabbits. By 1876 most of Otago was overrun. The rabbits had no natural land predators, and the mild climate lengthened the breeding season. Rabbits ate the sheep pastures bare. Flocks starved. More than a million acres of Otago landholds were abandoned.

By 1881, government officials were moved to take action. They passed a Rabbit Nuisance Act, hired rabbit inspectors and "rabbiters" to shoot and poison. Stoats and ferrets were shipped in from Europe; nearly eight thousand of these "enemies of the rabbit" were released into the New Zealand countryside and protected by law.

But rabbits were only one of the species that wound up on the stoats' plate. The fierce tubular hunters quickly set to the avifauna: eggs, chicks, small adults. The march to genocide began. As of 2019, 79 percent of New Zealand's land-dwelling vertebrate species were classed as either threatened with, or at risk of, extinction. That includes 78 bird species and 89 reptile species.

In 2012, the New Zealand government again stepped in. Where once it had imported and protected stoats, now it endeavors to be rid of them. Predator Free 2050 (PF2050) is a Department of Conservation (DOC) effort to protect native biodiversity by eradicating the three invasive predators that most threaten it: stoats, rats, and brushtail possums. (The year 2050 is when the DOC aims to have it done.) Citizens have been brought to the cause by energetic public outreach. Wander through any national park visitor center and you will see the Predator Free 2050 exhibit, with brochures and worrisome statistics and the requisite taxidermied stoat, needly teeth bared, one paw proprietarily placed on some savaged bird or violated egg.

Entire towns have taken the bait. Neighbors and farmers set out DOC traps by the hundreds. Success stories and tips are shared in a monthly newsletter that closes with the sign-off "Happy Trapping!"

Today has been a tour not just of Otago's birds but of the various models of DOC traps used to protect them. Each predator has its own challenges. "Cats like fresh meat," says Shaun, "so you need to rebait a lot."

Most places on earth have some feral cats, but this is well beyond *some feral cats*. Because along with stoats and ferrets, cats galore were set loose in the countryside to kill rabbits. Were in fact farmed for the job. And when demand outstripped supply, Dunedin lads were conscripted to prowl the city, stealing housecats.

So now, alongside the leering stoats in visitor centers and museums, you'll often see a taxidermied feral cat posed beside its stomach and the contents thereof: tiny paws, and feathers and bones, preserved in acrylic like some horrid paperweight that I kind of want on my desk.

On the walk back up the hill to the van, we pass a newer addition to the DOC arsenal, a self-resetting Goodnature A24. The bait sits at the end of a tube, behind a thin, flexible metal bar. When the head pushes the bar, it triggers the firing of a piston propelled by a carbon dioxide canister. Maybe you remember the unusual murder weapon used by Javier Bardem's character in *No Country for Old Men*. It's that sort of mechanism.

The stoat is famously hard to trap. "They don't like to stick their heads in things," Shaun says. Why then make a trap that's baited at the end of a tube? Because that way the head—the brain, more specifically—is positioned for a lethal strike and an instantaneous passing.

New Zealand is committed to eliminating invasive predators, but it is also committed to doing it humanely. No one person has contributed more to this effort than Bruce Warburton, whom I'm driving up to Christchurch tomorrow to see. Warburton helps design more humane traps, drafts animal welfare standards for traps, and tests traps to see whether they meet those standards. He is certainly the only New Zealander to hold professional affiliations with both the National Pest Control Agencies and the National Animal Welfare Advisory Committee.

What happened, in the end, to the rabbits? Through persistence, resistance, and multiplication, they're doing just fine. They survived the stoats and cats and rabbiters, and even the rabbit hemorrhagic virus smuggled in from Australia. Shaun says he's seen more than ever this year. With plenty of rabbits to eat, the stoats are thriving, too. "Stoats are through the roof," he says over his shoulder, from the driver's seat of the van that is taking us away from this beautiful and heartbreaking place.

Samantha Brown is a young biologist with freckles across her nose and an ear-pleasing Kiwi accent and extensive knowledge of how to kill quickly. Sam works with Bruce Warburton at Landcare Research, one of New Zealand's Crown Research Institutes. Their branch, on the outskirts of Christchurch, focuses on biodiversity and sustainability. That means lots of appealing job prospects for conservation majors. Though sometimes less so when they hear about the trap testing.

While we wait for Warburton to arrive, Sam cues up a video shot at the test facility, a few miles down the road. No tests are scheduled this week, which is, for me, less a disappointment than a relief. I appreciate and respect the good and emotionally wringing work the team does, but I wasn't sure I wanted to watch it live. There are things I too don't like to stick my head inside.

The setup is simple. A video camera on a tripod faces an observation pen. A person waits in the dark with a stopwatch and feelings it is hard for me to guess at. Sam points to a readout below the video frame. "You can see the time elapsing down here." Once the trap is sprung, the observer clicks the stopwatch. Humaneness, in the context of a lethal trap, is a function of speed: speed of death, yes, but more critically, speed to unconsciousness—to feeling and knowing nothing.

"This is a test of a Northland Regional Council cat trap," says Sam. The SA2 Kat trap is used not only on feral cats but also brushtail possums. The possums, native to Australia, were released in New Zealand in the nineteenth century to establish a fur trade. They have thrived in their new home and multiplied and spread, feeding on and destroying vast numbers of trees that support native birds. On top of the estimated 21,000 tons of leaves and shoots they consume nightly, the brushtail possum also enjoys an egg.

On the screen, the time is displayed in tenths of seconds, increments that speed by unreadably fast, unless of course it's you in the trap. Sam and I sit silently through a half minute of don't-go-in-there, and then the terrible in-there. A colleague of Sam's rushes into the frame. Some part of me hopes he's been compelled to set the animal free. Sam narrates. "Here Grant is trying to look 'round the side of the head and touch the side of the eye." He's checking for the palpebral* blink reflex. When the reflex disappears, the animal is unconscious and the stopwatch is stopped.

---

* *Palpebra* means eyelid. Medicine has its own word for everything. The back of the knee? The popliteus. Earwax? Cerumen. *Hallux* means "big toe," and *mental protuberance* refers to "a prominent cranial bulge caused by memorizing too much Latin," oh, sorry, "chin."

The trap being tested in the video employs a metal bar that, when sprung, clamps down on the neck. While it's possible to kill a mouse or even a rat by snapping the neck in this manner, a larger animal in a trap like this one is killed by strangulation: the bar clamping shut the carotid arteries, depriving the brain of blood and thus oxygen. Suffocation plays a supporting role, as the bar may also shut the windpipe. Suffocation achieves the same end point as strangulation, but it takes longer, because it is cutting off air intake instead of blood flow. Thus the blood keeps circulating, and it takes a while for the oxygen already in the blood to be depleted. With a well-designed trap of this variety, oblivion arrives in 40 to 50 seconds. (Most film directors, though not Tarantino, accelerate strangulation to five or ten seconds, because, really, who wants to see more than that?)

A faster, kinder death could be delivered through a blow to the head. As a humane means of ending life, few things rival a bullet to the brain; thus "properly placed gunshot" is included as an acceptable mode of euthanasia in the guidelines issued by the American Veterinary Medical Association (AVMA).* The world's first humane kill trap featured, in fact, a gun to the head. U.S. Patent No. 269,766, granted to James Alexander Williams of Fredonia, Texas, in 1882, featured a revolver set upright in a frame. In the illustration accompanying the patent, the barrel is pointed at some variety of rodent emerging from a burrow. Varmint steps on rod, rod pushes trigger. You don't get the sense, in reading the patent details, that Mr. Williams cared much about safety or

---

*   As is decapitation. Humaneness was in fact Monsieur Guillotine's motivation. Scientific supply catalogs used to list an item called the "small animal guillotine." These days, at least online, you can only buy them used, perhaps because they attracted negative attention. For regardless of how humane a guillotine may be, it entails cutting off a head, and that is hard to stomach. And now I'm going to say this: if you want to sell a used rodent guillotine on eBay, for god's sake clean the blade before you take the picture.

humaneness or even rodent control: "This invention may also be used in connection with a door or window, so as to kill any person or thing opening the door or window to which it is attached."

So why are the makers of the SA2 Kat trap not trying to hit the skull? Because a humane head strike needs to be carefully placed. Carotid occlusion affords more leeway with body size and position. Also, Sam is saying, for the bar to hit with force sufficient to deliver a lethal, humane blow, one starts running into (human) safety issues. It would take some good strength to pull back the bar to set the thing, and if it got away and snapped shut on someone's fingers, bones would break.

Humane traps for stoats are among those designed to impact the head. This is because these animals have unusually strong, muscular necks, and the thick muscles protect the arteries. Also, the strength those muscles supply can enable the stoat to pull free.

Sam clicks a link for a video of a test of a stoat trap called a "modified Victor." This is the classic wooden Victor brand snap trap for rats, upgraded (by Warburton) for humanely killing stoats. A molded plastic hood, or shroud, is screwed in place over the bait to guide the head in from the proper direction and to the proper depth. The bar strikes at the latitude of the ears, and unconsciousness, Sam is saying, is almost instant.

"You can see it's started to go limp already." My brain offers a more benign interpretation of what it's taking in. *Look, it's one of those fur boas ladies wore to be classy in the nineteen forties.*

When I was twenty-two, I lived in a mouse-infested apartment building. The Mouse Death Count—lettered in German Gothic and stuck on the refrigerator by my graphic designer roommate—was up to thirty-two hash marks by the time I moved out. The landlord didn't allow cats, so we bought traps, the classic, cheap wooden Victor snap trap, the setting and disposal of which were my chore. I set them without much thought, assuming that they kill instantly because they hit the head or the neck. I was always taken aback by a mouse struck elsewhere: the one who came in from the side and was pinned by the shoulder. Another that changed its mind and got pinched by the snout while backing away.

According to University of Guelph bioscientist Georgia Mason's thorough, unblinking comparison of the humaneness of various rodent control methods, this sort of thing happens 4 percent of the time with a Victor snap trap. That's actually quite good, certainly compared to a competitor that, according to Mason, caught mice by the legs or tail 57 percent of the time. Mason studies behavioral biology at the university's Campbell Centre for the Study of Animal Welfare, and her comparison study ran in the journal *Animal Welfare*. She is Canada's Bruce Warburton. She's outstanding.

These days the Victor company sells its own version of Warburton's "modified Victor," a snap trap called the Quick-Kill.* I commend them for this, but am disappointed to see that they continue to sell glue traps in their 2019 product catalog.† On top of the protracted torment of being glued in place, the animal may, Mason writes, rip off pieces of skin or gnaw through its limbs trying to escape. A professional pest control person should be checking the traps daily and humanely killing any rodent that's been caught, but Victor and others sell glue traps online to anyone, and what homeowner is going to tackle that? So millions of mice and rats are left, adhered, to die slowly of dehydration. Luckier, Mason says, are the ones who get their snout stuck to the tray in the initial struggle and suffocate.

---

* Not to be confused with Victor Clean-Kill, Fast-Kill, Smart-Kill, or Power-Kill traps. All of them trademarked. Victor legal staff have basically trademarked the entire universe of small rodent killing: Kill Bar, Kill Gate, Kill Vault, Kill-Point, Multi-Kill, and even, if their catalog copy is to be believed, Mulit-Kill are all Victor trademarks.

† A lushly photographed 44-pager called Rodent Free Living. A number of pest control companies have adopted this term, as though mice and rats were a lifestyle choice, or something you go to rehab for. *I've been rodent-free for six years now.*

Glue traps are illegal in New Zealand and parts of Europe. I emailed the Victor product manager, asking whether the company had plans to stop selling glue traps. You won't believe this, but she didn't answer.

Sam's office door swings open. It's Warburton with a box of traps. A couple decades of his career in there. It clanks as he sets it down to shake my hand. Warburton has a genial, wry manner. He doesn't work at making everyone like him, but I imagine everyone does. He's an interesting hybrid: animal ethicist and hunter. I ask him how he got involved in the business of humaneness. It began, he says, when New Zealand's Society for the Prevention of Cruelty to Animals took an interest in a new, humane possum trap that they hoped could replace a particularly unkind leghold trap—key words: *serrated jaws*. (Possums are still commercially trapped for their fur, which is spun into wool.)[*] "They took it to our minister of forestry, and he said, 'Well, how good is this trap?' and he came to us." *Us* meaning Landcare.

I guess I was less interested in the how than the why. I rephrase my question.

"I mean, they are pests," Warburton says, "but they're sentient animals.[†] They have the capacity to suffer. We have a duty of care

---

[*] Possum merino is a heavenly soft wool blend. When I first arrived in New Zealand, I bought a pair of wonderful green possum merino gloves. I imagined peaceful possum flocks being sheared like sheep. Then Warburton explained that possum pile is too short to shear and is typically sliped instead. Sliping involves some kind of postmortem chemical depilatory. I still wear the gloves, though with diminished happiness.

[†] By definition of New Zealand's Animal Welfare Act, "sentient" applies to any animal with a nervous system that can relay stimuli from sensors around the body to the brain, plus a brain advanced enough to translate those signals into perceived sensations. That means all vertebrates, plus octopus, squid, crab, and lobster, but not, I'm relieved to report, oysters.

to think about it, to minimize their suffering. That's just been my philosophy."

The traps in Warburton's box are mostly mechanical: trigger and thwack. Where's the newfangled stuff I've been reading about? Before I arrived in New Zealand I spoke to a Predator Free 2050 researcher about a humane carbon dioxide trap under development for stoats and rats. The animal enters a tunnel and crosses an infrared beam, triggering doors at either end to close and carbon dioxide to be expelled.

At the right concentration and flow rate, carbon dioxide is believed to kill humanely. The gas is one of the AVMA's acceptable means of euthanasia. When a wildlife control operator live-traps Canada geese, say, or a raccoon in someone's attic, a CO_2 chamber may be the next stop. The operator may not mention this to the homeowner, and the homeowner may not ask, preferring to believe that the word *humane* on the operator's website means that the animal will be driven to some sunny woodland and set free. (Which may be less humane than the gas. See "The Fuzzy Trespasser: Resources for Homeowners," page 297, for advice on this.)

The humaneness of carbon dioxide was a subject of renewed debate at the 2018 AVMA Humane Endings Symposium. This is an animal euthanasia research conference held regularly near Chicago in November. Appealing! Here is the dilemma. It is elevated carbon dioxide levels in the blood, not low oxygen levels, that cause an organism to breathe harder. To avoid causing breathlessness and panic, you want the end to come quickly. But when the concentration and flow rate are high enough to achieve that, the CO_2 may start to form an acid on contact with mucous membranes, and the animal may experience burning and choking sensations. Getting it right is tricky.

Another newcomer to the humane endings arena is the electronic trap. These are box traps with double-decker electrified floor plates. When a visitor steps inside, the top plate tilts into contact with the lower one, completing the circuit and introducing the current into the visitor. Warburton tested one such device designed for possums. It was, he says, a mixed success in terms of animal welfare. "It works okay when the electrode plates are clean, but once they got a bit dirty, it cooked the wrists of the animal. So that wasn't very good at all." Warburton shares what seems to me to be a national distaste for euphemism and double-talk. His words are not offensive, just direct. Delivered flat, and at low volume. I won't be needing any exclamation points for the next few pages.

Georgia Mason weighed in on a commercially available version called the Rat Zapper. Two thousand volts for two minutes. (As this gets underway, the homeowner receives a text: "RODENT CAUGHT." ROMANTIC DINNER RUINED.) Electrocution kills by disrupting the normal movements of the heart and diaphragm muscles. It's a death by ventricular fibrillation and respiratory distress, both of which starve the brain of oxygen. So painful are the muscle constrictions thought to be that humane livestock electrocution requires the shock to the body be preceded—or at least accompanied—by passing a current through the brain to induce unconsciousness. Whether these traps also do this is not known.

Mason ranked a well-designed electronic trap alongside a good snap trap as most humane of all the lethal options for rodents, largely because both kill quickly. The cutoff in New Zealand for an allowably humane kill trap is three minutes to irreversible unconsciousness. Having watched test videos of two traps that came in under a minute, three minutes strikes me as an eternity. Warburton overhears me saying this to Sam.

"Three minutes is not too bad," he says, putting traps back in the box. "The poisons are the tough ones." Here is the ugly reality of keeping stoats and possums and rats from causing extinctions.

Predator Free 2050 relies more on aerial drops of poison bait than they do on the happy trapping of motivated citizens. Given the numbers and remoteness of some of the creatures that prey on native birds, traps are not going to get them all. And if you don't get them all, they will quickly bounce back—the rats especially.

New Zealand's DOC has been looking into better poisons: poisons that kill humanely, poisons that kill the targeted invasive species and no other creatures, poisons that don't build up on the land and in the animals that live off the land. And Landcare dutifully tests them. "It's quite hard on the staff," Warburton says. Because it's hard on the testees. Because now you're talking about hours or days, not minutes or seconds.

The tests are run near the facility where the traps are tested, a short drive from Landcare's offices. Warburton is holding car keys. "You wanted to see it?"

If you do an image search for "L pill," Google will present you with many close-up photographs of low-dose aspirin tablets marked with an *L*. This made me laugh out loud, because the L pill I was interested in was the kind issued to World War II operatives facing torture and possible spillage of top-secret beans. The *L* standing for *lethal*. L-pills, as the name was styled, contained potassium cyanide. The OSS, forerunner of the CIA, chose cyanide because it's a fast, easily concealed way to end a person's life.

In her consideration of the various rodent poisons, Georgia Mason ranks cyanide as the most humane. It suppresses central nervous system activity, and it interferes with the blood's ability to ferry oxygen to the cells, causing a sort of chemical asphyxiation. Mason cites two New Zealand studies of ingested cyanide, one indicating that possums lost consciousness after a minute to a minute and a half, the other finding that it took around five minutes.

In the span of time before oblivion, there were muscle spasms, which would be painful, and some convulsions. Because the convulsions occurred after the EEG indicated a loss of consciousness, the animal wouldn't have been aware of them.

However, an onlooker would. Appearances matter. When states execute prisoners via a "cocktail" of drugs, they usually opt to include something to paralyze the muscles. Are they wishing to paralyze the breathing muscles or are they more concerned with the clenching, grimacing, spasming, convulsing muscles? I ran this by longtime death row assistant federal public defender Robin Konrad, former director of research and special projects with the Death Penalty Information Center. State officials have given her both reasons, she said, but it was her belief that, yes, they were seeking to avoid unpleasant visuals and the likely outcry that could result.

That convulsions are visually disturbing is the operative principle behind an unusual frightening agent called Avitrol. As used by farmers, the chemical is applied to a tiny percentage of a bait scattered in the fields, maybe one in a hundred pieces. The lottery loser who eats it takes to the air, flapping and squawking and soon thereafter convulsing grandly and dropping dead. The idea is that the rest of the flock, witnessing the spectacle, would be spooked and flee the farmer's field.

In 1975, Ontario's minister of the environment declared that only pesticides shown to be humane could be applied in the province. Though Avitrol is used as a frightening agent, it is a poison and so was included in the testing. The convulsions, said the University of Ottawa report on the results, happened after the point at which an EEG indicated a lack of awareness similar to that caused by dissociative anesthesia. The team deemed Avitrol humane but warned that the scientific evidence would "never change the opinion of those observing the effects."

About that, they were correct. On YouTube, there is, or was, a video of a pigeon on Avitrol. Commenters assumed the bird was suffering and posted messages of pity and outrage. (At least the women did. Men's posts were more along the lines of: "Where can I buy this stuff?" And: "Maybe they shouldn't shit on everything and this wouldn't happen to them.") It seems the theatrical death throes of Avitrol victims are more upsetting to people than to other birds. In studies comparing its effectiveness with common frightening devices, the chemical placed last.

USDA Wildlife Services deploys a cyanide device on behalf of ranchers who complain about coyotes. The original was developed in the 1930s, and was known as the Humane Coyote Getter. It's a buried cyanide ejector, triggered by tugging on a bait that protrudes from the ground. Assuming it is jaws that do the tugging, the poison ejects directly into the mouth, à la an OSS cyanide pill.

You kind of knew there was going to be trouble. In a U.S. Fish and Wildlife Service study spanning 1940 to 1941, the Getter compared favorably to leghold traps in terms of humaneness—*what wouldn't?*—and the toll on nontarget wildlife. But the Getter got seven cows and twenty-four pet dogs over the course of that year, and so began its undoing. Between 2013 and 2016, the updated version, called the M-44, killed twenty-two pets and farm animals, and it has, over the years, injured a number of humans, including a man who mistook it for a geodetic survey marker he wanted to steal. At least one lawsuit is pending, and four states have banned M-44s. Sayonara, cyanide.

Warburton and I are walking along rows of enclosures. Most are empty. A few possums are asleep in hanging burlap sacks, which, he says, remind them of the mother's pouch. We're talking about a common and inexpensive rat poison, an anticoagulant.

In small doses, anticoagulants are used to prevent blood clots—in, say, patients laid up in bed after surgery. (When my brother, Rip, was prescribed warfarin, he texted, "I'm taking rat poison," and he was.) In larger doses, anticoagulants interfere with the kind of clotting that patches tiny breaches in the capillaries—the normal wear and tear of the circulatory system. Animals that consume anticoagulants bleed to death internally, and that, Warburton says, is an unpleasant way to die.

"Depending on where the bleeding is, it can be associated with a lot of pain," he adds. Moreover, it's a slow death, occurring over a period of one to three days in rodents, and up to a week in possums. In the United States, some classes of anticoagulants have been restricted for use by professional exterminators or for island eradications of rodents threatening native wildlife.

Predator Free 2050 doesn't use anticoagulants. New Zealand's invasive predators are served up aerial drops of 1080-infused bait. Given 1080's legacy from World War II and chapter 8, I was surprised to learn this. Warburton says the poison's effects—on possums, specifically—fall midpoint on the humaneness spectrum. "They get nauseous in the last few hours, but it's not too bad." (The Denver Wildlife Research Laboratory, back in the day, found 1080's effects to vary widely by species, from "progressive depression" to—in dogs and other extremely sensitive species—"most violent epileptiform convulsions.")

With differing degrees of nastiness, 1080 kills a broad range of mammals and birds. PF2050 baits are dyed to make them unappealing to birds, and New Zealand has almost no native mammals—just two bat species, neither interested in bait pellets. A body count published in the *New Zealand Journal of Ecology* determined the number of native birds killed in 1080 drops to be "negligible"—at any rate, not high enough to outweigh the benefits the birds derive from having their predators offed en masse.

New Zealand may have few native mammals, but it has seven introduced deer species—hunting stock shipped over by acclimitization societies and still hunted today. And 1080 kills the bejesus out of them. "The people who run 1080 operations get death threats from hunters," Warburton says, maneuvering a round-about. "They say 1080 is cruel to the deer, and then they go and stick arrows in them." Warburton half smiles. "I can say that, because I'm a hunter." To appease the hunters, deer repellents were developed and are added to 1080 bait in some areas.

Another concern with 1080 is secondary poisoning—the killing or sickening of creatures that might feed on the carcasses of stoats, possums, and rats. Pet dogs are the obvious example, but keas—the mountain parrots we encountered in chapter 11—also feed on carrion and have been killed as well. The kea is one of the native species Predator Free 2050 seeks to protect.

So now the DOC needs a kea repellent, too. "These guys in here are actually next week going to be getting some 1080 with kea repellent," says Sam, who is running the trial and has caught up with us. "We need to make sure the possums and rats will still eat the bait." Anyone would agree that adding a kea repellent to your poison bait is an excellent thing to do, unless perhaps they, like me, had just made the acquaintance of one of the taste-testers.

Brushtail possums are fluffy, and they don't have the naked pink tail or the long snout of the American opossum. Their eyes are more forward in their face, like human eyes, or kitten eyes or the eyes of pretty much anything you'd find incredibly adorable.

"I know." Sam acknowledges my pout. "You just hope it will reduce kea deaths."

We move to another set of enclosures. "So this guy," Warburton says, "has been given a form of vitamin D, called chole-calciferol, that's under consideration as a replacement for 1080." Possums are particularly sensitive to it, and birds are not. "But it's not a very nice toxin, you see. It calcifies things." Soft tissues. The heart. "And it takes quite a long time, and they stop eating. We were talking yesterday about whether we should just say, 'No, we are not doing any more trials.'"

Warburton and I say goodbye to Sam and head to the parking lot to drive back to the Landcare office building. Given what I've just seen and learned, the extent of public buy-in for Predator Free 2050 surprises me. A large 1080 operation may blanket 80,000 hectares (close to 200,000 acres). A government brochure illustrates the dispersal regime, using an image of a tennis court with five evenly spaced baits, as though someone had it in for Federer. At about five possums per hectare, that's 400,000 dead possums. And who knows how many dead stoats and rats. Plus all the deer and the occasional endangered bird. "Dead forest" is a term I've seen used—not by activists but by someone with the USDA—to describe the aftermath of a 1080 aerial campaign.

In light of New Zealand's widely touted commitment to the environment, I would have expected more widespread resistance.

"No one blinks an eye, because it's in the forest and it's at night," Warburton says. "If it was during the day and it was on pastureland, like this"—he tilts his head toward the car window—"we wouldn't be allowed to do it. I think it allows us to get a bit more social license than we deserve. That, and the fact that these animals are invasive. There's a pervasive media presence telling us that these things are pests and they're eating our forests, killing our birds, and there's just a general acceptance."

It's not just the media. Anti-predator propaganda turns up everywhere. Gift shops in New Zealand national parks sell jokey roadkill-shaped Squashed Possum chocolates. A popular children's book pits a scrappy band of endangered birds against a Grinch-faced stoat. ("There's no way any of us could fight him! We're doomed!")

Warburton rejects the stoat hate. "Stoats are amazing wee animals. They're incredible climbers, and they're incredible predators. They'll take on an animal bigger than themselves." Especially irksome to Warburton is the double standard applied to pets that

kill endangered birds. Mostly cats. "I detest cats."* Says the winner of the Royal Society of New Zealand Bronze Medal for significant contributions to animal welfare research. Warburton
would like to see the keeping of pet cats made illegal.

"And good luck with that, Bruce."

"I mean, you keep the one you have, but you don't get to
replace it." Otherwise Predator Free 2050 is actually Predator
Free Except for the Housecats That Decimate Endangered Bird
Populations and the Dogs That Kill Adult Kiwis Unless You Give
Them "Kiwi-Aversion Training" 2050. Which seems unfair to
the possums and stoats. I keep recalling my husband, Ed, imagin-
ing what the possums say to each other. *Why do they hate us? Why?
We give them the lovely wool for their gloves . . .*

Are there New Zealanders who advocate doing nothing?

"There are," Warburton says. "There are people who argue
that if you leave it long enough, things will find a new balance.
You'll lose some species, but others will adapt. Other people say
we can manage [the predators] at certain sites." Meaning an island
or a fenced area or something like the Otago Peninsula, with a
lot of endangered wildlife (and wildlife tourism) to protect. "And
then there's a suite of people who think that we can eradicate from
the whole island."

Warburton stands somewhere in the middle. "From a practical
standpoint, we can't afford to do the whole island. It's five hundred to a thousand New Zealand dollars per hectare. You can't roll
that out over 26 million hectares. And the rats may come back."
As they always do, from ships in port.

It strikes me that the PF2050 movement shares some DNA
with that of the old acclimitization societies. There's some of that

---

* New Zealand says it like it is. On the brink of a cliff above the surf at Pan-
cake Rocks is a sign warning tourists not to climb over the fence. It closes with
"Don't be an idiot."

same desire to have the land around you look the way you have always known it to look, a belief in an ideal, static ecosystem. But ecosystems are always evolving. "There are botanists," Warburton says, "who don't like the deer because they eat the understory and modify the forest. But we used to have moas that ate the understory." Moas were similar to emus, but larger, and long ago hunted to extinction. "So they're trying to restore a forest that's sort of post-moa extinction and pre-deer."

In parts of New Zealand's South Island you run across signs warning of the dangers of "wilding conifers." Pine trees! A threat to lands and lifestyles! Using up scarce water! Altering iconic landscapes! From what I could tell, it's something of a lost cause. They were planted as windbreaks, and now they're all over. And quite lovely! Forgive me, National Wilding Conifer Control Programme. It's just hard to know where to draw lines. What to save, and at what cost. Out on the beach yesterday, I was ready to support whatever it took to prevent the extinction of yellow-eyed penguins. Today I'm less sure. It's hard to feel peaceful about the killing of some species in order to preserve others.

In part, it's the way it's done. Poison seems so 1945. Isn't there, by now, something better?

15

THE DISAPPEARING MOUSE

The Scary Magic of Gene Drives

Everyone wants to eat a mouse. Hawks want to, coyotes want to, skunks, foxes, rats. A mouse is a nutritious morsel with no anatomical defenses: no venom or noxious exudations, no spines, no shell. A mouse's best hope is to get itself someplace safe, and do it fast. At that, the mouse excels. It can squeeze through a hole no bigger than its head. A motivated mouse can jump straight up to a height four times its body length. If I were a mouse, I could leap a wall twenty feet high with no running start. I could pass through the opening of my own mailbox.

I've seen the studies* and the video footage. It's on the com-

---

* Will Pitt, of the Smithsonian Conservation Biology Institute, documented it when he was with the NWRC. The house mice tested were all able to access food by passing through 13-millimeter holes—the average width of the local house mouse head. Less formal proof takes the form of my friend Steph, who, one hot day, upon returning from a walk with her dog, grabbed a water bottle from the floor of her truck. She took a swig and discovered, in order, a fetid taste and a dead mouse inside the bottle. "I don't think you'll need a doctor," an advice nurse said upon hearing that she'd spat the water out. "But you may need a psychiatrist." For me, Steph dutifully went back and measured the head. It was exactly the size of the water bottle's opening.

puter of Aaron Shiels, a wildlife biologist who has been working
to create an escape-proof habitat for mice at the National Wildlife Research Center headquarters in Fort Collins, Colorado. The
main feature is a room-sized simulated natural environment, or
SNE ("the snee"), which Aaron is showing me this morning. Each
mouse is implanted with an ID chip, and beneath the SNE's raised
floor is a chip reader, ensuring every rodent is accounted for. The
walls are smooth (unclimbable) plastic plates, screwed in place
from the outside to keep mice from springing up and securing
a clawhold on the screwheads, like rock climbers holding on by
their fingertips to barely discernible protrusions in the rock. Metal
flashing covers the seams of the SNE and the caulking in the corners of the outer room that contains it. Because a mouse can gnaw
a hole anywhere it can wedge its teeth—in wood, plastic, cinder
block, aluminum. The front teeth of a mouse are self-sharpening
chisels. The back sides are softer than the fronts, so each time a
mouse brings its teeth together, the hard enamel on the front of its
lower incisors wears away the softer backs of the uppers, honing
the edges. *Rodent* comes from the Latin for "gnaw." They do it
well and fast, so well and so fast that they need to suck the insides
of their cheeks through a gap in their teeth to close off their windpipe and keep the sawdust out.

The current residents of the SNE are mice of no great value or
concern. The security features are being designed for a group of
future occupants: mice genetically modified such that they produce
only male offspring. And further engineered, through a process
called a gene drive, to spread this trait far faster than it would spread
naturally. Gene drives are one possible future of invasive species
control, a potential alternative to strewing poison all over an island.

Like any genetic modification, gene drives make some people
uncomfortable—not just among the public but among the scientific community. Most prominently, Jane Goodall. And so, before
the engineering of the mice must come the engineering of the

habitat that will keep them in. Part of Aaron's work has been to create such a place and to demonstrate to a high level of certainty that escape is impossible. Even for a mouse.

"Otherwise, I'm all over the news." In 2017, an elk with a highly contagious bacterial infection escaped from a USDA Veterinary Services facility next door to the NWRC. "People thought it was ours." Aaron has hazel eyes and shoulder-length red-brown hair that is, today, gathered in an elastic band. I just realized it was Aaron I saw earlier at the front gate. He had sped past on foot, ponytail aloft, while the guard checked my ID. He looked like a man crashing the gate, but now I know he's just a man who runs to work in the morning.

Of all the invasive species harassing and killing endangered island natives, the house mouse is not high on the fret list. They can be a problem for seabirds on islands that evolved without seabird predators, for instance the endlessly beleaguered albatrosses of Midway Atoll. (Wildlife cameras in 2015 began capturing ghastly scenes of mice feeding on albatrosses as they sat incubating their eggs.)

But seabird conservation is not the reason mice were chosen as the gene-drive test species. Mice were chosen because science knows mice. You can't fiddle with an animal's genome without first decoding it. Additionally, mice can crank out a litter in a few weeks. Gene drives take effect over generations, so researchers who wish to have data before they reach retirement age prefer speedy breeders.

To date, only one mouse has escaped the SNE. It went down like a movie prison break. The mouse had burrowed deep into the sawdust shavings provided as bedding, and when staff came in to change out the shavings, the mouse was scooped up in the dustpan along with the bedding. He escaped in the laundry cart! (And was caught the next day in a trap outside the enclosure.)

The SNE is quiet at the moment. The mice are sleeping. They're part of a trial of the Goodnature A24, to see whether it also works, and works humanely, on mice. So far the challenge seems to be the bait. In a lush tropical environment, a bait has to be extremely tempting* to compete with the natural food sources. Aaron hands me a bottle of Goodnature bait, a chocolatey-coconutty goo that smells, but does not taste, delicious. It's like eating suntan lotion, I tell Aaron.

"You *tried* it?" His look combines horror, confusion, and pity. "You want some gum?"

---

* To eradicate invasive, bird-decimating brown tree snakes on Guam, the U.S. government has used a bait of dead neonatal mice (or "pinkies," in the pet snake community) laced with acetaminophen (or Tylenol, in the pain-relief community). But pinkies aren't cheap, and within three days they start to become unpalatable, even to a snake ("green coloration . . . followed by swelling, seepage, odor, and eventual rupture"). So began a fifteen-year search for a replacement bait. Because snakes don't chew and savor their food as we do, the thinking was that you could coat a cheap core material—sea sponge was tried, and gum rubber—with an irresistible attractant. A dozen or more wrappings and coatings failed to seduce captive snakes at Monell Chemical Senses Center in Philadelphia: Roquefort cheese, termite attractant, poultry fat, fetal pig skin, evening primrose oil, baby formula. (The researcher, Bruce Kimball, had expected resistance from the baby formula company, but they were all in. The Tylenol people have maintained a cool distance.) Eventually a winner emerged: "mouse butter" on a core of potted meat product. "We reverse-engineered Spam," Kimball told me with understandable pride. Spam is cheap, keeps for at least a week, and doesn't attract ants—and the snakes, like the humans of Guam, are inexplicably fond of it.

The other challenge was to get the baits where the tree snakes were (and other creatures weren't). The pinkies had been strung with plastic parachutes, like toy army men, and dropped from helicopters into the canopy, where the parachute cords would catch on branches. Your taxpayer dollars making my day. However, hand-attaching six thread-thin parachute cords to each bait proved "tedious," and to cover all of Guam, you'd need two million baits. Now they use an "aerial broadcast system," a sort of helicopter-mounted machine gun that shoots potted-meat baits with biodegradable cornstarch streamers to entangle them in the trees.

The upshot? While brown tree snakes have quickly eradicated nine out of twelve native Guam bird species, the combined might of the U.S. Navy, the USDA, and U.S. Department of Fish and Wildlife has thus far failed to return the favor.

Like many people, I have some qualms about genetic engineering and its possible future. Also like many people, I know diddly about how it works. My plan for this afternoon is to become someone who knows a little more than diddly. I'm scheduled to meet with the Center's wildlife genetics staff, upstairs in the Long Speak Room, which is an amusingly apt name for a government conference room (except that it isn't—a realization that will dawn when I take note of the plaque by the door, which reads: Longs Peak Room).

I'm in the main lobby now, waiting for my escort. There is, yes of course there is, taxidermy. A family of monk parakeets is posed in and around a nest in the top segment of a power pole. The diorama takes up most of the surface of the small table beside my chair, forcing me to place my coffee directly beneath the birds and fostering a vague sense of unease.

My escort arrives and we make our way toward the stairwell. The corridors are hung with mounted research posters and color blowups of photogenic "nuisance" species: cormorants, ground squirrels, beavers. Wildlife management agencies and pest control websites do this too, and it always hits me as slightly off—it would be like the FBI decorating the hallways with headshots of good-looking federal criminals.

Up in the Longs Peak Room, I take a seat next to Toni Piaggio, a conservation genetics specialist. Her own personal genetics have provided well. She received: graceful cheekbones, blinding intelligence, glossy black curls, deep reserves of patience. Toni introduces a young colleague, Kevin Oh, also a geneticist.

Before we get to the *should* and *could* of gene drives, a wobbly stab at the *how*. A gene drive is two unique manipulations. First, there is the one people are familiar with, at least in a general way: genetic modification, the GM in GMO. This is done using a technology called CRISPR-Cas (or CRISPER, for short). A target gene is selected—say, a gene for a trait that enables a mosquito to carry malaria—and then what Kevin calls the "molecular scissors" of CRISPR cut out and/or replace the target gene. The edit is done, in this case, early in the embryo's growth—when it's just a few dozen cells—so that the modified genome will be copied into every new cell going forward.

CRISPR-Cas is a natural element of bacteria, a part of the mechanism that serves as their defense against viruses called phages. This defense system includes an enzyme that cuts up the virus's DNA, and as it does so, it retains a memory of it—a "molecular bar code," as Kevin puts it. So if the virus returns, its specific genetic sequence will be recognized and cut up. Geneticists have harnessed the CRISPR scan-and-snip system as a way of precision-targeting and editing DNA.

"But how does the enzyme get *in* there?" I'm whining.

"They're literally injecting mouse embryos," says Kevin.

"Using, like, super-tiny dollhouse hypodermics?" I want to see these.

Toni steps in to move things along. "There are different methods. We sort of flood the embryo in its petri dish."

The enzyme gets in. It scans, it finds a match, and *beep*, it does its thing.

So let's say that you've managed, with your scissors and your bar code reader and your dollhouse drug paraphernalia, to manipulate the genome of a group of mice. And that now these mice cannot produce female offspring. If you set enough of them free on an invasive mouse-plagued island, the population will start to dwindle.

"Enough" is the challenge. This is where the gene drive element comes in. With normal Mendelian inheritance, this new trait would show up in 50 percent of the offspring, because half the offspring's genetics are contributed by the male. What a gene drive aims to do is deliver the machinery to make that gene 100 percent heritable. So that now all mice born to the gene-drive mouse will carry the trait. A successful gene drive would speed the time it takes for a trait to spread through a population.

Here's a hitch. In order to swamp an island population to a point where the gene-drive animals make sufficient inroads, scientists will have to release large numbers of the very species they seek to be rid of. Depending on the size of the island's invasive population, it may take a considerable number of lab-born gene-drive animals to tip the balance. So before a gene drive can improve the situation, it will temporarily make it worse. Thus the first step, with rodents, would likely be an aerial drop of a rodenticide—the practice we're trying to get away from. The gene-drive rodents would be released afterward as a sort of mop-up and maintenance program. So that any survivors, and any newly introduced future rodents, will peter out—rather than building up their numbers again and necessitating another hail of poison.

So far, the gene-drive element is proving tricky. The gene-drive mechanism isn't always copying correctly. And there seems to be a narrow window within the embryo's development during which the manipulation will take. Too early and it kills it; too late and it doesn't transform. The sexual practices of wild mice may also pose a challenge. Polyandry—whereby one litter may include offspring fertilized by multiple males—was recently shown to be more common among wild mice than previously thought. Thus it may take longer and require more gene-drive males to overtake the population.

On a more basic level, there's a possibility that the invasive residents won't breed with the genetically modified newcomers. In nature, mice from different islands or regions of the world begin to evolve separately. Not necessarily to the extent that they're different subspecies, but different enough that they may not breed with one another. "You create these mice in the lab and then you have to make sure the wild mice on the island find them sexy," says Katherine Horak, a toxicologist sitting a few seats to my right. Laboratory *Mus musculus* are surprisingly different from wild *Mus musculus*. "Lab mice want to just sit there and hang out in your hand. The first time I worked with a wild mouse," Horak recalls, "I was like, *What is that?* It wanted to jump up and bite my face off." (Thus the need for a high-security habitat.) One of the first things that will happen in the SNE are mating trials—back-crossing the gene-drive lab mice with wild mice to create a strain that's sufficiently sexy to the invasives one wishes to control.

The big concern with gene-drive organisms is that they'll make their way beyond the area—and the population—they're intended to control. And that the locals wherever they end up will *not* hesitate to breed with them. Say you create a gene-drive "daughterless" feral pig and one of them mates with a domestic pig. Pig farmers would be undelighted. This is one reason scientists would plan to start with a physically isolated population—invasive rodents on a remote, uninhabited island, for instance. (Preferably an island no ships visit, because mice and rats are notorious stowaways.)

There is a way to avoid the concerning scenario. The same genetic drift that could prevent mice from different land masses from mating can be harnessed as a safety feature. Geneticists can target a genetic bar code unique to the population of the particular island or region. "So CRISPR can sit down and cut at this place in the genome that we only find in this population of mice," Toni says. "So even if someone nefariously transported those mice somewhere"—or they transported themselves as stowaways—"we wouldn't have to worry about it migrating into the local population." In other reassuring news, research out of the University of California, San Diego, suggests it may be possible to halt or even reverse a gene drive. In a paper published in the fall of 2020, two new gene-drive control mechanisms were shown to work in fruit flies.

Katherine Horak has been working on something altogether different and less anxiety-provoking than gene drives. It's called interfering RNA, or RNAi. Riddle me this: It's a bait that kills, yet it contains no poison. It's a species-specific genetic solution, yet it doesn't modify the target species' genome. That makes it an appealing package: doesn't effect nontarget animals, safe for the environment, can't go rogue. RNAi is based on a mechanism all organisms have: enzymes that patrol for viral RNA and destroy it. So you would choose a protein critical to your target animal's life processes, dress it up like viral RNA, and let the interference mechanism destroy it. There are of course hundreds of proteins critical to life. Horak will look for one that would end it swiftly and without suffering, probably something neurological or cardiac.

The challenge with RNAi bait is that you're sending delicate strings of genetic code through the acids and enzymes of a digestive tract. Horak is working with biochemists to design a carrier molecule. That will take a while, as will getting RNAi registered with the EPA. It's at least a decade away from being something you'll see at Home Depot.

The novelty of the approach may trip it up. "There's a lot of talk in my world about perceived risk and actual risk," Horak says. "People are more comfortable with the actual risk of anticoagulant rodenticides, which will kill anything if they eat enough. But because we've done it for so many years, that level of risk is somehow acceptable." (In some places, and for some people, anyway.) "But the risks of RNAi are new risks, and so there's hesitation around those risks."

RNAi will face the inevitable challenge of any island eradication effort that relies on bait. The holdouts. The rodents that never encountered any bait. (Or, with poison bait, the ones that nibbled it, ate enough to feel sick but not die, and from then on steer clear.) Agencies can end up spending as much money tracking down the last ten of an invasive species—and monitoring for eleven, twelve, thirteen—than they spent eradicating the first ten thousand. This is going on right now in California's Sacramento–San Joaquin River Delta, where nutria have been proliferating. Nutria are similar to beavers—big swimming rodents that like to manipulate the landscape in ways that can make them unpopular. But nutria breed faster, and they're invasive. To find the holdouts, the California Department of Fish and Wildlife has been releasing neutered "Judas nutrias"—radio-collared individuals turned loose to betray their hidden kin.

A gene drive would have the rodents eradicate themselves. Without death or pain, and without killing any nontarget species.

And yet.

Here are some species the EPA, the USDA, and the Department of Health and Human Services consider "pests": chipmunks, bears, raccoons, foxes, coyotes, skunks, flying squirrels, tree squirrels, little brown bats, rattlesnakes, coral snakes, cliff swallows, crows, house finches, turkey vultures, black vultures, and mute swans.

This is what troubles me, I say to Aaron. We're back in his building now, watching rows of mice in stacked Plexiglas habitats, a Hollywood Squares of mice. Paul Lynde is doing backflips off one wall. What if a government agency eventually decides to go forward with gene drives on these other "pests"? What if economic considerations start to determine which species are next? What then? So long, pocket gophers? Toodle-oo, "nuisance beaver"? Right now, the focus is island conservation. It's a more appealing and less worrisome application: saving endangered species in a geographically isolated location. So you try it out there, and it works well, and the native species recover and there's good press. Now what? Where does the line get drawn, and who draws it? Let's remember: The National Wildlife Research Center is part of the USDA. It's not a conservation organization. "Aaron, the end point here, the ultimate target, is agricultural pests? Right?"

"That has been part of the discussion," he allows.

This is where it gets scary to me. We've seen what happens when the deciding factor is agriculture's bottom line. Will gene drive be a tidier rendition of the poisoning, shooting, trapping, bombing, wipe-'em-out campaigns of past centuries?

Aaron agrees that the decisions can't be just financial. "It has to be tied to ethics. We feel like we've taken a lot of steps to be sure this is accepted on a lot of fronts and we're not trying to go to third world countries to test it out." But the United States is surely not the only country working on gene drive in mammals. If we're on it, China is too. And China has not demonstrated a comforting abundance of oversight in the realm of genetic engineering.

Aaron was at the GBIRd meeting where Jane Goodall called for a moratorium on gene-drive research. GBIRd stands for Genetic Biocontrol of Invasive Rodents; it's a consortium of five U.S. and Australian government agencies and universities, plus the nonprofit Island Conservation. I ask him what Goodall's stated objection was. (Efforts to communicate directly with her were unsuccessful.)

"I think the concern is that the technology and the ability of people to experiment with it is moving way too fast, and the only way you're going to slow it down is to shut it down completely," he says. "And I think it's good. If someone reputable like Jane takes a stand, people will stop and think, *Maybe we need to establish some guidelines that we'll all follow.*"

Yes, please. Guidelines. Imagine if, rather than reducing the population or geographic distribution of a species, a gene drive wiped a species off a whole continent. Or an ecosystem was changed in some unanticipated and catastrophic way. It's the unknown unknowns that trouble some biologists. I spoke with

Will Pitt, a former project leader at NWRC and now deputy director of the Smithsonian Conservation Biology Institute. Rather than any specific fallout scenario, he expressed a general wariness. "People always say, 'We've thought of everything that might be a problem.' Well, it's probably one of the things you haven't thought of that's going to be a problem."

In a corner habitat on the top row, Charles Nelson Reilly is pirouetting on his hind legs. *Look how clever we are. See how we dance! Don't wipe us out!* Personally I would hate for mice to disappear. As would species for whom they're a common bill of fare. Who knows what defenseless small meat they'd turn to instead? But I suspect I'm not in the majority opinion as regards small rodents. I think plenty of people would be a-okay with, say, a global *Mus musculus* extinction.

"Right?"

"You mean if you asked a farmer or rancher who has a mouse problem?" Aaron chews his gum and considers this. "And you said, 'What if mice were eliminated from the planet, even from places where they have some important function?'"

"Yeah."

"Yeah, they might be like, 'I don't care.'"

Aaron knows a Big Agriculture guy with a lot of mice on his property. His name is Roger and he runs a feed lot where beef and dairy operations send their cattle to be raised. They're fed different diets depending on what they're being raised for, milk or beef or breeding more beef. The mice enjoy all of the diets. When Aaron needs wild mice for the SNE, he drives out to Roger's place. Roger surely has opinions about bothersome rodents and what their fate should be. Aaron agrees to drive out there with me after lunch.

Roger arrives to greet us driving a bulldozer-sized forklift. His cowboy hat is white felt, and the rest is mostly denim. He steps down and extends a hand. His grip is strong, but not in the manner of a person who's been coached on the importance of a firm handshake. More in the manner of a person who uses hand tools a lot. "Glad to meet you," Roger says.

Aaron hasn't been out here in a while, so he reintroduces himself. "I know who you are," Roger says. "You guys are the ones that let that elk out." Aaron lets this drift.

We follow the hat into the interior of Roger's grain elevator. As our eyes adjust to the dim, we start to see them. Every half minute or so, a mouse races along the base of a wall or shoots across the floor and disappears beneath a pile of metal machine parts. They say if you see twenty mice, there are two hundred more you're not seeing.

We go back out into the sun to continue the conversation. Above us is a silo of cracked corn and, I'm guessing, those other two hundred mice.

"Nah, that's pretty much mouseproof," Roger says. The top button on his shirt is open, and a long white chest hair quivers when the wind rises. "They could go up there, but why? They don't have to go very far to find something to eat." He scuffs the ground with one boot. There's enough spilled corn that the driveway crunched like gravel when we pulled up.

Other feed ingredients are stored out in the open. At the end of the drive is a low mountain range of brewer's grain and barley hops. I ask Roger to estimate the percentage that's lost to mice.

"Well, it comes in twenty-five-ton lots. How do you know if mice ate fifty pounds of that?" He removes his hat with one hand and with the other, wipes away sweat. His face is tanned up to where the hat begins, then not. "In the grand scheme of things, the wind probably blows away more than that. You know, so. I'm not sure that's a huge problem."

Roger's quibble with the mice is that they like to nest in the engines of his vehicles and sometimes they chew the wiring. But he doesn't set out traps or poison. "I try to keep barn cats. Though they keep going out there on the yellow line and getting run over. Or the barn owls get them."

I ask if he puts up nest boxes to encourage the barn owls, which also eat mice. This is a dumb question. Roger has barns. He doesn't need boxes. Though he has heard about the practice. "They're doing that out in California. Man, they eat a lot of mice." He surmises the reason he has no rat problem is that the foxes around his farm keep the population down. Probably so. In the late 1950s, overexuberant slaughter of foxes and coyotes in Oregon contributed to a massive mouse infestation. In California, circa 1918, one bounty program begat a second: three cents per ground squirrel tail or, in some counties, scalp.*

In the sky over a corral of Holsteins, twenty or so black birds wheel east. Aaron asks about bird-hazing strategies. "There's guys that'll come in and shoot at the starlings," Roger says. He doesn't use them, he adds, because it's not effective. The birds take off, circle around, and quickly come back. "It's more of a psychological benefit. To feel like you're doing something." He watches the birds disappear behind a stand of trees. "It's not a huge problem."

I'd like to end this book right here in Roger's smelly, baking cattle feedlot. The man in the big white hat gives me hope. To me, he represents a possible future where people may be frustrated by wild animals that get up in their business but they're living with them. In that possible future, people's reaction to the damage brought about by wildlife is something akin to acceptance. Or maybe res-

---

*   Unscrupulous bounty collectors would cash in twice: scalps in one county and tails in another. Some would fashion faux tails by wrapping portions of the hide around a skinny stick. Still others would cut off the tails and set the rodent free to breed yet more tails. In considering a bounty program for invasive brown tree snakes in Guam, officials worried that people might be tempted to release them on islands where they did not yet exist, in order to create a novel income source.

ignation is closer. Anyway, something far short of the conscience-free rush to annihilation that characterized previous decades and centuries. If people are able to step outside the anger, they may find that more humane approaches are also more effective.

There are plenty of farmers and ranchers more progressive than Roger, and that's precisely why he makes me hopeful. He's big agriculture, not small organic, and yet he gets it. Without ever uttering the words, he's practicing coexistence and biocontrol. The feed he loses to rodents and birds is part of the cost of doing business. Perhaps the model should be shoplifting. Supermarkets and chain stores don't poison shoplifters; they come up with better ways to outsmart them.

Before we leave, Roger shows us around the feed lots. The breeding bulls are on a maintenance ration. Roger takes a handful from a trough and holds it out for me to smell. We walk on. "Across here are the commercial beef cattle." They fatten up on corn. "Real high-calorie, high-carb."

The fat cattle stand at the fence, swishing their tails and staring. *You're all on about the mice, but what about us?*

"They'll go to JDA or Cargill for slaughter," Roger adds casually. "Probably sixty days from now." Because people like me want to have their hamburgers. *Only once or twice a year,* I want to say. But I know that's a lame defense. It's not the quantity that matters, it's the statement you make or don't make. When you tell people you don't eat beef—or would never use a glue trap—you make the alternative a little less comfortable for them. You keep it from being a thing they give no thought to.

For centuries, people have killed trespassing wildlife—or brought in someone to do it for them—without compunction and with scant thought to whether it's done humanely. We have detailed protocols for the ethical treatment and humane "euthanizing" of laboratory rats and mice, but no formal standards exist for the rodents or raccoons in our homes and yards. We leave the details to the exterminators and the "wildlife control operators," the latter a profession that got rolling when the bottom dropped out of the fur market and trappers realized they could make better money getting squirrels out of people's attics.

Rodents are a good bellwether. If people can be less cruel to rats, if it even crosses their minds to be less cruel to a rat, then things are heading in a good direction. Not just good for rats, maybe also good for humans. "If man can be taught to respect the home of the worm," wrote the nineteenth-century historian Léon Ménabréa, "how much more ought he to regard that of his fellow man."

A few months after I returned from Colorado, I was reading a book outside. I happened to look up as a roof rat ran across the end of the deck. My immediate impulse was to drive to the hardware store and buy a snap trap. But I didn't. How could I? Little Miss Coexistence. Put your money where your yap is. Besides, as I knew, removing one raises a VACANCY sign for another.* My neighbor livetraps the squirrels that raid her peach tree and lets them go in a neighborhood park. She has been doing this, Sisyphus-like, the entire decade we've lived next door to each other.

A few days later, heading down the stairs from the deck, I saw the rat again. He was running down a tree branch, with a loquat in his jaws. Our eyes met. He froze. I froze. He dropped his fruit. Seen head-on, the naked tail hidden from view, the roof rat is certifiably cute. The species is smaller than the Norway rat, with fur of a warmer, prettier brown. It's a squirrel without tail fluff. This

---

* Or on a more magisterial scale, "I know from my own estate in Yorkshire. . . . as soon as one kills 300 or 400 grey squirrels in any given month a similar number come . . . and take their places."—the Earl of Feversham, parliamentary secretary of the Ministry of Agriculture and Fisheries, at a meeting of the House of Lords to discuss the Grey Squirrels (Prohibition of Importation and Keeping) Order of 1937.

guy was fully as adorable as the ground squirrels that run around the bayside park where I hike. (And if history is any indication, less likely to pass on disease.) I went on down the steps and put my load in the washer and forgot about the rat.

A week later, I heard something moving inside a wall. "Your little friend is going to chew through the wiring and set the house on fire," said Ed. I told him I wanted to figure out how it was getting in and practice "exclusion." He gave me a week to figure it out.

I set up my wildlife camera in different places around the outside of the house, and we figured out where the rat was getting in. Ed filled the gap, and that was that. The noises stopped. I continued to see the rat around, mostly on the camera, but once or twice our paths would cross. I'd nod hello, and we'd go about our days.

Acknowledgments

While working on this project, I would occasionally encounter the term "vertebrate pest." I don't much care for it, because it reduces an animal to its role in the context of human enterprise. However, there is one mammal for whom the term seems fair and apt, and that is me. For kindness and forbearance in the face of ceaseless pestering, a few people must be singled out. Bowed down to. Carried high by the cheering crowd. They are: Stewart Breck, Justin Dellinger, Travis DeVault, André Frijters, Joel Kline, Dipanjan Naha, Aaron Shiels, Bruce Warburton, and Dazy Weymer. You made this book, as much as I did. And in return I have those two words that never really capture the feelings they represent. Thank you.

Though they are less visible on the page, the following people also welcomed me and shared their time and knowledge, often with little or no advance warning. Giant neon thanks to Samantha Brown, Father Carlo Casalone, Aaron Koss-Young, Charlie Martin, Dean McGeough, Nico Nijenhuis, Toni Piaggio, Qamar

Qureshi, Saroj Raj, Tom Seamans, Shaun Templeton, Kurtis Tesch, Rafael Tornini, R. B. S Tyagi, and Tina White.

Kim Annis, Jonathan Clemente, Bradley Cohen, Sarah Courchesne, Doug Eckery, Julie Carol Ellis, Esteban Fernandez-Juricic, Dave Garshelis, Katherine Horak, John Humphrey, Bruce Kimball, Mario Klip, Page Klug, Tim Manley, Stella McMillin, Vicky Monroe, Julie Oakes, Seth Pincus, William Pitt, Samantha Pollak, Heather Reich, Virginia Roxas-Duncan, Shane Siers, Steve Smith, Peter Tira, Catherine VandeVoort, Harry Wetherbee, Kate Wilmot, and Bonnie Yates: if I did not achieve full pest status, I was, at least for an hour or two, a thing buzzing in your ear. Thank you for not waving me away.

Tim Bibby, Paul Deckers, Carol Glatz, Thaddeus Jones, Gail Keirn, Kirsten Macintyre, Fabrizio Mastrofini, Heather Steere, Kevin Van Damme, and Brian Wakeling: your help made possible entire chapters that would otherwise not exist. My thanks to you. Keli Hendricks, John Griffin, John Hadidian, and Kellie Nicholas, you put things in context both historical and political, and I deeply appreciate your insights. For allowing me to infest your inbox, thank you, John Anderson, Meera Bhatia, Johan Elmberg, Ann Filmer, Robin Konrad, Georgia Mason, Christina Meister, Sanath Muliya, and George Smith.

My reporting took me to places where I did not understand the language and culture. My gratitude to Raffaella Buschiazzo and Charles Lansdorp for help with translation and interpreting. Nilanjana Bhowmick, Aritra Naha, and Shweta Singh, to each of you: your skillful ear and quick mind added nuance to my reporting, and your companionship made me feel at home when home was far away.

To Jill Bialosky and Jay Mandel: twenty years and seven books down the line, here I am, thanking you again. Yet it never feels like enough. Because *everything always goes well.* How often, in publishing—in any business? in life!—does that happen? Contributing in other immeasurable ways to the ongoing W. W. Norton miracle are the following excellent humans: Steve Attardo, Louise Brockett, Steve Colca, Brendan Curry, Ingsu Liu, Erin Lovett, Meredith McGinnis, Stephanie Romeo, and Drew Weitman.

I am indebted to Janet Byrne for her graceful and expert copy-editing and for the patience, tact, and zeal with which she checked my work. Few are as good as she (as *her?* Janet! Help!).

A nod to Carlton Engelhardt for the emus and to Andy Karam for the stilettos. Cynthia, thank you for introducing me to Nila. Jeff, thank you for listening and for thinking it was a good idea, and Jesse, thank you for the New Zealand connections and hospitality. Steph, thank you for adding monkey-time to the itinerary. And Ed, always Ed, thank you for everything.

The Fuzzy Trespasser

The Humane Society of the United States (HSUS) has a helpful "What to Do About" series on its website. Included are strategies for resolving—or, better, preventing—problems with urban and suburban wildlife species, among them, bats, bears, Canada geese, chipmunks, coyotes, crows, deer, foxes, mice, opossums, pigeons, rabbits, raccoons, rats, skunks, snakes, squirrels, sparrows, starlings, wild turkeys, and woodchucks. https://www.humanesociety.org/resource/wildlife-management-solutions

More good advice can be found in the "Living in Harmony with Wildlife" series on the PETA website. Included are bats, geese, mice, chipmunks, pigeons, raccoons, skunks, squirrels, rabbits, and rats. https://www.peta.org/issues/wildlife/living-harmony-wildlife/

If wild animals have begun raising a family in an attic or crawl space, you will need professional help to undertake a humane eviction of both mother and young. Before you make any calls,

I recommend reading the HSUS "Choosing a Wildlife Control Company" web page. Getting them out requires specialized know-how.

As does releasing them afterward. Best practice these days is called "on-site" release. Once the operator has helped you seal any entry points and eliminate or close off other appealing nesting sites, the animals are released inside their home range—that is, your property. Driving them to a nearby woods or park sounds humane but likely isn't. "The squirrels did not fare well," concluded a study by University of Maryland and HSUS researchers who radio-collared thirty-eight gray squirrels and relocated them inside nearby Patuxent Research Refuge. Seventeen wound up as carcasses, or a skull or piece of fluffy tail lying by a collar, or just collars—two "with tooth marks on them" and one inside a fox den. The remaining eighteen squirrels disappeared, on average within eleven days, fate unknown. Raccoons, by one study, fare better, but some states prohibit the practice, because rabies virus could be relocated at the same time.

Good news for homeowners and rodents: more and more pest control companies are offering "exclusion" as an alternative to poison bait boxes or trapping. It's a matter of locating all the (surprisingly small) crevices and gaps where mice, rats, or squirrels could gain entry and then filling them with something rustproof that the animals can't easily gnaw through: steel wool, typically. Xcluder Rodent and Pest Defense makes "cut-and-stuff" stainless steel fiber products. In a seven-day NWRC test, none of 10 Xcluder-blocked gaps were breached by rats and mice attempting to reach their preferred "lure foods" (peanut butter oatmeal balls and, for the house mice, hot dogs and cheese).

Bibliography

A QUICK WORD OF INTRODUCTION

Evans, E. P. *The Criminal Prosecution and Capital Punishment of Animals.* New York: E. P. Dutton and Company, 1906.

1. MAUL COPS

Conover, Michael R. *Resolving Human Wildlife Conflicts: The Science of Wildlife Damage Management.* Boca Raton: Lewis Publishers, 2002. Table 3.1: Studies of Nonfatal and Fatal Injuries to Humans by Wildlife in Different Parts of the U.S. and Canada.

Floyd, Timothy. "Bear-Inflicted Human Injury and Fatality." *Wilderness and Environmental Medicine* 10 (1999): 75–87.

U.S. Consumer Product Safety Commission. "Product Instability or Tip-Over Injuries and Fatalities Associated with Televisions, Furniture, and Appliances: 2012 Report." Graph, p. 17.

Young, Stanley Paul, and Edward Alphonso Goldman. *The Puma.* Washington, DC: American Wildlife Institute, 1946.

2. BREAKING AND ENTERING AND EATING

Alldredge, Mat W., et al. "Evaluation of Translocation of Black Bears Involved in Human-Bear Conflicts in South-Central Colorado." *Wildlife Society Bulletin* 39, no. 2 (June 2015): 334–40.

Beckmann, Jon P., Carl W. Lackey, and Joel Berger. "Evaluation of Deterrent Techniques and Dogs to Alter Behavior of 'Nuisance' Black Bears." *Wildlife Society Bulletin* 32, no. 4 (2004): 1141–46.

Breck, Stewart W. "Selective Foraging for Anthropogenic Resources by Black Bears: Minivans in Yosemite National Park." *Journal of Mammalogy* 90, no. 5 October 2009): 1041–44.

George, Kelly A., et al. "Changes in Attitude Toward Animals in the United States from 1978 to 2014." *Biological Conservation* 201 (2016): 237–42.

Johnson, Heather E., et al. "Human Development and Climate Affect Hibernation in a Large Carnivore with Implications for Human-Carnivore Conflicts." *Journal of Applied Ecology* 55, no. 2 (March 2018): 663–72.

Johnson, Heather E., et al. "Assessing Ecological and Social Outcomes of a Bear-Proofing Experiment." *Journal of Wildlife Management* 82, no. 6 (2018): 1102–14.

Linnell, John D. C., et al. "Translocation of Carnivores as a Method for Managing Problem Animals: A Review." *Biodiversity and Conservation* 6, no. 9 (September 1997): 1245–57.

Manning, Elizabeth. "Tasers for Moose and Bears: Alaska Explores Law Enforcement Tool for Wildlife." *Alaska Fish & Wildlife News*, March 2010.

Nelson, Ralph A., et al. "Behavior, Biochemistry, and Hibernation in Black, Grizzly, and Polar Bears." *Proceedings of the International Conference on Bear Research and Management* 5 (1983): 284–90.

Roenigk, Adolph. *Pioneer History of Kansas.* Transcribed by his great-grandniece L. Ann Bowler. Denver, CO, 1933. https://www.kancoll.org/books/roenigk/index.html.

Rogers, Lynn L. "Homing by Radio-Collared Black Bears, *Ursus americanus*, in Minnesota." *Canadian Field Naturalist* 100, January 1986.

Spencer, Rocky D., Richard A. Beausoleil, and Donald A. Martorello. "How Agencies Respond to Human–Black Bear Conflicts: A Survey of Wildlife Agencies in North America." *Ursus* 18, no. 2 (2007): 217–29.

3. THE ELEPHANT IN THE ROOM

The Asian Elephant (Elephas maximus) of Nagaland: Landscape & Human-Elephant Conflict Management. Dimapur, Nagaland: Government of Nagaland, Wildlife Wing, Department of Forests, Environment and Wildlife.

Gopalakrishnan, Shankar, Terpan Singh Chauhan, and M. S. Selvaraj. "It Is Not Just About Fences: Dynamics of Human-Wildlife Conflict in Tamil Nadu and Uttarakhand." *Economic & Political Weekly* 52: 97–104.

Hindustan Times. "Appetite for Money: Elephants Who Entered a Shop Gorge on Rs 2,000, 500 Notes." April 25, 2017.

———. "Drunken Man Challenges Elephants' Herd, Trampled to Death in Jharkhand." December 19, 2018.

Jayewardene, Jayantha. *The Elephant in Sri Lanka.* Colombo, Sri Lanka: Wildlife Heritage Trust of Sri Lanka, 1994.

Lahiri-Choudhury, Dhriti K. "History of Elephants in Captivity in India and Their Use: An Overview." *Gajah* 14 (June 1995): 28–31.

McKay, George M. *Behavior and Ecology of the Asiatic Elephant* (Smithsonian Contributions to Zoology, Number 125). Washington, DC: Smithsonian Institution Press, 1973.

Naha, Dipanjan, et al. "Assessment and Prediction of Spatial Patterns of Human-

Elephant Conflicts in Changing Land Cover Scenarios of a Human-Dominated Landscape in North Bengal." *PLOS ONE*, February 1, 2019.

Outlook India. "928 Elephants Died Unnaturally Since 2009 Including 565 Due to Electrocution Alone." March 25, 2019.

————. "Delhi Planning to Club Old Age Home with Cow Shelter." January 9, 2019.

Siegel, Ronald K., and Mark Brodie. "Alcohol Self-Administration by Elephants." *Bulletin of the Psychonomic Society* 22, no. 1 (1984): 49–52.

U.S. House of Representatives, Committee on the Judiciary, Hearing before the Subcommittee on Crime. *Captive Elephant Accident Prevention Act of 1999*. 106th Cong., 2d sess., 2000. H.R. 2273.

4. A SPOT OF TROUBLE

Athreya, Vidya. "Is Relocation a Viable Option for Unwanted Animals? The Case of the Leopard in India." *Conservation and Society* 4, no. 3 (2006): 419–23.

Athreya, Vidya, et al. "Translocation as a Tool for Mitigating Conflict with Leopards in Human-Dominated Landscapes of India." *Conservation Biology* 25, no. 1 (November 2010): 133–41.

Corbett, Jim. *The Man-Eating Leopard of Rudraprayag*. New Delhi: Rupa, 2016.

Naha, Dipanjan, S. Sathyakumar, and G. S. Rawat. "Understanding Drivers of Human-Leopard Conflicts in the Indian Himalayan Region: Spatio-Temporal Patterns of Conflicts and Perception of Local Communities Towards Conserving Large Carnivores." *PLOS ONE*, October 2018.

Singh, H. S. *Leopards in the Changing Landscapes*. Dehra Dun: Bishen Singh Mahendra Pal Singh, 2014.

Times of India. "Leopard Enters Hema Malini's House." May 28, 2011.

5. THE MONKEY FIX

Chauhan, Arvind. "Monkey Snatches Baby from Mom, Kills It." *Times of India*, November 14, 2018.

————. "UP: After Infant's Death, 2 More Toddlers Attacked by Monkeys." *Times of India*, November 17, 2018.

Colagross-Schouten, A., et al. "The Contraceptive Efficacy of Intravas Injection of Vasalgel™ for Adult Male Rhesus Monkeys." *Basic Clinical Andrology* 27, no. 1 (2017), article no. 4.

Gandhiok, Jasjeev, and Paras Singh. "Delhi: Simians Wreak Havoc; Forest Department, Corporations Pass Buck." *Times of India*, January 19, 2019.

Harris, Gardiner. "Indians Feed the Monkeys, Which Bite the Hand." *New York Times*, May 22, 2012.

Killian, G., D. Wagner, and L. Miller. "Observations on the Use of the GnRH Vaccine Gonacon™ in Male White-Tailed Deer (*Odocoileus virginianus*)." *Proceedings of the 11th Wildlife Damage Management Conference*, 2005.

Miller, Lowell A., Kathleen A. Fagerstone, and Douglas C. Eckery. "Twenty Years of Immunocontraceptive Research: Lessons Learned." *Journal of Zoo and Wildlife Medicine* 44, Supplement 4 (December 2013): S84–S96.

Mohan, Vishwa. "Order to Cull HP's 'Vermin' Monkeys Draws Activists' Ire." *Times of India*, July 19, 2019.

Mohapatra, Bijayeeni, et al. "Snakebite Mortality in India: A Nationally Repre-
    sentative Mortality Survey." *PLOS Neglected Tropical Diseases* 5, no 4 (April
    2011): e1018.
Singh, Paras. "Delhi: South Corporation Finally Nets Eight Monkey Catchers."
    Times of India, October 8, 2018.
Times of India. "Teen Killed in Monkey Attack in Kasganj; 5th Death in a Month."
    December 3, 2018.
———. "Simians Lay Siege to Agra." November 16, 2018.
———. "70-Year-Old Allegedly Stoned to Death by Monkeys; Kin Demands
    FIR." October 20, 2018.

6. MERCURIAL COUGARS

Beier, Paul, Seth P. D. Riley, and Raymond M. Sauvajot. "Mountain Lions." In
    Urban Carnivores: Ecology, Conflict, and Conservation, edited by Stanley D. Gehrt,
    Seth P. D. Riley, and Brian L Cypher. Baltimore: Johns Hopkins University
    Press, 2010.
Brewster, R. Kyle, et al. "Do You Hear What I Hear? Human Perception of Coyote
    Group Size." *Human–Wildlife Interaction* 11, no. 2 (Fall 2017): 167–74.
Clemente, Jonathan D. "CIA's Medical and Psychological Analysis Center (MPAC)
    and the Health of Foreign Leaders." *International Journal of Intelligence and Coun-
    terintelligence* 19, no. 3 (2006): 385–423.
Fisher, A. K. "The Hawks and Owls of the United States in Their Relation to
    Agriculture." Washington, DC: U.S. Department of Agriculture, Division of
    Ornithology and Mammalogy, Bulletin No. 3, 1893.
Hunter, J. S. "The Mountain Lion." Article manuscript, undated. Joseph S. Hunter
    Papers, F3735:618. California State Archives: Records of the Division of Fish
    and Game. [Jay Bruce statement]
"Mountain Lion." Letter from Jay C. Bruce to J. S. Hunter, March 23, 1941. Joseph
    S. Hunter Papers, F3735:618. California State Archives: Records of the Divi-
    sion of Fish and Game.
Peirce, E. R. "A Method of Determining the Prevalence of Rats in Ships." *The
    Medical Officer* 43 (1930): 222–24.
Todd, Kim. "Coyote Tracker." *Bay Nature*, January–March 2018.
Welch, David. "Dung Properties and Defecation Characteristics in Some Scottish
    Herbivores, with an Evaluation of the Dung-Volume Method of Assessing
    Occupance." *Acta Theriologica* 27, no. 15 (October 1982): 191–212.
Yiakoulaki, M. D., and A. S. Nastis. "A Modified Faecal Harness for Grazing Goats
    on Mediterranean Shrublands." *Journal of Range Management* 51, no. 5 (Septem-
    ber 1998): 545–46.
Young, Stanley Paul, and Edward Alphonso Goldman. *The Puma.* Washington, DC:
    The American Wildlife Institute, 1946.

7. WHEN THE WOOD COMES DOWN

BC Parks. Wildlife/Danger Tree Assessor's Course Workbook. Revised edition,
    March 2012. https://www2.gov.bc.ca/assets/gov/environment/plants-animals

-and-ecosystems/conservation-habitat-management/wildlife-conservation/
wildlife-tree-committee/parks-handbook.pdf.

Brookes, Andrew. "Preventing Death and Serious Injury from Falling Trees and Branches." *Australian Journal of Outdoor Education* 11, no. 2 (2007): 50–59.

Mulford, J. S., H. Oberli, and S. Tovosia. "Coconut Palm-Related Injuries in the Pacific Islands." *ANZ Journal of Surgery* 71, no. 1 (2001): 32–34.

Oregon Fatality Assessment and Control Evaluation. *Fallers Logging Safety* (manual). 2007. https://www.ohsu.edu/sites/default/files/2019-02/ORFACE -SafetyBooklet-FallersLoggingSafety-Eng.pdf.

Schmidlin, Thomas. "Human Fatalities from Wind-Related Tree Failures in the United States, 1995–2007." *Natural Hazards* 50, no. 1 (2009): 13–25.

Tribun-Bali. "Falling Durian Possibly Killed Man in West Bali" (via Google Translate). January 28, 2015.

—. "To the Durian Garden Without Head Shield, Kusman Found Dead" (via Google Translate). March 26, 2015.

Walsh, Raoul A., and Lara Ryan. "Hospital Admissions in the Hunter Region from Trees and Other Falling Objects, 2008–2012." *Australian and New Zealand Journal of Public Health* 41, no. 2 (2017): 121–24.

8. THE TERROR BEANS

Arianti, V. "Biological Terrorism in Indonesia." *The Diplomat*, November 20, 2019. https://thediplomat.com/2019/11/biological-terrorism-in-indonesia/.

"Compound 1080—Powerful New Rat Killer." Press release, n.d. Fort Collins, CO: National Wildlife Research Center Archive.

Dymock, William, C. J. H. Warden, and David Hooper. *Pharmacographia Indica: A History of the Principal Drugs of Vegetable Origin, Met with in British India.* London: Kegan Paul, Trench, Trübner & Co., 1891.

Eisemann, John D., Patricia A. Pipas, and John L. Cummings. "Acute and Chronic Toxicity of Compound DRC-1339 (3-Chloro-4-Methylaniline Hydrochloride) to Birds." *USDA National Wildlife Research Center Staff Publications* 211, November 2003.

Filmer, Ann. "Safe and Poisonous Garden Plants." University of California, Davis, October 2012. https://ucanr.edu/sites/poisonous_safe_plants/files/154528 .pdf.

Jacobsen, W. C., and S. V. Christierson, eds., Rodent Control Division. "California Ground Squirrels: A Bulletin Dealing with Life Histories, Habits and Control of the Ground Squirrels in California." *Monthly Bulletin of the California State Commission of Horticulture* VII, nos. 11 and 12, November–December 1918.

Jain, Ankita, et al. "Foreign Body (Kidney Beans) in Urinary Bladder: An Unusual Case Report." *Annals of Medicine and Surgery* 32 (August 2018): 22–25.

Karthikeyan, Aishwarya, and S. Deepak Amalnath. "*Abrus precatorius* Poisoning: A Retrospective Study of 112 Patients." *Indian Journal of Critical Care* 21, no. 4 (April 2017): 224–25.

Linz, George M., and H. Jeffrey Honan. "Tracing the History of Blackbird Research Through an Industry's Looking Glass: *The Sunflower Magazine*." *Proceedings of the 18th Vertebrate Pest Conference*, 1998.

Malik, Balwant Singh. "Punishment of Transportation for Life." *Journal of the Indian Law Institute* 36, no. 1 (1994): 111–20.

Nicholson, Blake. "Debate Rises over Blackbirds." *Bismarck Tribune*, March 18, 2007.

Ogawa, Haruko, and Kimie Date. "The 'White Kidney Bean Incident' in Japan." In *Lectins: Methods and Protocols*. Part of Methods in Molecular Biology book series, volume 1200. New York: Humana Press, 2014.

Ormsbee, R. A. "A Summary of Field Reports on 1080 (Sodium Fluoroacetate)." National Research Council, Insect Control Committee, Technical Report No. 163. December 17, 1945.

Pincus, Seth H., et al. "Passive and Active Vaccination Strategies to Prevent Ricin Poisoning." *Toxins* 3, no. 9 (September 2011): 1163–84.

Pitschmann, Vladimír, and Zdeněk Hon. "Military Importance of Natural Toxins and Their Analogs." *Molecules* 21, no. 5 (April 2016): 556–78.

Renshaw, Birdsey, to Dr. W. R. Kirner. Memorandum regarding "Animal Poisons" sent from the Office for Emergency Management National Defense Research Committee of the Office of Scientific Research and Development. Washington, DC (1530 P Street, NW), December 30, 1943. Fort Collins, CO: National Wildlife Research Center Archive.

Roxas-Duncan, Virginia I., and Leonard A. Smith. "Of Beans and Beads: Ricin and Abrin in Bioterrorism and Biocrime." *Journal of Bioterrorism & Biodefense* S2:002, January 2012.

Smith, George, and Dick Destiny. "Great WMD Failures: Casey the Castor Oil Killer." *The Register*, October 18, 2006. http://www.theregister.com/2006/10/18/dd_castor_oil_wmd/.

The Sunflower. "Blackbird Project Focuses on Population Reduction." December 1, 1996.

Thornton, S. L., et al. "Castor Bean Seed Ingestions: A State-Wide Poison Control System's Experience." *Clinical Toxicology* 52, no. 4 (March 2014): 265–68.

Ward, Justus C. "Rodent Control with 1080, ANTU, and Other War-Developed Toxic Agents." *American Journal of Public Health* 36, no. 12 (December 1946): 1427–31.

Wildlife Research Laboratory, Division of Wildlife Research, U.S. Fish & Wildlife Services. "Compound 1080—A New Agent for the Control of Noxious Mammals." Denver, CO, n.d. Fort Collins, CO: National Wildlife Research Center Archive.

9. OKAY, BOOMER

Blackwell, Bradley F., Eric Huszar, George M. Linz, and Richard A. Dolbeer. "Lethal Control of Red-Winged Blackbirds to Manage Damage to Sunflower: An Economic Evaluation." *Journal of Wildlife Management* 67, no. 4 (October 2003): 818–28.

Daily News (Perth). "Emus Outwit Gunners." November 4, 1932, p. 1.

———. "Campion Evacuated: Emus Flourish Unharried." November 10, 1932, p. 5.

Daily Telegraph (Sydney). "Not Easy to Kill Emus: A Thousand Rounds Fired, 12 Dead." November 5, 1932, p. 3.

Fisher, Harvey I. "Airplane-Albatross Collisions on Midway Atoll." *The Condor* 68 (May 1966): 229–42.

Flying Magazine. "MATS Versus the Gooney Bird." September 1958.

Frings, Hubert. *The Scientific Scobberlotching of Hubert and Mable Frings.* BTcurlew Press, 2015.

Frings, Hubert, and Mable Frings. "Problems of Albatrosses and Men on Midway Islands." *The Elepaio: Journal of the Hawaii Audubon Society* 20, no. 5 (1959).

Kenyon, Karl W., Dale W. Rice, Chandler S. Robbins, and John W. Aldrich. "Birds and Aircraft on Midway Islands: 1956–57 Investigations." *Special Scientific Report—Wildlife* No. 38. Washington, DC: United States Department of the Interior, Fish and Wildlife Service, January 1958.

Mail (Adelaide). "Request to Use Bombs to Kill Emus." July 3, 1943, p. 12.

Rice, Dale W. "Birds and Aircraft on Midway Islands: 1957–58 Investigations." *Special Scientific Report—Wildlife* No. 44. Washington, DC: United States Department of the Interior, Fish and Wildlife Service, 1959.

Robbins, Chandler S. "Birds and Aircraft on Midway Islands: 1959–63 Investigations. *Special Scientific Report—Wildlife* No. 85. Washington, DC: United States Department of the Interior, Fish and Wildlife Service, 1966.

Sydney Morning Herald. "War on Emus." October 12, 1932, p. 11.

USFWS—Pacific Region. "Night Vision Trail Cameras Capture Mouse Attacks on Albatross." Video, October 31, 2017.

West Australian (Perth). "War on Emus: Ambush at a Dam." November 8, 1932, p. 8.

Western Mail (Perth). "A Thousand Birds in Luck: Machine Guns Jam." November 10, 1932, p. 28.

10. ON THE ROAD AGAIN

Ansari, S. A., et al. "Dorsal Spine Injuries in Saudi Arabia—An Unusual Cause." *Surgical Neurology* 56, no. 3 (2001): 181–84.

Biondi, Kristin M. "White-Tailed Deer Incidents with U.S. Civil Aircraft." *Wildlife Society Bulletin* 35, no. 3 (September 2011): 303–9.

Blackwell, Bradley F., and Thomas W. Seamans. "Enhancing the Perceived Threat of Vehicle Approach to Deer." *Journal of Wildlife Management* 73, no. 1 (2009): 128–35.

Cohen, Bradley S., et al. "Behavioral Measure of the Light-Adapted Visual Sensitivity of the White-Tailed Deer." *Wildlife Society Bulletin* 38, no. 3 (September 2014): 480–85.

D'Angelo, Gino, et al. "Development and Evaluation of Devices Designed to Minimize Deer-Vehicle Collisions." Final Project Report, Daniel B. Warnell School of Forestry and Natural Resources. July 2, 2007.

DeVault, Travis, et al. "Effects of Vehicle Speed on Flight Initiation by Turkey Vultures: Implications for Bird-Vehicle Collisions." *PLOS ONE*, February 4, 2014.

————. "Speed Kills: Ineffective Avian Escape Responses to Oncoming Vehicles." *Proceedings of the Royal Society B: Biological Sciences*, February 22, 2015.

DeVault, Travis, Bradley F. Blackwell, and Jerrold L. Belant, eds. *Wildlife in Airport Environments: Preventing Animal-Aircraft Collisions through Science-Based Management.* Baltimore: Johns Hopkins University Press, 2013.

DeVault, Travis, Thomas W. Seamans, and Brad Blackwell. "Frontal Vehicle Illumination via Rear-Facing Lighting Reduces Potential for Collisions with White-Tailed Deer." *Ecosphere* (manuscript accepted).

Dolbeer, Richard A., et al. "Wildlife Strikes to Civil Aircraft in the United States, 1990–2015." *National Wildlife Strike Database Serial Report Number 21*, July 2015.

Gens, Magnus. "Moose Crash Test Dummy." Master's thesis, Royal Institute of Technology, Stockholm, Sweden. *VTI särtryck 342*, 2001.

Hughson, Debra L., and Neal Darby. "Desert Tortoise Road Mortality in Mojave National Preserve, California." *California Fish and Game* 99, no. 4 (September 2013): 222–32.

Kennedy Space Center Status Report. "Roadkill Roundup." April 26, 2006.

Kim, Sharon, and A. Robertson Harrop. "Maxillofacial Injuries in Moose–Motor Vehicle Collisions Versus Other High-Speed Motor Vehicle Collisions." *Canadian Journal of Plastic Surgery* 13, no. 4 (December 2005): 191–94.

Knapp, Keith K. "Deer-Vehicle Crash Countermeasure Toolbox." Iowa State University Institute for Transportation, Deer-Vehicle Crash Information Clearinghouse, 2001.

Pynn, Tania P., and Bruce R. Pynn. "Moose and Other Large Animal Wildlife Collisions: Implications for Prevention and Emergency Care." *Journal of Emergency Nursing* 30, no. 6 (2004): 542–47.

Riginos, Corinna, et al. "Wildlife Warning Reflectors and White Canvas Reduce Deer-Vehicle Collisions and Risky Road-Crossing Behavior." *Wildlife Society Bulletin* 42, no. 4 (March 2018): 1–11.

Al-Sebai, M. W., and S. Al-Zahrani. "Cervical Spinal Injuries Caused by Collision of Cars with Camels." *Injury* 28, no. 3 (April 1997): 191–94.

Simmons, James Raymond. *Feathers and Fur on the Turnpike*. Boston: The Christopher Publishing House, 1938.

Williams, Allan F., and Joann K. Wells. "Characteristics of Vehicle-Animal Crashes in Which Vehicle Occupants Are Killed." *Traffic Injury Prevention* 6, no. 1 (2005): 56–59.

11. TO SCARE A THIEF

Avery, Michael L., et al. "Dispersing Vulture Roosts on Communication Towers." *Journal of Raptor Research* 36, no. 1 (February 2002): 45–50.

Bildstein, Keith. *Raptors: The Curious Nature of Diurnal Birds of Prey*. Ithaca, NY: Cornell University Press, 2017.

Blackwell, Bradley, Thomas W. Seamans, Morgan B. Pfeiffer, and Bruce N. Buckingham. "European Starling (*Sturnus vulgaris*) Reproduction Undeterred by Predator Scent Inside Nest Boxes." *Canadian Journal of Zoology* 96, no. 9 (2018): 980–86.

Chipman, Richard B., et al. "Emergency Wildlife Management Response to Protect Evidence Associated with the Terrorist Attack on the World Trade Center, New York City." *Proceedings of the 21st Vertebrate Pest Conference*, 2004.

Mauldin, Richard E., et al. "Development of a Synthetic Materials Mimic for Vulture Olfaction Research." *Proceedings of the 10th Damage Management Conference*, 2003.

Seamans, Thomas W. "Response of Roosting Turkey Vultures to a Vulture Effigy." *Ohio Journal of Science* 104, no. 5 (December 2004): 136–38.

Tillman, Eric A., John S. Humphrey, and Michael L. Avery. "Use of Vulture Carcasses and Effigies to Reduce Vulture Damage to Property and Agriculture." *Proceedings of the 20th Vertebrate Pest Conference*, 2002, pp. 123–28.

12. THE GULLS OF ST. PETER'S

Glahn, James F., Greg Ellis, Paul Fioranelli, and Brian Dorr. "Evaluation of Moderate and Low-Powered Lasers for Dispersing Double-Crested Cormorants from Their Night Roosts." *Proceedings of the Ninth Wildlife Damage Management Conference*, January 2001.

Glatz, Carol. "Feathery Fiascos: The Unfortunate Prey for Peace." Catholic News Service Blog, January 27, 2014.

Graham, Frank, Jr. *Gulls: An Ecological History*. New York: Van Nostrand Reinhold, 1975.

Linton, E., et al. "Retinal Burns from Laser Pointers: A Risk in Children with Behavioral Problems." *Eye* 33, no. 3 (March 2019): 492–504.

Markham, Gervase. *Markham's Farewell to Husbandry*. London: Nicholas Oakes for John Harrison, 1631.

Parsons, Jasper. "Cannibalism in Herring Gulls." British Birds (newsletter), December 1, 1971.

Vickery, Juliet A., and Ronald W. Summers. "Cost-Effectiveness of Scaring Brent Geese (*Branta b. bernicla*) from Fields of Arable Crops by a Human Bird Scarer." *Crop Protection* 11, no. 5 (October 1992): 480–84.

13. THE JESUIT AND THE RAT

Francis (pope). "'Laudato Si'": On Care for Our Common Home." Encyclical of the Holy Father on Climate Change and Inequality. http://w2.vatican.va/content/francesco/en/encyclicals/documents/papa-francesco_20150524_enciclica-laudato-si.html.

Philippi, Dieter. "Campagi—The Footwear of the Pope and the Clergy." http://www.dieter-philippi.de/en/ecclesiastical-fineries/campagi-the-footgear-of-the-pope-and-the-clergy.

14. KILLING WITH KINDNESS

Adams, Lowell W., J. Hadidian, and V. Flyger. "Movement and Mortality of Translocated Urban-Suburban Grey Squirrels." *Animal Welfare* 13, no. 1 (February 2004): 45–50.

American Veterinary Medical Association. AVMA Guidelines for the Euthanasia of Animals. American Veterinary Medical Association: 2013 edition.

———. "AVMA May Change Guidance for CO_2 Euthanasia in Rodents." *JAVMA News*, January 1, 2019.

Egerton, Rachael. "'Unconquerable Enemy or Bountiful Resource?' A New Perspective on the Rabbit in Central Otago." Bachelor's thesis, University of Otago, Dunedin, New Zealand, March 18, 2014. Australian & New Zealand Environmental History Network, https://www.environmentalhistory-au-nz.org/publications/.

King, Carolyn M. "Liberation and Spread of Stoats (*Mustela erminea*) and Weasels (*M. nivalis*) in New Zealand, 1883–1920." *New Zealand Journal of Ecology* 41, no. 2 (2017): 163–76.

Littin, Kate E., et al. "Behavior and Time to Unconsciousness of Brushtail Pos-
    sums (*Trichosurus vulpecula*) After a Lethal or Sublethal Dose of 1080." *Wildlife
    Research* 36, no. 8 (2009): 709–20.
Mason, G., and K. E. Littin. "The Humaneness of Rodent Pest Control." *Animal
    Welfare* 12, no. 1 (February 2003): 1–37.
Morriss, Grant A., Graham Nugent, and Jackie Whitford. "Dead Birds Found After
    Aerial Poisoning Operations Targeting Small Mammal Pests in New Zealand
    2003–14." *New Zealand Journal of Ecology* 40, no. 3 (January 2016): 361–70.
Robinson, Weldon B. "The 'Humane Coyote-Getter' vs. the Steel Trap in Control
    of Predatory Animals." *Journal of Wildlife Management* 7, no. 2 (April 1943):
    179–89.
Stats NZ. "Conservation Status of Indigenous Land Species." April 17, 2019. https://
    www.stats.govt.nz/indicators/conservation-status-of-indigenous-land-species.
Warburton, Bruce, Nick Poutu, and Ian Domigan. "Effectiveness of the Victor
    Snapback Trap for Killing Stoats." *DOC Science Internal Series* 83. Wellington:
    New Zealand Department of Conservation. October 2002.
Warburton, Bruce, Neville G. Gregory, and Grant Morriss. "Effect of Jaw Shape in
    Kill-Traps on Time to Loss of Palpebral Reflexes in Brushtail Possums." *Journal
    of Wildlife Diseases* 36, no. 1 (2000): 92–96.

15. THE DISAPPEARING MOUSE

Kimball, Bruce, et al. "Development of Artificial Bait for Brown Treesnake Sup-
    pression." *Biological Invasions* 18 (2016): 359–69.
Pitt, William C., et al. "Physical and Behavioral Abilities of Commensal Rodents
    Related to the Design of Selective Rodenticide Bait Stations." *International
    Journal of Pest Management* 57, no. 3 (July–September 2011): 189–93.

THE PALE
HORSEMAN

BOOKS BY BERNARD CORNWELL

The King Alfred Novels

THE LAST KINGDOM

The Sharpe Novels (in chronological order)

SHARPE'S TIGER*
Richard Sharpe and the Siege of Seringapatam, 1799

SHARPE'S TRIUMPH*
Richard Sharpe and the Battle of Assaye, September 1803

SHARPE'S FORTRESS*
Richard Sharpe and the Siege of Gawilghur, December 1803

SHARPE'S TRAFALGAR*
Richard Sharpe and the Battle of Trafalgar, 21 October 1805

SHARPE'S PREY*
Richard Sharpe and the Expedition to Copenhagen, 1807

SHARPE'S RIFLES
Richard Sharpe and the French Invasion of Galicia, January 1809

SHARPE'S HAVOC*
Richard Sharpe and the Campaign in Northern Portugal, Spring 1809

SHARPE'S EAGLE
Richard Sharpe and the Talavera Campaign, July 1809

SHARPE'S GOLD
Richard Sharpe and the Destruction of Almeida, August 1810

SHARPE'S ESCAPE*
Richard Sharpe and the Bussaco Campaign, 1810

SHARPE'S BATTLE*
Richard Sharpe and the Battle of Fuentes de Onoro, May 1811

SHARPE'S COMPANY
Richard Sharpe and the Siege of Badajoz, January to April 1812

SHARPE'S SWORD
Richard Sharpe and the Salamanca Campaign, June and July 1812

SHARPE'S ENEMY
Richard Sharpe and the Defense of Portugal, Christmas 1812

SHARPE'S HONOUR
Richard Sharpe and the Vitoria Campaign, February to June 1813

* Published by HarperCollins*Publishers*

THE PALE
HORSEMAN

Bernard Cornwell

HarperCollins*Publishers*

HarperCollins books may be purchased for educational, business, or sales promotional use. For information, please write: Special Markets Department, HarperCollins Publishers, 10 East 53rd Street, New York, NY 10022.

FIRST EDITION

Printed on acid-free paper

Library of Congress Cataloging-in-Publication Data
Cornwell, Bernard.
    The pale horseman / Bernard Cornwell.—1st ed.
      p. cm.
    Sequel to: The last kingdom.
    ISBN 10: 0-06-078712-0
    ISBN 13: 978-0-06-078712-7
    1. Great Britain—History—Alfred, 871–899—Fiction. 2. Alfred, King of England, 849–899—Fiction. 3. Vikings—Fiction. I. Title.
    PR6053.O75P35   2006
    823'.914—dc22   2005046290

06 07 08 09 10   ❖/RRD   10 9 8 7 6 5 4 3 2 1

THE PALE HORSEMAN

is for
George MacDonald Fraser,
in admiration.

Ac her forþ beraŏ; fugelas singaŏ, gylleŏ græghama.

For here starts war, carrion birds sing,
and gray wolves howl.
(From *The Fight at Finnsburh*)

PLACE-NAMES

The spelling of place-names in Anglo-Saxon England was an uncertain business, with no consistency and no agreement even about the name itself. Thus London was variously rendered as Lundonia, Lundenberg, Lundenne, Lundene, Lundenwic, Lundenceaster, and Lundres. Doubtless some readers will prefer other versions of the names listed below, but I have usually employed whichever spelling is cited in the *Oxford Dictionary of English Place-Names* for the years nearest or contained within Alfred's reign, A.D. 871–899, but even that solution is not foolproof. Hayling Island, in 956, was written as both Heilincigae and Hæglingaiggæ. Nor have I been consistent myself; I use England instead of Englaland, and have preferred the modern form Northumbria to Norðhymbralond to avoid the suggestion that the boundaries of the ancient kingdom coincide with those of the modern county. So this list, like the spellings themselves, is capricious.

Æsc's Hill	Ashdown, Berkshire
Ætheling Aeg	Athelney, Somerset
Afen	River Avon, Wiltshire
Andefera	Andover, Wiltshire
Baðum (pronounced Bathum)	Bath, Avon
Bebbanburg	Bamburgh Castle, Northumberland

Brant	Brent Knoll, Somerset
Bru	River Brue, Somerset
Cippanhamm	Chippenham, Wiltshire
Cracgelad	Cricklade, Wiltshire
Cridianton	Crediton, Devon
Cynuit	Cynuit Hillfort, near Cannington, Somerset
Contwaraburg	Canterbury, Kent
Cornwalum	Cornwall
Dærentmora	Dartmoor, Devon
Defereal	Kingston Deverill, Wiltshire
Defnascir	Devonshire
Dornwaraceaster	Dorchester, Dorset
Dreyndynas	"Fort of thorns," fictional, set in Cornwall
Dunholm	Durham, County Durham
Dyfed	Southwest Wales, mostly now Pembrokeshire
Dyflin	Dublin, Eire
Eoferwic	York (also the Danish Jorvic, pronounced Yorvik)
Ethandun	Edington, Wiltshire
Exanceaster	Exeter, Devon
Exanmynster	Exminster, Devon
Gewæsc	The Wash
Gifle	Yeovil, Somerset
Gleawecestre	Gloucester, Gloucestershire
Glwysing	Welsh kingdom, approximately Glamorgan and Gwent
Hamptonscir	Hampshire
Hamtun	Southampton, Hampshire
Lindisfarena	Lindisfarne (Holy Island), Northumberland

Lundene	London
Lundi	Lundy Island, Devon
Mærlebeorg	Marlborough, Wiltshire
Ocmundtun	Okehampton, Devon
Palfleot	Pawlett, Somerset
Pedredan	River Parrett
Penwith	Land's End, Cornwall
Readingum	Reading, Berkshire
Sæfern	River Severn
Sceapig	Isle of Sheppey, Kent
Scireburnan	Sherborne, Dorset
Sillans	The Scilly Isles
Soppan Byrg	Chipping Sodbury, Gloucestershire
Sumorsæte	Somerset
Suth Seaxa	Sussex (South Saxons)
Tamur	River Tamar
Temes	River Thames
Thon	River Tone, Somerset
Thornsæta	Dorset
Uisc	River Exe
Werham	Wareham, Dorset
Wilig	River Wylye
Wiltunscir	Wiltshire
Winburnan	Wimborne Minster, Dorset
Wintanceaster	Winchester, Hampshire

THE PALE
HORSEMAN

ONE

These days I look at twenty-year-olds and think they are pathetically young, scarcely weaned from their mothers' tits, but when I was twenty I considered myself a full-grown man. I had fathered a child, fought in the shield wall, and was loath to take advice from anyone. In short I was arrogant, stupid, and headstrong. Which is why, after our victory at Cynuit, I did the wrong thing.

We had fought the Danes beside the ocean, where the river runs from the great swamp and the Sæfern Sea slaps on a muddy shore, and there we had beaten them. We had made a great slaughter and I, Uhtred of Bebbanburg, had done my part. More than my part, for at the battle's end, when the great Ubba Lothbrokson, most feared of all the Danish leaders, had carved into our shield wall with his great war ax, I had faced him, beaten him, and sent him to join the *einherjar*, that army of the dead who feast and swive in Odin's corpse hall.

What I should have done then, what Leofric told me to do, was ride hard to Exanceaster where Alfred, King of the West Saxons, was besieging Guthrum. I should have arrived deep in the night, woken the king from his sleep, and laid Ubba's battle banner of the black raven and Ubba's great war ax, its blade still crusted with blood, at Alfred's feet. I should have given the king the good news that the Danish army was beaten, that the few survivors had taken to their dragon-headed ships, that Wessex was safe, and that I, Uhtred of Beb-banburg, had achieved all of those things.

Instead I rode to find my wife and child.

At twenty years old I would rather have been plowing Mildrith than reaping the rewards of my good fortune, and that is what I did wrong, but, looking back, I have few regrets. Fate is inexorable, and Mildrith, though I had not wanted to marry her and though I came to detest her, was a lovely field to plow.

So, in that late spring of the year 877, I spent the Saturday riding to Cridianton instead of going to Alfred. I took twenty men with me and I promised Leofric that we would be at Exanceaster by midday on Sunday and I would make certain Alfred knew we had won his battle and saved his kingdom.

"Odda the Younger will be there by now," Leofric warned me. Leofric was almost twice my age, a warrior hardened by years of fight-ing the Danes. "Did you hear me?" he asked when I said nothing. "Odda the Younger will be there by now," he said again, "and he's a piece of goose shit who'll take all the credit."

"The truth cannot be hidden," I said loftily.

Leofric mocked that. He was a bearded squat brute of a man who should have been the commander of Alfred's fleet, but he was not well born and Alfred had reluctantly given me charge of the twelve ships because I was an ealdorman, a noble, and it was only fitting that a high-born man should command the West Saxon fleet even though it had been much too puny to confront the massive array of Danish ships that had come to Wessex's south coast. "There are times," Leofric grumbled, "when you are an earsling." An earsling was something that had dropped out of a creature's backside and was one of Leofric's favorite insults. We were friends.

"We'll see Alfred tomorrow," I said.

"And Odda the Younger," Leofric said patiently, "has seen him today."

Odda the Younger was the son of Odda the Elder who had given my wife shelter, and the son did not like me. He did not like me because he wanted to plow Mildrith, which was reason enough for him to dislike me. He was also, as Leofric said, a piece of goose shit, slippery and slick, which was reason enough for me to dislike him.

"We shall see Alfred tomorrow," I said again, and next morning we all rode to Exanceaster, my men escorting Mildrith, our son, and his nurse, and we found Alfred on the northern side of Exanceaster where his green-and-white dragon banner flew above his tents. Other banners snapped in the damp wind, a colorful array of beasts, crosses, saints, and weapons announcing that the great men of Wessex were with their king. One of those banners showed a black stag, which confirmed that Leofric had been right and that Odda the Younger was here in south Defnascir. Outside the camp, between its southern margin and the city walls, was a great pavilion made of sailcloth stretched across guyed poles, and that told me that Alfred, instead of fighting Guthrum, was talking to him. They were negotiating a truce, though not on that day, for it was a Sunday and Alfred would do no work on a Sunday if he could help it. I found him on his knees in a makeshift church made from another poled sailcloth, and all his nobles and thegns were arrayed behind him, and some of those men turned as they heard our horses' hooves. Odda the Younger was one of those who turned and I saw the apprehension show on his narrow face.

The bishop who was conducting the service paused to let the congregation make a response, and that gave Odda an excuse to look away from me. He was kneeling close to Alfred, very close, suggesting that he was high in the king's favor, and I did not doubt that he had brought the dead Ubba's raven banner and war ax to Exanceaster and claimed the credit for the fight beside the sea. "One day," I said to Leofric, "I shall slit that bastard from the crotch to the gullet and dance on his offal."

"You should have done it yesterday."

A priest had been kneeling close to the altar, one of the many priests who always accompanied Alfred, and he saw me and slid backward as unobtrusively as he could until he was able to stand and hurry toward me. He had red hair, a squint, a palsied left hand, and

an expression of astonished joy on his ugly face. "Uhtred!" he called
as he ran toward our horses. "Uhtred! We thought you were dead!"

"Me?" I grinned at the priest. "Dead?"

"You were a hostage!"

I had been one of the dozen English hostages in Werham, but
while the others had been murdered by Guthrum, I had been spared
because of Earl Ragnar who was a Danish war-chief and as close to me
as a brother. "I didn't die, father," I said to the priest, whose name
was Beocca, "and I'm surprised you did not know that."

"How could I know it?"

"Because I was at Cynuit, father, and Odda the Younger could
have told you that I was there and that I lived."

I was staring at Odda as I spoke and Beocca caught the grimness
in my voice. "You were at Cynuit?" he asked nervously.

"Odda the Younger didn't tell you?"

"He said nothing."

"Nothing!" I kicked my horse forward, forcing it between the
kneeling men and thus closer to Odda. Beocca tried to stop me, but I
pushed his hand away from my bridle. Leofric, wiser than me, held
back, but I pushed the horse into the back rows of the congregation
until the press of worshippers made it impossible to advance farther,
and then I stared at Odda as I spoke to Beocca. "He didn't describe
Ubba's death?" I asked.

"He says Ubba died in the shield wall," Beocca said, his voice a
hiss so that he did not disturb the liturgy, "and that many men contributed to his death."

"Is that all he told you?"

"He says he faced Ubba himself," Beocca said.

"So who do men think killed Ubba Lothbrokson?" I asked.

Beocca could sense trouble coming and he tried to calm me. "We
can talk of these things later," he said, "but for now, Uhtred, join us
in prayer." He used my name rather than calling me lord because he
had known me since I was a child. Beocca, like me, was a Northumbrian, and he had been my father's priest, but when the Danes took

our country he had come to Wessex to join those Saxons who still resisted the invaders. "This is a time for prayer," he insisted, "not for quarrels."

But I was in a mood for quarrels. "Who do men say killed Ubba Lothbrokson?" I asked again.

"They give thanks to God that the pagan is dead." Beocca evaded my question and tried to hush my voice with frantic gestures from his palsied left hand.

"Who do you think killed Ubba?" I asked, and when Beocca did not answer, I provided the answer for him. "You think Odda the Younger killed him?" I could see that Beocca did believe that, and the anger surged in me. "Ubba fought me man on man," I said, too loudly now, "one on one, just me and him. My sword against his ax. And he was unwounded when the fight began, father, and at the end of it he was dead. He had gone to his brothers in the corpse hall." I was furious now and my voice had risen until I was shouting, and the distracted congregation all turned to stare at me. The bishop, whom I recognized as the bishop of Exanceaster, the same man who had married me to Mildrith, frowned nervously. Only Alfred seemed unmoved by the interruption, but then, reluctantly, he stood and turned toward me as his wife, the pinch-faced Ælswith, hissed into his ear.

"Is there any man here," I was still shouting, "who will deny that I, Uhtred of Bebbanburg, killed Ubba Lothbrokson in single combat?"

There was silence. I had not intended to disrupt the service, but monstrous pride and ungovernable rage had driven me to defiance. The faces gazed at me, the banners flapped in the desultory wind, and the small rain dripped from the edges of the sailcloth awning. Still no one answered me, but men saw that I was staring at Odda the Younger and some looked to him for a response, but he was struck dumb. "Who killed Ubba?" I shouted at him.

"This is not seemly," Alfred said angrily.

"This killed Ubba!" I declared, and I drew Serpent-Breath.

And that was my next mistake.

In the winter, while I was mewed up in Werham as one of the hostages given to Guthrum, a new law had been passed in Wessex, a law which decreed that no man other than the royal bodyguards was to draw a weapon in the presence of the king. The law was not just to protect Alfred, but also to prevent the quarrels between his great men becoming lethal and, by drawing Serpent-Breath, I had unwittingly broken the law so that his household troops were suddenly converging on me with spears and drawn swords until Alfred, red-cloaked and bare-headed, shouted for every man to be still.

Then he walked toward me and I could see the anger on his face. He had a narrow face with a long nose and chin, a high forehead, and a thin-lipped mouth. He normally went clean-shaven, but he had grown a short beard that made him look older. He was not lived thirty years yet, but looked closer to forty. He was painfully thin, and his frequent illnesses had given his face a crabbed look. He looked more like a priest than the king of the West Saxons, for he had the irritated, pale face of a man who spends too much time out of the sun and poring over books, but there was an undoubted authority in his eyes. They were very light eyes, as gray as mail, unforgiving. "You have broken my peace," he said, "and offended the peace of Christ."

I sheathed Serpent-Breath, mainly because Beocca had muttered at me to stop being a damned fool and to put my sword away, and now the priest was tugging my right leg, trying to make me dismount and kneel to Alfred, whom he adored. Ælswith, Alfred's wife, was staring at me with pure scorn. "He should be punished," she called out.

"You will go there," the king said, pointing toward one of his tents, "and wait for my judgment."

I had no choice but to obey, for his household troops, all of them in mail and helmets, pressed about me and so I was taken to the tent where I dismounted and ducked inside. The air smelled of yellowed, crushed grass. The rain pattered on the linen roof and some leaked through onto an altar that held a crucifix and two empty candleholders. The tent was evidently the king's private chapel and Alfred made me wait there a long time. The congregation dispersed, the rain ended, and a watery sunlight emerged between the clouds. A harp played somewhere, perhaps serenading Alfred and his wife as they ate. A dog came into the tent, looked at me, lifted its leg against the altar, and went out again. The sun vanished behind a cloud and more rain pattered on the canvas, then there was a flurry at the tent's opening and two men entered. One was Æthelwold, the king's nephew, and the man who should have inherited Wessex's throne from his father except he had been reckoned too young and so the crown had gone to his uncle instead. He gave me a sheepish grin, deferring to the second man who was heavyset, full-bearded, and ten years older than Æthelwold. He introduced himself by sneezing, then blew his nose into his hand and wiped the snot onto his leather coat. "Call it springtime," he grumbled, then stared at me with a truculent expression. "Damned rain never stops. You know who I am?"

"Wulfhere," I said, "Ealdorman of Wiltunscir." He was a cousin to the king and a leading power in Wessex.

He nodded. "And you know who this damn fool is?" he asked, gesturing at Æthelwold who was holding a bundle of white cloth.

"We know each other," I said. Æthelwold was only a month or so younger than I, and he was fortunate, I suppose, that his uncle Alfred was such a good Christian or else he could have expected a knife in the night. He was much better looking than Alfred, but foolish, flippant, and usually drunk, though he appeared sober enough on that Sunday morning.

"I'm in charge of Æthelwold now," Wulfhere said, "and of you. And the king sent me to punish you." He brooded on that for a heartbeat. "What his wife wants me to do," he went on, "is pull the guts out of your smelly arse and feed them to the pigs." He glared at me. "You know what the penalty is for drawing a sword in the king's presence?"

"A fine?" I guessed.

"Death, you fool, death. They made a new law a month ago."

"How was I supposed to know?"

"But Alfred's feeling merciful." Wulfhere ignored my question.

"So you're not to dangle off a gallows. Not today, anyhow. But he wants your assurance you'll keep the peace."

"What peace?"

"His damned peace, you fool. He wants us to fight the Danes, not slice each other up. So for the moment you have to swear to keep the peace."

"For the moment?"

"For the moment," he said tonelessly, and I just shrugged. He took that for acceptance. "So you killed Ubba?" he asked.

"I did."

"That's what I hear." He sneezed again. "You know Edor?"

"I know him," I said. Edor was one of Ealdorman Odda's battle chiefs, a warrior of the men of Defnascir, and he had fought beside us at Cynuit.

"Edor told me what happened," Wulfhere said, "but only because he trusts me. For God's sake stop fidgeting!" This last shout was directed at Æthelwold who was poking beneath the altar's linen cover, presumably in search of something valuable. Alfred, rather than murder his nephew, seemed intent on boring him to death. Æthelwold had never been allowed to fight, lest he make a reputation for himself; instead he had been forced to learn his letters, which he hated, and so he idled his time away, hunting, drinking, whoring, and filled with resentment that he was not the king. "Just stand still, boy," Wulfhere snarled.

"Edor told you," I said, unable to keep the outrage from my voice, "because he trusts you? You mean what happened at Cynuit is a secret? A thousand men saw me kill Ubba!"

"But Odda the Younger took the credit," Wulfhere said, "and his father is badly wounded and if he dies then Odda the Younger will become one of the richest men in Wessex, and he'll lead more troops and pay more priests than you can ever hope to do, so men won't want to offend him, will they? They'll pretend to believe him, to keep him generous. And the king already believes him, and why shouldn't he? Odda arrived here with Ubba Lothbrokson's banner and war ax.

He dropped them at Alfred's feet, then knelt and gave the praise to God, and promised to build a church and monastery at Cynuit, and what did you do? Ride a damned horse into the middle of mass and wave your sword about. Not a clever thing to do with Alfred."

I half smiled at that, for Wulfhere was right. Alfred was uncommonly pious, and a sure way to succeed in Wessex was to flatter that piety, imitate it, and ascribe all good fortune to God.

"Odda's a prick," Wulfhere growled, surprising me, "but he's Alfred's prick now, and you're not going to change that."

"But I killed . . ."

"I know what you did!" Wulfhere interrupted me. "And Alfred probably suspects you're telling the truth, but he believes Odda made it possible. He thinks Odda and you both fought Ubba. He may not even care if neither of you did, except that Ubba's dead and that's good news, and Odda brought that news and so the sun shines out of Odda's arse, and if you want the king's troops to hang you off a high branch, then you'll make a feud with Odda. Do you understand me?"

"Yes."

Wulfhere sighed. "Leofric said you'd see sense if I beat you over the head long enough."

"I want to see Leofric," I said.

"You can't," Wulfhere said sharply. "He's being sent back to Hamtun where he belongs. But you're not going back. The fleet will be put in someone else's charge. You're to do penance."

For a moment I thought I had misheard. "I'm to do what?" I asked.

"You're to grovel." Æthelwold spoke for the first time. He grinned at me. We were not exactly friends, but we had drunk together often enough and he seemed to like me. "You're to dress like a girl," Æthelwold continued, "go on your knees and be humiliated."

"And you're to do it right now," Wulfhere added.

"I'll be damned . . ."

"You'll be damned anyway," Wulfhere snarled at me, then snatched the white bundle from Æthelwold's grasp and tossed it at my feet. It was a penitent's robe, and I left it on the ground. "For

God's sake, lad," Wulfhere said, "have some sense. You've got a wife and land here, don't you? So what happens if you don't do the king's bidding? You want to be outlawed? You want your wife in a nunnery? You want the church to take your land?"

I stared at him. "All I did was kill Ubba and tell the truth."

Wulfhere sighed. "You're a Northumbrian," he said, "and I don't know how they did things up there, but this is Alfred's Wessex. You can do anything in Wessex except piss all over his church, and that's what you just did. You pissed, son, and now the church is going to piss all over you." He grimaced as the rain beat harder on the tent. Then he frowned, staring at the puddle spreading just outside the entrance. He was silent a long time, before turning and giving me a strange look. "You think any of this is important?"

I did, but I was so astonished by his question that had been asked in a soft, bitter voice, that I had nothing to say.

"You think Ubba's death makes any difference?" he asked, and again I thought I had misheard. "And even if Guthrum makes peace," he went on, "you think we've won?" His heavy face was suddenly savage. "How long will Alfred be king? How long before the Danes rule here?"

I still had nothing to say. Æthelwold, I saw, was listening intently. He longed to be king, but he had no following, and Wulfhere had plainly been appointed as his guardian to keep him from making trouble. But Wulfhere's words suggested the trouble would come anyway. "Just do what Alfred wants," the ealdorman advised me, "and afterward find a way to keep living. That's all any of us can do. If Wessex falls we'll all be looking for a way to stay alive, but in the meantime put on that damned robe and get it over with."

"Both of us," Æthelwold said, and he picked up the robe and I saw he had fetched two of them, folded together.

""You?" Wulfhere snarled at him. "Are you drunk?"

"I'm penitent for being drunk. Or I was drunk, now I'm penitent." He grinned at me, then pulled the robe over his head. "I shall go to the altar with Uhtred," he said, his voice muffled by the linen.

Wulfhere could not stop him, but Wulfhere knew, as I knew, that Æthelwold was making a mockery of the rite. And I knew Æthelwold was doing it as a favor to me, though as far as I knew he owed me no favor. But I was grateful to him, so I put on the damned frock and, side by side with the king's nephew, went to my humiliation.

I meant little to Alfred. He had a score of great lords in Wessex while across the frontier in Mercia there were other lords and thegns who lived under Danish rule but who would fight for Wessex if Alfred gave them an opportunity. All of those great men could bring him soldiers, could rally swords and spears to the dragon banner of Wessex, while I could bring him nothing except my sword, Serpent-Breath. True, I was a lord, but I was from far-off Northumbria and I led no men and so my only value to him was far in the future. I did not understand that yet. In time, as the rule of Wessex spread northward, my value grew, but back then, in 877, when I was an angry twenty-year-old, I knew nothing except my own ambitions.

And I learned humiliation. Even today, a lifetime later, I remember the bitterness of that penitential grovel. Why did Alfred make me do it? I had won him a great victory, yet he insisted on shaming me, and for what? Because I had disturbed a church service? It was partly that, but only part. He loved his god, loved the church, and passionately believed that the survival of Wessex lay in obedience to the church and so he would protect the church as fiercely as he would fight for his country. And he loved order. There was a place for everything and I did not fit and he genuinely believed that if I could be brought to God's heel then I would become part of his beloved order. In short he saw me as an unruly young hound that needed a good whipping before it could join the disciplined pack.

So he made me grovel.

And Æthelwold made a fool of himself.

Not at first. At first it was all solemnity. Every man in Alfred's army was there to watch, and they made two lines in the rain. The lines stretched to the altar under the guyed sailcloth where Alfred and his wife waited with the bishop and a gaggle of priests. "On your knees," Wulfhere said to me. "You have to go on your knees," he insisted tonelessly, "and crawl up to the altar. Kiss the altar cloth, then lie flat."

"Then what?"

"Then God and the king forgive you," he said, and waited. "Just do it," he snarled.

So I did it. I went down on my knees and I shuffled through the mud, and the silent lines of men watched me, and then Æthelwold, close beside me, began to wail that he was a sinner. He threw his arms in the air, fell flat on his face, howled that he was penitent, shrieked that he was a sinner, and at first men were embarrassed and then they were amused. "I've known women!" Æthelwold shouted at the rain. "And they were bad women! Forgive me!"

Alfred was furious, but he could not stop a man making a fool of himself before God. Perhaps he thought Æthelwold's remorse was genuine? "I've lost count of the women!" Æthelwold shouted, then beat his fists in the mud. "Oh God, I love tits! God, I love naked women. God, forgive me for that!" The laughter spread, and every man must have remembered that Alfred, before piety caught him in its clammy grip, had been notorious for the women he had pursued. "You must help me, God!" Æthelwold cried as we shuffled a few feet farther. "Send me an angel!"

"So you can hump her?" a voice called from the crowd and the laughter became a roar.

Ælswith was hurried away, lest she hear something unseemly. The priests whispered together, but Æthelwold's penitence, though extravagant, seemed real enough. He was weeping. I knew he was really laughing, but he howled as though his soul was in agony. "No more tits, God!" he called. "No more tits!" He made a fool of himself, but, as men already thought him a fool, he did not mind. "Keep me from tits, God!" he shouted, and now Alfred left, knowing that the solemnity of the day was ruined, and most of the priests left with him, so that Æthelwold and I crawled to an abandoned altar where

Æthelwold turned in his mud-spattered robe and leaned against the table. "I hate him," he said softly, and I knew he referred to his uncle. "I hate him," he went on, "and now you owe me a favor, Uhtred."

"I do," I said.

"I'll think of one," he said.

Odda the Younger had not left with Alfred. He seemed bemused. My humiliation, which he had surely thought to enjoy, had turned into laughter and he was aware that men were watching him, judging his truthfulness, and he moved closer to a huge man who was evidently one of his bodyguards. That man was tall and very broad about the chest, but it was his face that commanded attention for it looked as though his skin had been stretched too tight across his skull, leaving his face incapable of making any expression other than one of pure hatred and wolfish hunger. Violence came off the man like the stench of a wet hound and when he looked at me it was like a beast's soulless stare, and I instinctively understood that this was the man who would kill me if Odda found a chance to commit murder. Odda was nothing, a rich man's spoiled son, but his money gave him the means to command men who were killers. Then Odda plucked at the tall man's sleeve and they both turned and walked away.

Father Beocca had stayed by the altar. "Kiss it," he ordered me, "then lie flat."

I stood up instead. "You can kiss my arse, father," I said. I was angry, and my anger frightened Beocca who backed away.

But I had done what the king wanted. I had been penitent.

The tall man beside Odda the Younger was named Steapa. Steapa Snotor, men called him, or Steapa the Clever. "It's a joke," Wulfhere told me as I ripped off the penitent's frock and pulled on my mail coat.

"A joke?"

"Because he's dumb as an ox," Wulfhere said. "He's got frog spawn instead of a brain. He's stupid, but he's not a stupid fighter. You didn't see him at Cynuit?"

"No," I said curtly.

"So what's Steapa to you?" Wulfhere asked.

"Nothing," I said. I had asked the ealdorman who Odda's body-guard was so that I could learn the name of the man who might try to kill me, but that possible murder was none of Wulfhere's business.

Wulfhere hesitated, wanting to ask more, then deciding he would fetch no better answer. "When the Danes come," he said, "you'll be welcome to join my men."

Æthelwold, Alfred's nephew, was holding my two swords and he drew Serpent-Breath from her scabbard and stared at the wispy patterns in her blade. "If the Danes come," he spoke to Wulfhere, "you must let me fight."

"You don't know how to fight."

"Then you must teach me." He slid Serpent-Breath back into the scabbard. "Wessex needs a king who can fight," he said, "instead of pray."

"You should watch your tongue, lad," Wulfhere said, "in case it gets cut out." He snatched the swords from Æthelwold and gave them to me. "The Danes will come," he said, "so join me when they do."

I nodded, but said nothing. When the Danes came, I thought, I planned to be with them. I had been raised by Danes after being captured at the age of ten and they could have killed me, but instead they had treated me well. I had learned their language and worshipped their gods until I no longer knew whether I was Danish or English. Had Earl Ragnar the Elder lived I would never have left them, but he had died, murdered in a night of treachery and fire, and I had fled south to Wessex. But now I would go back. Just as soon as the Danes left Exanceaster I would join Ragnar's son, Ragnar the Younger, if he lived. Ragnar the Younger's ship had been in the fleet that had been hammered in the great storm. Scores of ships had sunk, and the remnants of the fleet had limped to Exanceaster where the boats were now burned to ash on the riverbank beneath the town. I did not know if Ragnar lived. I hoped he lived, and I prayed he would escape Exanceaster and then I would go to him, offer him my sword, and carry that blade against Alfred of Wessex. Then, one day, I would dress Alfred in a frock and make him crawl on his knees to an altar of Thor. Then kill him.

Those were my thoughts as we rode to Oxton. That was the estate Mildrith had brought me in marriage and it was a beautiful place, but so saddled with debt that it was more of a burden than a pleasure. The farmland was on the slopes of hills facing east toward the broad sea-reach of the Uisc and above the house were thick woods of oak and ash from which flowed small clear streams that cut across the fields where rye, wheat, and barley grew. The house—it was not a hall—was a smoke-filled building made from mud, dung, oak, and rye straw, and so long and low that it looked like a green, moss-covered mound from which smoke escaped through the roof's central hole. In the attached yard were pigs, chickens, and mounds of manure as big as the house. Mildrith's father had farmed it, helped by a steward named Oswald who was a weasel, and he caused me still more trouble on that rainy Sunday as we rode back to the farm.

I was furious, resentful, and vengeful. Alfred had humiliated me, which made it unfortunate for Oswald that he had chosen that Sunday afternoon to drag an oak tree down from the high woods. I was brooding on the pleasures of revenge as I let my horse pick its way up the track through the trees and saw eight oxen hauling the great trunk toward the river. Three men were goading the oxen, while a fourth, Oswald, rode the trunk with a whip. He saw me and jumped off and, for a heartbeat, it looked as if he wanted to run into the trees, but then he realized he could not evade me and so he just stood and waited as I rode up to the great oak log.

"Lord," Oswald greeted me. He was surprised to see me. He probably thought I had been killed with the other hostages, and that belief had made him careless.

My horse was nervous because of the stink of blood from the oxen's flanks and he stepped backward and forward in small steps until I calmed him by patting his neck. Then I looked at the oak trunk that must have been forty feet long and as thick about as a man is tall. "A fine tree," I said to Oswald.

He glanced toward Mildrith who was twenty paces away. "Good

day, lady," he said, clawing off the woolen hat he wore over his springy red hair.

"A wet day, Oswald," she said. Her father had appointed the steward and Mildrith had an innocent faith in his reliability.

"I said," I spoke loudly, " a fine tree. So where was it felled?"

Oswald tucked the hat into his belt. "On the top ridge, lord," he said vaguely.

"The top ridge on my land?"

He hesitated. He was doubtless tempted to claim it came from a neighbor's land, but that lie could easily have been exposed and so he said nothing.

"From my land?" I asked again.

"Yes, lord," he admitted.

"And where is it going?"

He hesitated again, but had to answer. "Wigulf's mill."

"Wigulf buys it?"

"He'll split it, lord."

"I didn't ask what he will do with it," I said, "but whether he will buy it."

Mildrith, hearing the harshness in my voice, intervened to say that her father had sometimes sent timber to Wigulf's mill, but I waved her to silence. "Will he buy it?" I asked Oswald.

"We need the timber, lord, to make repairs," the steward said, "and Wigulf takes his fee in split wood."

"And you drag the tree on a Sunday?" He had nothing to say to that. "Tell me," I went on, "if we need planks for repairs, then why don't we split the trunk ourselves? Do we lack men? Or wedges? Or mauls?"

"Wigulf has always done it," Oswald said in a surly tone.

"Always?" I repeated and Oswald said nothing. "Wigulf lives in Exanmynster?" I guessed. Exanmynster lay a mile or so northward and was the nearest settlement to Oxton.

"Yes, lord," Oswald said.

"So if I ride to Exanmynster now," I said, "Wigulf will tell me how many similar trees you've delivered to him in the last year?"

There was silence, except for the rain dripping from leaves and the intermittent burst of birdsong. I edged my horse a few steps closer to Oswald, who gripped his whip's handle as if readying it to lash out at me. "How many?" I asked.

Oswald said nothing.

"How many?" I demanded, louder.

"Husband," Mildrith called.

"Quiet!" I shouted at her and Oswald looked from me to her and back to me. "And how much has Wigulf paid you?" I asked. "What does a tree like this fetch? Eight shillings? Nine?"

The anger that had made me act so impetuously at the king's church service rose again. It was plain that Oswald was stealing the timber and being paid for it, and what I should have done was charge him with theft and have him arraigned before a court where a jury of men would decide his guilt or innocence, but I was in no mood for such a process. I just drew Serpent-Breath and kicked my horse forward. Mildrith screamed a protest, but I ignored her. Oswald ran, and that was a mistake, because I caught him easily, and Serpent-Breath swung once and opened up the back of his skull so I could see brains and blood as he fell. He twisted in the leaf mold and I wheeled the horse back and stabbed down into his throat.

"That was murder!" Mildrith shouted at me.

"That was justice," I snarled at her, "something lacking in Wessex." I spat on Oswald's body that was still twitching. "The bastard's been stealing from us."

Mildrith kicked her horse, leading the nurse who carried our child uphill. I let her go. "Take the trunk up to the house," I ordered the slaves who had been goading the oxen. "If it's too big to drag uphill, then split it here and take the planks to the house."

I searched Oswald's house that evening and discovered fifty-three shillings buried in the floor. I took the silver, confiscated his cooking pots, spit, knives, buckles, and a deerskin cloak, then drove his wife and three children off my land. I had come home.

TWO

My anger was not slaked by Oswald's killing. The death of a
dishonest steward was no consolation for what I perceived
as a monstrous injustice. For the moment Wessex was safe from the
Danes, but it was only safe because I had killed Ubba Lothbrokson and
my reward had been humiliation.

Poor Mildrith. She was a peaceable woman who thought well of
everyone she met, and now she found herself married to a resentful,
angry warrior. She was frightened of Alfred's wrath, terrified that the
church would punish me for disturbing its peace, and worried that
Oswald's relatives would demand a wergild from me. And so they
would. A wergild was the blood price that every man, woman, and
child possessed. Kill a man and you must pay his price or else die
yourself, and I had no doubt that Oswald's family would go to Odda
the Younger, who had been named the Ealdorman of Defnascir
because his father was too badly wounded to continue as ealdorman,
and Odda would instruct the shire reeve to pursue me and place me
on trial, but I did not care. I hunted boar and deer, I brooded and
waited for news of the negotiations at Exanceaster. I was expecting
Alfred to do what he always did, which was to make peace with the
Danes and so release them, and when he did I would go to Ragnar.

And as I waited I found my first retainer. He was a slave and I discovered him in Exanmynster on a fine spring day. There was a hiring fair where men looked for employment through the busy days of hay-making and harvest, and like all fairs there were jugglers, storytellers, stilt walkers, musicians, and acrobats. There was also a tall, white-haired man with a lined, serious face, who was selling enchanted leather bags that turned iron into silver. He showed us how it was done, and I saw him place two common nails into the bag and a moment later they were pure silver. He said we had to place a silver crucifix in the bag and then sleep one night with it tied around our necks before the magic worked and I paid him three silver shillings for one bag, and it never worked. I spent months searching for the man, but never found him. Even these days I come across such men and women, selling sorcerous pouches or boxes, and now I have them whipped and run off my land, but I was only twenty then and I believed my own eyes. That man had attracted a large crowd, but there were even more people gathered by the church gate where shouts erupted every few minutes. I pushed my horse into their rear ranks, getting dirty looks from folk who knew I had killed Oswald, but none dared accuse me of the murder for I carried both Serpent-Breath and Wasp-Sting.

A young man was by the church gate. He was stripped to the waist, barefooted, and had a rope round his neck, and the rope was tied to the gatepost. In his hand was a short, stout stave. He had long unbound fair hair, blue eyes, a stubborn face, and blood all over his chest, belly, and arms. Three men guarded him. They, too, were fair-haired and blue-eyed, and they shouted in a strange accent. "Come and fight the heathen! Three pennies to make the bastard bleed! Come and fight!"

"Who is he?" I asked.

"A Dane, lord, a pagan Dane." The man tugged off his hat when he spoke to me, then turned back to the crowd. "Come and fight him! Get your revenge! Make a Dane bleed! Be a good Christian! Hurt a pagan!"

The three men were Frisians. I suspected they had been in Alfred's army and, now that he was talking to the Danes rather than fighting them, the three had deserted. Frisians come from across the sea and they come for one reason only, money, and this trio had somehow captured the young Dane and were profiting from him so long as he lasted. And that could have been some time, for he was good. A strong young Saxon paid his three pence and was given a sword with which he hacked wildly at the prisoner, but the Dane parried every blow, wood chips flying from his stave, and when he saw an opening he cracked his opponent around the head hard enough to draw blood from his ear. The Saxon staggered away, half stunned, and the Dane rammed the stave into his belly and, as the Saxon bent to gasp for breath, the stave whistled around in a blow that would have cracked his skull open like an egg, but the Frisians dragged on the rope so that the Dane fell backward. "Do we have another hero?" a Frisian shouted as the young Saxon was helped away. "Come on, lads! Show your strength! Beat a Dane bloody!"

"I'll beat him," I said. I dismounted and pushed through the crowd. I gave my horse's reins to a boy, then drew Serpent-Breath. "Three pence?" I asked the Frisians.

"No, lord," one of them said.

"Why not?"

"We don't want a dead Dane, do we?" The man answered.

"We do!" someone shouted from the crowd. The folk in the Uisc valley did not like me, but they liked the Danes even less and they relished the prospect of watching a prisoner being slaughtered.

"You can only wound him, lord," the Frisian said. "And you must use our sword." He held out the weapon. I glanced at it, saw its blunt edge, and spat.

"Must?" I asked.

The Frisian did not want to argue. "You can only draw blood, lord," he said.

The Dane flicked hair from his eyes and watched me. He held the stave low. I could see he was nervous, but there was no fear in his eyes. He had probably fought a hundred battles since the Frisians cap-

tured him, but those fights had been against men who were not soldiers, and he must have known, from my two swords, that I was a
warrior. His skin was blotched with bruises and laced by blood and
scars, and he surely expected another wound from Serpent-Breath,
but he was determined to give me a fight.

"What's your name?" I asked in Danish.

He blinked at me, surprised.

"Your name, boy," I said. I called him "boy," though he was not
much younger than I.

"Haesten," he said.

"Haesten who?"

"Haesten Storrison," he said, giving me his father's name.

"Fight him! Don't talk to him!" a voice shouted from the crowd.

I turned to stare at the man who had shouted and he could not
meet my gaze, then I turned fast, very fast, and whipped Serpent-
Breath in a quick sweep that Haesten instinctively parried so that
Serpent-Breath cut through the stave as if it was rotten. Haesten was
left with a stub of wood, while the rest of his weapon, a yard of thick
ash, lay on the ground.

"Kill him!" someone shouted.

"Just draw blood, lord," a Frisian said, "please, lord. He's not a
bad lad, for a Dane. Just make him bleed and we'll pay you."

I kicked the ash stave away from Haesten. "Pick it up," I said.

He looked at me nervously. To pick it up he would have to go to
the end of his tether, then stoop, and at that moment he would
expose his back to Serpent-Breath. He watched me, his eyes bitter
beneath the fringe of dirty hair, then decided I would not attack him
as he bent over. He went to the stave and, as he leaned down, I kicked
it a few inches farther away. "Pick it up," I ordered him again.

He still held the stub of ash and, as he took a further step, straining against the rope, he suddenly whipped around and tried to ram
the broken end into my belly. He was fast, but I had half expected the
move and caught his wrist in my left hand. I squeezed hard, hurting
him. "Pick it up," I said a third time.

This time he obeyed, stooping to the stave, and to reach it he stretched his tether tight and I slashed Serpent-Breath onto the taut rope, severing it. Haesten, who had been straining forward, fell onto his face as the hide rope was cut. I put my left foot onto his back and let the tip of Serpent-Breath rest on his spine. "Alfred," I said to the Frisians, "has ordered that all Danish prisoners are to be taken to him."

The three looked at me, said nothing.

"So why have you not taken this man to the king?" I demanded.

"We didn't know, lord," one of them said. "No one told us," which was not surprising because Alfred had given no such order.

"We'll take him to the king now, lord," another reassured me.

"I'll save you the trouble," I said. I took my foot off Haesten. "Get up," I told him in Danish. I threw a coin to the boy holding my horse and hauled myself into the saddle where I offered Haesten a hand. "Get up behind me," I ordered him.

The Frisians protested, coming at me with their swords drawn, so I pulled Wasp-Sting from her scabbard and gave it to Haesten who had still not mounted. Then I turned the horse toward the Frisians and smiled at them. "These people," I waved Serpent-Breath at the crowd, "already think I am a murderer. I'm also the man who met Ubba Lothbrokson beside the sea and killed him there. I tell you this so you may boast that you killed Uhtred of Bebbanburg."

I lowered the sword so it pointed at the nearest man and he backed away. The others, no more eager to fight than the first, went with him. Haesten then pulled himself up behind me and I spurred the horse into the crowd that parted reluctantly. Once free of them I made Haesten dismount and give me back Wasp-Sting. "How did you get captured?" I asked him.

He told me he had been on one of Guthrum's ships caught in the storm, and his ship had sunk, but he had clung to some wreckage and been washed ashore where the Frisians had found him. "There were two of us, lord," he said, "but the other died."

"You're a free man now," I told him.

"Free?"

"You're my man," I said, "and you'll give me an oath, and I'll give you a sword."

"Why?" he wanted to know.

"Because a Dane saved me once," I said, "and I like the Danes."

I also wanted Haesten because I needed men. I did not trust Odda the Younger, and I feared Steapa Snotor, Odda's warrior, and so I would have swords at Oxton. Mildrith, of course, did not want sword Danes at her house. She wanted plowmen and peasants, milkmaids and servants, but I told her I was a lord, and a lord has swords.

I am indeed a lord, a lord of Northumbria. I am Uhtred of Bebbanburg. My ancestors, who can trace their lineage back to the god Woden, the Danish Odin, were once kings in northern England, and if my uncle had not stolen Bebbanburg from me when I was just ten years old, I would have lived there still as a Northumbrian lord safe in his sea-washed fastness. The Danes had captured Northumbria, and their puppet king, Ricsig, ruled in Eoferwic, but Bebbanburg was too strong for any Dane and my uncle Ælfric ruled there, calling himself Ealdorman Ælfric, and the Danes left him in peace so long as he did not trouble them, and I often dreamed of going back to Northumbria to claim my birthright. But how? To capture Bebbanburg I would need an army, and all I had was one young Dane, Haesten.

And I had other enemies in Northumbria. There was Earl Kjartan and his son Sven, who had lost an eye because of me, and they would kill me gladly, and my uncle would pay them to do it, and so I had no future in Northumbria, not then. But I would go back. That was my soul's wish, and I would go back with Ragnar the Younger, my friend, who still lived because his ship had weathered the storm. I heard that from a priest who had listened to the negotiations outside Exanceaster and he was certain that Earl Ragnar had been one of the Danish lords in Guthrum's delegation. "A big man," the priest told me, "and very loud." That description convinced me that Ragnar lived and my heart was glad for it, for I knew that my future lay with him, not with Alfred. When the negotiations were finished and a truce made the Danes would doubtless leave Exanceaster and I would give my sword to Ragnar and carry it against Alfred, who hated me. And I hated him.

I told Mildrith that we would leave Defnascir and go to Ragnar, that I would be his man and that I would pursue my blood feud against Kjartan and against my uncle under Ragnar's eagle banner, and Mildrith responded with tears and more tears.

I cannot bear a woman's crying. Mildrith was hurt and she was confused and I was angry and we snarled at each other like wildcats and the rain kept falling and I raged like a beast in a cage and wished Alfred and Guthrum would finish their talking because everyone knew that Alfred would let Guthrum go, and once Guthrum left Exanceaster then I could join the Danes and I did not care whether Mildrith came or not, so long as my son, who bore my name, went with me. So by day I hunted, at night I drank and dreamed of revenge, and then one evening I came home to find Father Willibald waiting in the house.

Willibald was a good man. He had been chaplain to Alfred's fleet when I commanded those twelve ships, and he told me he was on his way back to Hamtun, but he thought I would like to know what had unfolded in the long talks between Alfred and Guthrum. "There is peace, lord," he told me. "Thanks be to God, there is peace."

"Thanks be to God," Mildrith echoed.

I was cleaning the blood from the blade of a boar spear and said nothing. I was thinking that Ragnar was released from the siege now and I could join him.

"The treaty was sealed with solemn oaths yesterday," Willibald said, "and so we have peace."

"They gave each other solemn oaths last year," I said sourly. Alfred and Guthrum had made peace at Werham, but Guthrum had broken the truce and murdered the hostages he had been holding. Eleven of the twelve had died, and only I had lived because Ragnar was there to protect me. "So what have they agreed?" I asked.

"The Danes are to give up all their horses," Willibald said, "and march back into Mercia."

Good, I thought, because that was where I would go. I did not say that to Willibald, but instead sneered that Alfred was just letting them march away. "Why doesn't he fight them?" I asked.

"Because there are too many, lord. Because too many men would die on both sides."

"He should kill them all."

"Peace is better than war," Willibald said.

"Amen," Mildrith said.

I began sharpening the spear, stroking the whetstone down the long blade. It seemed to me that Alfred had been absurdly generous. Guthrum, after all, was the one remaining leader of any stature on the Danish side, and he had been trapped, and if I had been Alfred there would have been no terms, only a siege, and at its end the Danish power in southern England would have been broken. Instead Guthrum was to be allowed to leave Exanceaster. "It is the hand of God," Willibald said.

I looked at him. He was a few years older than I was, but always seemed younger. He was earnest, enthusiastic, and kind. He had been a good chaplain to the twelve ships, though the poor man was ever seasick and blanched at the sight of blood. "God made the peace?" I asked skeptically.

"Who sent the storm that sank Guthrum's ships?" Willibald retorted fervently. "Who delivered Ubba into our hands?"

"I did," I said.

He ignored that. "We have a godly king, lord," he said, "and God rewards those who serve him faithfully. Alfred has defeated the Danes! And they see it! Guthrum can recognize divine intervention! He has been making enquiries about Christ."

I said nothing.

"Our king believes," the priest went on, "that Guthrum is not far from seeing the true light of Christ." He leaned forward and touched my knee. "We have fasted, lord," he said, "we have prayed, and the king believes that the Danes will be brought to Christ, and when that happens there will be a permanent peace."

He meant every word of that nonsense and, of course, it was sweet music to Mildrith's ears. She was a good Christian and had great faith in Alfred, and if the king believed that his god would bring victory, then she would believe it, too. It seemed madness to me, but I said nothing as a servant brought us barley ale, bread, smoked mackerel, and cheese. "We shall have a Christian peace," Willibald said, making the sign of the cross above the bread before he ate, "sealed by hostages."

"We've given Guthrum hostages again?" I asked, astonished.

"No," Willibald said. "But he has agreed to give us hostages. Including six earls!"

I stopped sharpening the spear and looked at Willibald. "Six earls?"

"Including your friend, Ragnar!" Willibald seemed pleased by this news, but I was appalled. If Ragnar was not with the Danes, then I could not go to them. He was my friend and his enemies were my enemies, but without Ragnar to protect me I would be horribly vulnerable to Kjartan and Sven, the father and son who had murdered Ragnar's father and who wanted me dead. Without Ragnar, I knew, I could not leave Wessex.

"Ragnar's one of the hostages?" I asked. "You're sure?"

"Of course I'm sure. He will be held by Ealdorman Wulfhere. All the hostages are to be held by Wulfhere."

"For how long?"

"For as long as Alfred wishes, or until Guthrum is baptized. And Guthrum has agreed that our priests can talk to his men." Willibald gave me a pleading look. "We must have faith in God," he said. "We must give God time to work on the hearts of the Danes. Guthrum understands now that our god has power!"

I stood and went to the door, pulling aside the leather curtain and staring down at the wide sea-reach of the Uisc. I was sick at heart. I hated Alfred, did not want to be in Wessex, but now it seemed I was doomed to stay there. "And what do I do?" I asked.

"The king will forgive you, lord," Willibald said nervously.

"Forgive me?" I turned on him. "And what does the king believe happened at Cynuit? You were there, father," I said, "so did you tell him?"

"I told him."

"And?"

"He knows you are a brave warrior, lord," Willibald said, "and that your sword is an asset to Wessex. He will receive you again, I'm sure, and he will receive you joyfully. Go to church, pay your debts, and show that you are a good man of Wessex."

"I'm not a West Saxon," I snarled at him. "I'm a Northumbrian!"

And that was part of the problem. I was an outsider. I spoke a different English. The men of Wessex were tied by family, and I came from the strange north and folk believed I was a pagan, and they called me a murderer because of Oswald's death, and sometimes, when I rode about the estate, men would make the sign of the cross to avert the evil they saw in me. They called me Uhtredærwe, which means Uhtred the Wicked, and I was not unhappy with the insult, but Mildrith was. She assured them I was a Christian, but she lied, and our unhappiness festered all that summer. She prayed for my soul, I fretted for my freedom, and when she begged me to go with her to the church at Exanmynster I growled at her that I would never set foot in another church all my days. She would weep when I said that and her tears drove me out of the house to hunt, and sometimes the chase would take me down to the water's edge where I would stare at *Heahengel*.

She lay canted on the muddy foreshore, lifted and dropped repeatedly by the tides, abandoned. She was one of Alfred's fleet, one of the twelve large warships he had built to harry the Danish boats that raided Wessex's coast, and Leofric and I had brought *Heahengel* up from Hamtun in pursuit of Guthrum's fleet and we had survived the storm that sent so many Danes to their deaths and we had beached *Heahengel* here, left her mastless and without a sail, and she was still on the Uisc's foreshore, rotting and apparently forgotten.

Archangel. That was what her name meant. Alfred had named her and I had always hated the name. A ship should have a proud name, not a sniveling religious word, and she should have a beast on her prow, high and defiant, a dragon's head to challenge the sea or a snarling wolf to terrify an enemy. I sometimes climbed on board *Heahengel* and saw how the local villagers had plundered some of her upper strakes, and how there was water in her belly, and I remembered her proud days at sea and the wind whipping through her sealhide rigging and the crash as we had rammed a Danish boat.

Now, like me, *Heahengel* had been left to decay, and sometimes I dreamed of repairing her, of finding new rigging and a new sail, of finding men and taking her long hull to sea. I wanted to be anywhere but where I was, I wanted to be with the Danes, and every time I said that Mildrith would weep again. "You can't make me live among the Danes!"

"Why not? I did."

"They're pagans! My son won't grow up a pagan!"

"He's my son, too," I said, "and he will worship the gods I worship." There would be more tears then, and I would storm out of the house and take the hounds up to the high woods and wonder why love soured like milk. After Cynuit I had so wanted to see Mildrith, yet now I could not abide her misery and piety and she could not endure my anger. All she wanted me to do was till my fields, milk my cows, and gather my harvest to pay the great debt she had brought me in marriage. That debt came from a pledge made by Mildrith's father, a pledge to give the church the yield of almost half his land. That pledge was for all time, binding on his heirs, but Danish raids and bad harvests had ruined him. Yet the church, venomous as serpents, still insisted that the debt be paid, and said that if I could not pay then our land would be taken by monks, and every time I went to Exanceaster I could sense the priests and monks watching me and enjoying the prospect of their enrichment. Exanceaster was English again for Guthrum had handed over the hostages and gone north so that peace of a sort had come to Wessex. The fyrds, the armies of each shire, had been disbanded and sent back to their farms. Psalms were being sung in all the churches and Alfred, to mark his victory, was sending gifts to every monastery and nunnery. Odda the Younger, who was being celebrated as the champion of Wessex, had been given all the land about the place where the battle had been fought at Cynuit and he had ordered a church to be built there, and it was rumored that the church would have an altar of gold as thanks to God for allowing Wessex to survive.

Though how long would it survive? Guthrum lived and I did not share the Christian belief that God had sent Wessex peace. Nor was I the only one, for in midsummer Alfred returned to Exanceaster where he summoned his witan, a council of the kingdom's leading thegns and churchmen, and Wulfhere of Wiltunscir was one of the men summoned and I went into the city one evening and was told the ealdorman and his followers had lodgings in The Swan, a tavern by the east gate. He was not there, but Æthelwold, Alfred's nephew, was doing his best to drain the tavern of ale. "Don't tell me the bastard summoned you to the witan?" he greeted me sourly. The "bastard" was Alfred who had snatched the throne from the young Æthelwold.

"No," I said. "I came to see Wulfhere."

"The ealdorman is in church," Æthelwold said, "and I am not." He grinned and waved to the bench opposite him. "Sit and drink. Get drunk. Then we'll find two girls. Three, if you like. Four, if you want?"

"You forget I'm married," I said.

"As if that ever stopped anyone."

I sat and one of the maids brought me ale. "Are you in the witan?" I asked Æthelwold.

"What do you think? You think that bastard wants my advice? 'Lord king,' I'd say, 'why don't you jump off a high cliff and pray that God gives you wings.'" He pushed a plate of pork ribs toward me. "I'm here so they can keep an eye on me. They're making sure I'm not plotting treason."

"Are you?"

"Of course I am." He grinned. "Are you going to join me? You do owe me a favor."

"You want my sword at your service?" I asked.

"Yes." He was serious.

"So it's you and me," I said, "against all Wessex. Who else will fight with us?"

He frowned, thinking, but came up with no names. He stared down at the table and I felt sorry for him. I had always liked Æthelwold, but no one would ever trust him for he was as careless as he was irresponsible. Alfred, I thought, had judged him right. Let him be free and he would drink and whore himself into irrelevance. "What I should do," he said, "is go and join Guthrum."

"Why don't you?"

He looked up at me, but had no answer. Maybe he knew the answer, that Guthrum would welcome him, honor him, use him, and eventually kill him. But maybe that was a better prospect than his present life. He shrugged and leaned back, pushing hair off his face. He was a startlingly handsome young man, and that, too, distracted him for girls were attracted to him like priests to gold. "What Wulfhere thinks," he said, his voice slurring slightly, "is that Guthrum is going to come and kill us all."

"Probably," I said.

"And if my uncle dies," he said, not bothering to lower his voice even though there were a score of men in the tavern, "his son is much too young to be king."

"True."

"So it'll be my turn!" He smiled.

"Or Guthrum's turn," I said.

"So drink, my friend," he said, "because we're all in the cesspit." He grinned at me, his charm suddenly evident. "So if you won't fight for me," he asked, "how do you propose to pay back the favor?"

"How would you like it paid?"

"You could kill Abbot Hewald? Very nastily? Slowly?"

"I could do that," I said. Hewald was abbot at Winburnan and famous for the harshness with which he taught boys to read.

"On the other hand," Æthelwold went on, "I'd like to kill that scrawny bastard myself, so don't do it for me. I'll think of something that won't make my uncle happy. You don't like him, do you?"

"No."

"Then we'll brew up some mischief. Oh God." This last impreca-
tion was because Wulfhere's voice was suddenly loud just outside the
door. "He's angry at me."

"Why?"

"One of the dairy maids is pregnant. I think he wanted to do it
himself, but I churned her first." He drained his ale. "I'm going to the
Three Bells. Want to come?"

"I have to speak to Wulfhere."

Æthelwold left by the back door as the ealdorman ducked
through the front. Wulfhere was accompanied by a dozen thegns, but
he saw me and crossed the room. "They've been reconsecrating the
bishop's church," he grumbled. "Hours upon damned hours! Nothing
but chanting and prayers, hours of prayers just to get the taint of the
Danes out of the place." He sat heavily. "Did I see Æthelwold here?"

"Yes."

"Wanted you to join his rebellion, did he?"

"Yes."

"Damned fool. So why are you here? Come to offer me your
sword?" He meant swear my allegiance to him and so become his
warrior.

"I want to see one of the hostages," I said, "so I seek your permission."

"Hostages." He sat down opposite me and snapped his fingers for
ale. "Damned hostages. I've had to make new buildings to house
them. And who pays for that?"

"You do?"

"Of course I do. And I'm supposed to feed them, too? Feed them?
Guard them? Wall them in? And does Alfred pay anything?"

"Tell him you're building a monastery," I suggested.

He looked at me as if I was mad, then saw the jest and laughed.
"True enough, he'd pay me then, wouldn't he? Have you heard about
the monastery they're building at Cynuit?"

"I hear it's to have an altar of gold."

He laughed again. "That's what I hear. I don't believe it, but I hear it." He watched one of the tavern girls cross the floor. "It's not my permission you need to see the hostages," he said, "but Alfred's, and he won't give it to you."

"Alfred's permission?" I asked.

"They're not just hostages," he said, "but prisoners. I have to wall them in and watch them day and night. Alfred's orders. He might think God brought us peace, but he's made damn sure he's got high-born hostages. Six earls! You know how many retainers they have? How many women? How many mouths to feed?"

"If I go to Wiltunscir," I said, "can I see Earl Ragnar?"

Wulfhere frowned at me. "Earl Ragnar? The noisy one? I like him. No, lad, you can't, because no one's allowed to see them except a damned priest who talks their language. Alfred sent him and he's trying to make them into Christians, and if you go without my permission then Alfred will hear you've been there and he'll want an explanation from me. No one can see the poor bastards." He paused to scratch at a louse under his collar. "I have to feed the priest, too, and Alfred doesn't pay for that, either. He doesn't even pay me to feed that lout Æthelwold!"

"When I was a hostage in Werham," I explained, "Earl Ragnar saved my life. Guthrum killed the others, but Ragnar guarded me. He said they'd have to kill him before they killed me."

"And he looks like a hard man to kill," Wulfhere said, "but if Guthrum attacks Wessex that's what I'm supposed to do. Kill the lot of them. Maybe not the women." He stared gloomily into the tavern's yard where a group of his men were playing dice in the moonlight. "And Guthrum will attack," he added in a low voice.

"That's not what I hear."

He looked at me suspiciously "And what do you hear, young man?"

"That God has sent us peace."

Wulfhere laughed at my mockery. "Guthrum's in Gleawecestre," he said, "and that's just a half day's march from our frontier. And they say more Danish ships arrive every day. They're in Lundene, they're in the Humber, they're in the Gewæsc." He scowled. "More ships, more men, and Alfred's building churches! And there's this fellow Svein."

"Svein?"

"Brought his ships from Ireland. Bastard's in Wales now, but he won't stay there, will he? He'll come to Wessex. And they say more Danes are joining him from Ireland." He brooded on this bad news. I did not know whether it was true, for such rumors were ever current, but Wulfhere plainly believed it. "We should march on Gleawecestre," he said, "and slaughter the lot of them before they slaughter us, but we've got a kingdom ruled by priests."

That was true, I thought, just as it was certain that Wulfhere would not make it easy for me to see Ragnar. "Will you give a message to Ragnar?" I asked.

"How? I don't speak Danish. I could ask the priest, but he'll tell Alfred."

"Does Ragnar have a woman with him?" I asked.

"They all do."

"A thin girl," I said, "black hair. Face like a hawk."

He nodded cautiously. "Sounds right. Has a dog, yes?"

"She has a dog," I said, "and its name is Nihtgenga."

He shrugged as if he did not care what the dog was called. Then he understood the significance of the name. "An English name?" he asked. "A Danish girl calls her dog Goblin?"

"She isn't Danish," I said. "Her name is Brida, and she's a Saxon."

He stared at me, then laughed. "The cunning little bitch. She's been listening to us, hasn't she?"

Brida was indeed cunning. She had been my first lover, an East Anglian girl who had been raised by Ragnar's father and who now slept with Ragnar. "Talk to her," I said, "and give her my greetings, and say that if it comes to war . . ." I paused, not sure what to say. There was no point in promising to do my best to rescue Ragnar, for if war came then the hostages would be slaughtered long before I could reach them.

"If it comes to war?" Wulfhere prompted me.

"If it comes to war," I said, repeating the words he had spoken to me before my penance, "we'll all be looking for a way to stay alive."

Wulfhere stared at me for a long time and his silence told me that though I had failed to find a message for Ragnar, I had given a message to Wulfhere. He drank ale. "So the bitch speaks English, does she?"

"She's a Saxon."

As was I, but I hated Alfred and I would join Ragnar when I could, if I could, whatever Mildrith wanted, or so I thought. But deep under the earth, where the corpse serpent gnaws at the roots of Yggdrasil, the tree of life, there are three spinners, three women who make our fate. We might believe we make choices, but in truth our lives are in the spinners' fingers. They make our lives, and destiny is everything. The Danes know that, and even the Christians know it. Wyrd bið ful āræd, we Saxons say, fate is inexorable, and the spinners had decided my fate because, a week after the witan had met, when Exanceaster was quiet again, they sent me a ship.

The first I knew of it was when a slave came running from Oxton's fields saying that there was a Danish ship in the estuary of the Uisc and I pulled on boots and mail, snatched my swords from their peg, shouted for a horse to be saddled, and rode to the foreshore where *Heahengel* rotted.

And where, standing in from the long sand spit that protects the Uisc from the greater sea, another ship approached. Her sail was furled on the long yard and her dripping oars rose and fell like wings and her long hull left a spreading wake that glittered silver under the rising sun. Her prow was high, and standing there was a man in full mail, a man with a helmet and spear, and behind me, where a few fisherfolk lived in hovels beside the mud, people were hurrying toward the hills and taking with them whatever few possessions they could snatch. I called to one of them. "It's not a Dane!"

"Lord?"

"It's a West Saxon ship," I called, though they did not believe me and hurried away with their livestock. For years they had done this. They would see a ship and they would run, for ships brought Danes and Danes brought death, but this ship had no dragon or wolf or eagle's head on its prow. I knew the ship. It was the *Eftwyrd*, the best named of all Alfred's ships that otherwise bore pious names like *Heahengel* or *Apostol* or *Cristenlic*. *Eftwyrd* meant judgment day, which, though Christian in inspiration, accurately described what she had brought to many Danes.

The man in the prow waved and, for the first time since I had crawled on my knees to Alfred's altar, my spirits lifted. It was Leofric, and then the *Eftwyrd*'s bow slid onto the mud and the long hull juddered to a halt. Leofric cupped his hands. "How deep is this mud?"

"It's nothing!" I shouted back. "A hand's depth, no more!"

"Can I walk on it?"

"Of course you can!" I shouted back.

He jumped and, as I had known he would, he sank up to his thighs in the thick black slime, and I bent over my saddle's pommel in laughter, and the *Eftwyrd*'s crew laughed with me as Leofric cursed, and it took ten minutes to extricate him from the muck, by which time a score of us were plastered in the stinking stuff, but then the crew, who were mostly my old oarsmen and warriors, brought ale ashore, and bread and salted pork, and we made a midday meal beside the rising tide.

"You're an earsling," Leofric grumbled, looking at the mud clogging up the links of his mail coat.

"I'm a bored earsling," I said.

"You're bored?" Leofric said. "So are we." It seemed the fleet was not sailing. It had been given into the charge of a man named Burgweard who was a dull, worthy soldier whose brother was Bishop of Scireburnan, and Burgweard had orders not to disturb the peace. "If the Danes aren't off the coast," Leofric said, "then we aren't."

"So what are you doing here?"

"He sent us to rescue that piece of shit," he nodded at *Heahengel*. "He wants twelve ships again, see?"

"I thought they were building more?"

"They were building more, only it all stopped because some thieving bastards stole the timber while we were fighting at Cynuit, and then someone remembered *Heahengel* and here we are. Burgweard can't manage with just eleven."

"If he isn't sailing," I asked, "why does he want another ship?"

"In case he has to sail," Leofric explained, "and if he does, then he wants twelve. Not eleven, twelve."

"Twelve? Why?"

"Because," Leofric paused to bite off a piece of bread, "because it says in the gospel book that Christ sent out his disciples two by two, and that's how we have to go, two ships together, all holy, and if we've only got eleven, then that means we've only got ten, if you follow me."

I stared at him, not sure whether he was jesting. "Burgweard insists you sail two by two?"

Leofric nodded. "Because it says so in Father Willibald's book."

"In the gospel book?"

"That's what Father Willibald tells us," Leofric said with a straight face, then saw my expression and shrugged. "Honest! And Alfred approves."

"Of course he does."

"And if you do what the gospel book tells you," Leofric said, still with a straight face, "then nothing can go wrong, can it?"

"Nothing," I said. "So you're here to rebuild *Heahengel*?"

"New mast," Leofric said, "new sail, new rigging, patch up those timbers, caulk her, then tow her back to Hamtun. It could take a month!"

"At least."

"And I never was much good at making things. Good at fighting, I am, and I can drink ale as well as any man, but I was never much good with a mallet and wedge or with adzes. They are." He nodded at a group of a dozen men who were strangers to me.

"Who are they?"

"Shipwrights."

"So they do the work?"

"Can't expect me to do it!" Leofric protested. "I'm in command of the *Eftwyrd*!"

"So," I said, "you're planning to drink my ale and eat my food for a month while those dozen men do the work?"

"You have any better ideas?"

I gazed at the *Eftwyrd*. She was a well-made ship, longer than most Danish boats and with high sides that made her a good fighting platform. "What did Burgweard tell you to do?" I asked.

"Pray," Leofric said sourly, "and help repair *Heahengel*."

"I hear there's a new Danish leader in the Sæfern Sea," I said, "and I'd like to know if it's true. A man called Svein. And I hear more ships are joining him from Ireland."

"He's in Wales, this Svein?"

"That's what I hear."

"He'll be coming to Wessex then," Leofric said.

"If it's true."

"So you're thinking . . ." Leofric said, then stopped when he realized just what I was thinking.

"I'm thinking that it doesn't do a ship or crew any good to sit around for a month," I said, "and I'm thinking that there might be plunder to be had in the Sæfern Sea."

"And if Alfred hears we've been fighting up there," Leofric said, "he'll gut us."

I nodded upriver toward Exanceaster. "They burned a hundred Danish ships up there," I said, "and their wreckage is still on the riverbank. We should be able to find at least one dragon's head to put on her prow."

Leofric stared at the *Eftwyrd*. "Disguise her?"

"Disguise her," I said, because if I put a dragon head on *Eftwyrd* no one would know she was a Saxon ship. She would be taken for a Danish boat, a sea raider, part of England's nightmare.

Leofric smiled. "I don't need orders to go on a patrol, do I?"

"Of course not."

"And we haven't fought since Cynuit," he said wistfully, "and no fighting means no plunder."

"What about the crew?" I asked.

He turned and looked at them. "Most of them are evil bastards," he said. "They won't mind. And they all need plunder."

"And between us and the Sæfern Sea," I said, "there are the Britons."

"And they're all thieving bastards, the lot of them," Leofric said. He looked at me and grinned. "So if Alfred won't go to war, we will?"

"You have any better ideas?" I asked.

Leofric did not answer for a long time. Instead, idly, as if he was just thinking, he tossed pebbles toward a puddle. I said nothing, just watched the small splashes, watched the pattern the fallen pebbles made, and knew he was seeking guidance from fate. The Danes cast runesticks, we all watched for the flight of birds, we tried to hear the whispers of the gods, and Leofric was watching the pebbles fall to find his fate. The last one clicked on another and skidded off into the mud and the trail it left pointed south toward the sea. "No," he said, "I don't have any better ideas."

And I was bored no longer, because we were going to be Vikings.

We found a score of carved beasts' heads beside the river beneath Exanceaster's walls, all of them part of the sodden, tangled wreckage that showed where Guthrum's fleet had been burned and we chose two of the least scorched carvings and carried them on board *Eftwyrd*. Her prow and stern culminated in simple posts and we had to cut the posts down until the sockets of the two carved heads fitted. The creature at the stern, the smaller of the two, was a gape-mouthed serpent, probably intended to represent Corpse-Ripper, the monster that tore at the dead in the Danish underworld, while the beast we placed at the bow was a dragon's head, though it was so blackened and disfigured by fire that it looked more like a horse's head. We dug into the scorched eyes until we found unburned wood, and did the same with the open mouth, and when we were finished the thing looked dramatic and fierce. "Looks like a fyrdraca now," Leofric said happily. A fire-dragon.

The Danes could always remove the dragon or beast heads from the bows and sterns of their ships because they did not want the horrid-looking creatures to frighten the spirits of friendly lands and so they only displayed the carved monsters when they were in enemy waters. We did the same, hiding our fyrdraca and serpent head in *Eftwyrd*'s bilges as we went back downriver to where the shipwrights were beginning their work on *Heahengel*. We hid the beast heads because Leofric did not want the shipwrights to know he planned mischief. "That one," he said as he jerked his head toward a tall, lean, gray-haired man who was in charge of the work, "is more Christian than the pope. He'd bleat to the local priests if he thought we were going off to fight someone, and the priests will tell Alfred and then Burgweard will take *Eftwyrd* away from me."

"You don't like Burgweard?"

Leofric spat for answer. "It's a good thing there are no Danes on the coast."

"He's a coward?"

"No coward. He just thinks God will fight the battles. We spend more time on our knees than at the oars. When you commanded the fleet we made money. Now even the rats on board are begging for crumbs."

We had made money by capturing Danish ships and taking their plunder, and though none of us had become rich we had all possessed silver to spare. I was still wealthy enough because I had a hoard hidden at Oxton, a hoard that was the legacy of Ragnar the Older, and a hoard that the church and Oswald's relatives would make their own if they could, but a man can never have enough silver. Silver buys land, it buys the loyalty of warriors, it is the power of a lord, and without silver a man must bend the knee or else become a slave. The Danes led men by the lure of silver, and we were no different. If I was to be a lord, if I was to storm the walls of Bebbanburg, then I would

need men and I would need a great hoard to buy the swords and
shields and spears and hearts of warriors, and so we would go to sea
and look for silver, though we told the shipwrights that we merely
planned to patrol the coast. We shipped barrels of ale, boxes of hard-
baked bread, cheeses, kegs of smoked mackerel, and flitches of bacon.
I told Mildrith the same story, that we would be sailing back and forth
along the shores of Defnascir and Thornsæta. "Which is what we
should be doing anyway," Leofric said, "just in case a Dane arrives."

"The Danes are lying low," I said.

Leofric nodded. "And when a Dane lies low you know there's
trouble coming."

I believed he was right. Guthrum was not far from Wessex, and
Svein, if he existed, was just a day's voyage from her north coast.
Alfred might believe his truce would hold and that the hostages would
secure it, but I knew from my childhood how land-hungry the Danes
were, and how they lusted after the lush fields and rich pastures of
Wessex. They would come, and if Guthrum did not lead them then
another Danish chieftain would gather ships and men and bring his
swords and axes to Alfred's kingdom. The Danes, after all, ruled the
other three English kingdoms. They held my own Northumbria, they
were bringing settlers to East Anglia, their language was spreading
southward through Mercia, and they would not want the last English
kingdom flourishing to their south. They were like wolves, shadow-
skulking for the moment, but watching a flock of sheep fatten.

I recruited eleven young men from my land and took them on
board *Eftwyrd*, and brought Haesten, too, and he was useful for he
had spent much of his young life at the oars. Then, one misty morning, as the strong tide ebbed westward, we slid *Eftwyrd* away from the
river's bank, rowed her past the low sand spit that guards the Uisc
and so out to the long swells of the sea. The oars creaked in their
leather-lined holes, the bow's breast split the waves to shatter water
white along the hull, and the steering oar fought against my touch. I
felt my spirits rise to the small wind and I looked up into the pearly
sky and said a prayer of thanks to Thor, Odin, Njord, and Hoder.

A few small fishing boats dotted the inshore waters, but as we went south and west, away from the land, the sea emptied. I looked back at the low dun hills slashed brighter green where rivers pierced the coast, and then the green faded to gray, the land became a shadow, and we were alone with the white birds crying. It was then that we heaved the serpent's head and the fyrdraca from the bilge and slotted them over the posts at stem and stern, pegged them into place, and turned our bow westward.

The *Eftwyrd* was no more. Now the *Fyrdraca* sailed, and she was hunting trouble.

THREE

The crew of the *Eftwyrd* turned *Fyrdraca* had been at Cynuit with me. They were fighting men and they were offended that Odda the Younger had taken credit for a battle they had won. They had also been bored since the battle. Once in a while, Leofric told me, Burgweard exercised his fleet by taking it to sea, but most of the time they waited in Hamtun. "We did go fishing once, though," Leofric admitted.

"Fishing?"

"Father Willibald preached a sermon about feeding five thousand folk with two scraps of bread and a basket of herring," he said, "so Burgweard said we should take nets out and fish. He wanted to feed the town, see? Lots of hungry folk."

"Did you catch anything?"

"Mackerel. Lots of mackerel."

"But no Danes?"

"No Danes," Leofric said, "and no herring, only mackerel. The bastard Danes have vanished."

We learned later that Guthrum had given orders that no Danish ships were to raid the Wessex coast and so break the truce. Alfred was to be lulled into a conviction that peace had come, and that meant there were no pirates roaming the seas between Kent and Cornwalum and their absence encouraged traders to come from the lands to the south to sell wine or to buy fleeces. The *Fyrdraca* took two such ships in the first four days. They were both Frankish ships, tubby in their hulls, and neither with more than six oars a side, and both believed the *Fyrdraca* was a Viking ship for they saw her beast heads, and they heard Haesten and me speaking Danish and they saw my arm rings. We did not kill the crews, but just stole their coins, weapons, and as much of their cargo as we could carry. One ship was heaped with bales of wool, for the folk across the water prized Saxon fleeces, but we could only take three of the bales for fear of cluttering the *Fyrdraca's* benches.

At night we found a cove or river's mouth, and by day we rowed to sea and looked for prey, and each day we went farther westward until I was sure we were off the coast of Cornwalum, and that was enemy country. It was the old enemy that had confronted our ancestors when they first came across the North Sea to make England. That enemy spoke a strange language, and some Britons lived north of Northumbria and others lived in Wales or in Cornwalum, all places on the wild edges of the isle of Britain where they had been pushed by our coming. They were Christians. Indeed, Father Beocca had told me they had been Christians before we were and he claimed that no one who was a Christian could be a real enemy of another Christian. But nevertheless the Britons hated us. Sometimes they allied themselves with the Northmen to attack us. Sometimes the Northmen raided them. Sometimes they made war against us on their own. In the past the men of Cornwalum had made much trouble for Wessex, although Leofric claimed they had been punished so badly that they now pissed themselves whenever they saw a Saxon.

Not that we saw any Britons at first. The places we sheltered were deserted, all except one river mouth where a skin boat pushed offshore and a half-naked man paddled out to us and held up some crabs, which he wanted to sell to us. We took a basketful of the beasts and paid him two pennies. Next night we grounded *Fyrdraca* on a rising tide and collected fresh water from a stream, and Leofric and I climbed a hill and stared inland. Smoke rose from distant valleys, but there was no one in sight, not even a shepherd. "What are you expecting," Leofric asked, "enemies?"

"A monastery," I said.

"A monastery!" He was amused. "You want to pray?"

"Monasteries have silver," I said.

"Not down here, they don't. They're poor as stoats. Besides."

"Besides what?"

He jerked his head toward the crew. "You've got a dozen good Christians aboard. Lots of bad ones, too, of course, but at least a dozen good ones. They won't raid a monastery with you." He was right. A few of the men had showed some scruples about piracy, but I assured them that the Danes used trading ships to spy on their enemies. That was true enough, though I doubt either of our victims had been serving the Danes. But both ships had been crewed by foreigners and, like all Saxons, the crew of the *Fyrdraca* had a healthy dislike of foreigners, though they made an exception for Haesten and the dozen crewmen who were Frisians. The Frisians were natural pirates, bad as the Danes, and these twelve had come to Wessex to get rich from war and so were glad that the *Fyrdraca* was seeking plunder.

As we went west we began to see coastal settlements, and some were surprisingly large. Cenwulf, who had fought with us at Cynuit and was a good man, told us that the Britons of Cornwalum dug tin out of the ground and sold it to strangers. He knew that because his father had been a trader and had frequently sailed this coast. "If they sell tin," I said, "then they must have money."

"And men to guard it," Cenwulf said drily.

"Do they have a king?"

No one knew. It seemed probable, though where the king lived or who he was we could not know, and perhaps, as Haesten suggested, there was more than one king. They did have weapons because, one night, as the *Fyrdraca* crept into a bay, an arrow flew from a cliff top to be swallowed in the sea beside our oars. We might never have known that arrow had been shot except I happened to be looking up and saw it, fledged with dirty gray feathers, flickering down from the sky to vanish with a plop. One arrow, and no others followed, so perhaps it was a warning, and that night we let the ship lie at its anchor and in the dawn we saw two cows grazing close to a stream and Leofric fetched his ax. "The cows are there to kill us," Haesten warned us in his new and not very good English.

"The cows will kill us?" I asked in amusement.

"I have seen it before, lord. They put cows to bring us on land. Then they attack."

We granted the cows mercy, hauled the anchor, and pulled toward the bay's mouth. A howl sounded behind us and I saw a crowd of men appear from behind bushes and trees. I took one of the silver rings from my left arm and gave it to Haesten. It was his first arm ring and, being a Dane, he was inordinately proud of it. He polished it all morning.

The coast became wilder and refuge more difficult to find, but the weather was placid. We captured a small eight-oared ship that was returning to Ireland and relieved it of sixteen pieces of silver, three knives, a heap of tin ingots, a sack of goose feathers, and six goatskins. We were hardly becoming rich, though *Fyrdraca*'s belly was becoming cluttered with pelts, fleeces, and ingots of tin. "We need to sell it all," Leofric said.

But to whom? We knew no one who traded here. What I needed to do, I thought, was land close to one of the larger settlements and steal everything. Burn the houses, kill the men, plunder the headman's hall, and go back to sea. But the Britons kept lookouts on the headlands and they always saw us coming, and whenever we were close to one of their towns we would see armed men waiting. They had learned how to deal with Vikings, which was why, Haesten told me, the Northmen now sailed in fleets of five or six ships.

"Things will be better," I said, "when we turn the coast." I knew Cornwalum ended somewhere to the west and we could then sail up into the Sæfern Sea where we might find a Danish ship on its voyage from Ireland, but Cornwalum seemed to be without end. Whenever we saw a headland that I thought must mark the end of the land, it turned out to be a false hope, for another cliff would lie beyond, and then another, and sometimes the tide flowed so strongly that even when we sailed due west we were driven back east. Being a Viking was more difficult than I thought, and then one day the wind freshened from the west and the waves heaved higher and their tops were torn ragged and rain squalls hissed dark from a low sky and we ran north-ward to seek shelter in the lee of a headland. We dropped our anchor there and felt *Fyrdraca* jerk and tug like a fretful horse to her long rope of twisted hide.

All night and all the next day the weather raged past the head-land. Water shattered white on high cliffs. We were safe enough, but our food was getting low, and I had half decided we must abandon our plans to make ourselves rich and sail back to the Uisc where we could pretend we had only been patrolling the coast, but on our second dawn under the lee of that high cliff, as the wind subsided and the rain dropped to a chill drizzle, a ship appeared about the eastern spit of land.

"Shields!" Leofric shouted, and the men, cold and unhappy, found their weapons and lined the ship's side.

The ship was smaller than ours, much smaller. She was squat, high-bowed, with a stumpy mast holding a wide yard on which a dirty sail was furled. A half dozen oarsmen manned her, and the steersman was bringing her directly toward *Fyrdraca*, and then, as she came closer and as her small bow broke the water white, I saw a green bough had been tied to her short mast. "They want to talk," I said.

"Let's hope they want to buy," Leofric grumbled.

There was a priest in the small ship. I did not know he was a priest at first, for he looked as ragged as any of the crewmen, but he shouted that he wished to speak with us, and he spoke Danish, though not well, and I let the boat come up on the flank protected from the wind where her crewmen gazed up at a row of armed men holding shields. Cenwulf and I pulled the priest over our side. Two other men wanted to follow, but Leofric threatened them with a spear and they dropped back and the smaller ship drew away to wait while the priest spoke with us.

He was called Father Mardoc and, once he was aboard and sitting wetly on one of *Fyrdraca*'s rowing benches, I saw the crucifix about his neck. "I hate Christians," I said, "so why should we not feed you to Njord?"

He ignored that, or perhaps he did not know that Njord was one of the sea gods. "I bring you a gift," he said, "from my master," and he produced, from beneath his cloak, two battered arm rings.

I took them. They were poor things, mere ringlets of copper, old, filthy with verdigris, and of almost no value, and for a moment I was tempted to toss them scornfully into the sea, but reckoned our voyage had made such small profit that even those scabby treasures must be kept. "Who is your master?" I asked.

"King Peredur."

I almost laughed. King Peredur? A man can expect a king to be famous, but I had never heard of Peredur, which suggested he was little more than a local chieftain with a high-sounding title. "And why does this Peredur," I asked, "send me miserable gifts?"

Father Mardoc still did not know my name and was too frightened to ask it. He was surrounded by men in leather, by men in mail, and by shields and swords, axes and spears, and he believed all of us were Danes for I had ordered any of *Fyrdraca*'s crew who wore crosses or crucifixes to hide them beneath their clothes. Only Haesten and I spoke, and if Father Mardoc thought that strange, he did not say anything of it. Instead he told me how his lord, King Peredur, had been treacherously attacked by a neighbor called Callyn, and Callyn's forces had taken a high fort close to the sea and Peredur would pay us well if we were to help him recapture the fort that was called Dreyndynas.

I sent Father Mardoc to sit in the *Fyrdraca*'s bow while we talked about his request. Some things were obvious. Being paid well did not mean we would become rich, but that Peredur would try to fob us off with as little as possible and, most likely, having given it to us he would then try to take it back by killing us all. "What we should do," Leofric advised, "is find this man Callyn and see what he'll pay us."

Which was good enough advice except none of us knew how to find Callyn, whom we later learned was King Callyn, which did not mean much, for any man with a following of more than fifty armed men called himself a king in Cornwalum, and so I went to the *Fyr-draca*'s bow and talked with Father Mardoc again, and he told me that Dreyndynas was a high fort, built by the old people, and that it guarded the road eastward, and so long as Callyn held the fort, so long were Peredur's people trapped in their lands.

"You have ships," I pointed out.

"And Callyn has ships," he said, "and we cannot take cattle in ships."

"Cattle?"

"We need to sell cattle to live," he said.

So Callyn had surrounded Peredur and we represented a chance to tip the balance in this little war. "So how much will your king pay us?" I asked.

"A hundred pieces of silver," he said.

I drew Serpent-Breath. "I worship the real gods," I told him, "and I am a particular servant of Hoder, and Hoder likes blood, and I have given him none in many days."

Father Mardoc looked terrified, which was sensible of him. He was a young man, though it was hard to tell for his hair and beard were so thick that most of the time he was just a broken nose and pair of eyes surrounded by a greasy black tangle. He told me he had learned to speak Danish when he had been enslaved by a chieftain called Godfred, but that he had managed to escape when Godfred raided the Sillans, islands that lay well out in the western sea-wastes. "Is there any wealth in the Sillans?" I asked him. I had heard of the islands, though some men claimed they were mythical and others said the islands came and went with the moons, but Father Mardoc said they existed and were called the Isles of the Dead.

"So no one lives there?" I asked.

"Some folk do," he said, "but the dead have their houses there."

"Do they have wealth as well?"

"Your ships have taken it all," he said. This was after he had promised me that Peredur would be more generous, though he did not know how generous. He said the king was willing to pay far more than a hundred silver coins for our help, and so we had him shout to his ship that they were to lead us around the coast to Peredur's settlement. I did not let Father Mardoc go back to his ship for he would serve as a hostage if the tale he had told us was false and Peredur was merely luring us to an ambush.

He was not. Peredur's home was a huddle of buildings built on a steep hill beside a bay and protected by a wall of thorn bushes. His people lived within the wall. Some were fishermen and some were cattle herders and none was wealthy. The king himself, however, had a high hall where he welcomed us, though not before we had taken more hostages. Three young men, all of whom we were assured were Peredur's sons, were delivered to *Fyrdraca* and I gave the crew orders that the three were to be killed if I did not return, and then I went ashore with Haesten and Cenwulf. I went dressed for war, with mail coat and helmet polished, and Peredur's folk watched with frightened eyes as the three of us passed. The place stank of fish and shit. The people were ragged and their houses mere hovels that were built up the side of the steep hill that was crowned with Peredur's hall. There was a church beside the hall, its thatch thick with moss and its gable decorated with a cross made from sea-whitened driftwood.

Peredur was twice my age, a squat man with a sly face and a forked black beard. He greeted us from a throne, which was just a chair with a high back, and he waited for us to bow to him, but none of us did and that made him scowl. A dozen men were with him, evidently his courtiers, though none looked wealthy and all were elderly except for one much younger man who was in the robes of a Christian monk, and he stood out in that smoke-darkened hall like a raven in a clutch of gulls for his black robes were clean, his face close shaven, and his hair and tonsure neatly trimmed. He was scarcely older than I, was thin and stern-faced, and that face looked clever. It also carried an expression of marked distaste for us. We were pagans, or at least Haesten and I were pagans, and I had told Cenwulf to keep his mouth shut and his crucifix hidden, and so the monk assumed all three of us were heathen Danes. The monk spoke Danish, far better Danish than Father Mardoc. "The king greets you," he said. He had a voice as thin as his lips and as unfriendly as his pale green eyes. "He greets you and would know who you are."

"My name is Uhtred Ragnarson," I said.

"Why are you here, Uhtred Ragnarson?" the priest asked.

I contemplated him. I did not just look at him, but I studied him as a man might study an ox before killing it. I gave him a look that suggested I was wondering where to make the cuts, and he got my meaning and did not wait for an answer to his question, an answer that was obvious if we were Danes. We were here to thieve and kill, of course. What else did he think a Viking ship would be doing?

Peredur spoke to the monk and they muttered for some time and I looked around the hall, searching for any evidence of wealth. I saw almost nothing except for three whalebones stacked in a corner, but Peredur plainly had some treasure for he wore a great heavy torque of bronze about his neck and there were silver rings on his grubby fingers, an amber brooch at the neck of his cloak, and a golden crucifix hidden in the cloak's lice-ridden folds. He would keep his hoard buried, I thought, but I doubted any of us would become rich from this alliance; in truth we were not becoming rich from our voyage either, and at least Peredur would have to feed us while we haggled.

"The king," the monk interrupted my thoughts, "wishes to know how many men you can lead against the enemy."

"Enough," I said flatly.

"Does that not depend," the monk observed slyly, "on how many enemy there are?"

"No," I said. "It depends on this," and I slapped Serpent-Breath's hilt. It was a good, arrogant reply, and probably what the monk expected. And, in truth, it was convincing for I was broad in the chest and a giant in this hall where I was a full head taller than any other man. "And who are you, monk?" I demanded.

"My name is Asser," he said. It was a British name, of course, and in the English tongue it meant a he-ass, and ever after I thought of him as the Ass. And there was to be a lot of the ever after for, though I did not know it, I had just met a man who would haunt my life like a louse. I had met another enemy, though on that day in Peredur's hall he was just a strange British monk who stood out from his companions because he washed. He invited me to follow him to a small door at the side of the hall and, motioning Haesten and Cenwulf to stay where they were, I ducked through the door to find myself standing beside a dung heap, but the point of taking me outside had been to show me the view eastward.

I stared across a valley. On the nearer slope were the smoke-blackened roofs of Peredur's settlement. Then came the thorn fence that had been made along the stream that flowed to the sea. On the stream's far side the hills rose gently to a far-off crest and there, breaking the skyline like a boil, was Dreyndynas. "The enemy," Asser said.

A small fort, I noted. "How many men are there?"

"Does it matter to you?" Asser asked sourly, paying me back for my refusal to tell him how many men I led, though I assumed Father Mardoc had made a count of the crew while he was on board *Fyrdraca* so my defiance had been pointless.

"You Christians," I said, "believe that at death you go to heaven. Isn't that right?"

"What of it?"

"You must surely welcome such a fate?" I asked. "To be near your god?"

"Are you threatening me?"

"I don't threaten vermin," I said, enjoying myself. "How many men are in that fort?"

"Forty? Fifty?" He plainly did not know. "We can assemble forty."

"So tomorrow," I said, "your king can have his fort back."

"He is not my king," Asser said, irritated by the assumption.

"Your king or not," I said, "he can have his fort back so long as he pays us properly."

That negotiation lasted until dark. Peredur, as Father Mardoc had said, was willing to pay more than a hundred shillings, but he feared we would take the money and leave without fighting and so he wanted some kind of surety from me. He wanted hostages, which I refused to give, and after an hour or more of argument we had still not reached an agreement, and it was then that Peredur summoned his queen. That meant nothing to me, but I saw the Ass stiffen as though he were offended, then sensed that every other man in the hall was strangely apprehensive. Asser made a protest, but the king cut him off with an abrupt slice of his hand and then a door at the back of the hall was opened and Iseult came to my life.

Iseult. Finding her there was like discovering a jewel of gold in a midden. I saw her and I forgot Mildrith. Dark Iseult, black-haired Iseult, huge-eyed Iseult. She was small, thin as an elf, with a luminous face and hair as black as a raven's feathers. She wore a black cloak and had silver bands about her neck and silver bracelets at her wrists and silver rings at her ankles and the jewelry clinked gently as she walked toward us. She was maybe two or three years younger than I, but somehow, despite her youth, she managed to scare Peredur's courtiers, who backed away from her. The king looked nervous, while Asser, standing beside me, made the sign of the cross, then spat to ward off evil.

I just stared at her, entranced. There was pain on her face, as if she found life unbearable, and there was fear on her husband's face when he spoke to her in a quiet, respectful voice. She shuddered when he talked and I thought that perhaps she was mad, for the grimace on her face was awful, disfiguring her beauty, but then she calmed and looked at me and the king spoke to Asser.

"You will tell the queen who you are and what you will do for King Peredur," Asser told me in a distant, disapproving voice.

"She speaks Danish?" I asked.

"Of course not," he snapped. "Just tell her and get this farce over."

I looked into her eyes, those big, dark eyes, and had the uncanny suspicion that she could see right through my gaze and decipher my innermost thoughts. But at least she did not grimace when she saw me, as she had when her husband spoke. "My name is Uhtred Ragnarson," I said, "and I am here to fight for your husband if he pays what I am worth. And if he doesn't pay, we go."

I thought Asser would translate, but the monk stayed silent.

Iseult still stared at me and I stared back. She had flawless skin, untouched by illness, and a strong face, but sad. Sad and beautiful. Fierce and beautiful. She reminded me of Brida, the East Anglian who had been my lover and who was now with Ragnar, my friend. Brida was as full of fury as a scabbard is filled with blade, and I sensed the same in this queen who was so young and strange and dark and lovely.

"I am Uhtred Ragnarson," I heard myself say as I spoke again, though I had scarcely been aware of any urge to talk, "and I work miracles."

Why I said that I do not know. I later learned that she had no idea what I had said, for at that time the only tongue she spoke was that of the Britons, but nevertheless she seemed to understand me and she smiled. Asser caught his breath. "Be careful, Dane," he hissed. "She is a queen."

"A queen," I asked, still staring at her, "or the queen?"

"The king is blessed with three wives," the monk said disapprovingly.

Iseult turned away and spoke to the king. He nodded, then gestured respectfully toward the door through which Iseult had come. She was evidently dismissed and she obediently went to the door, but paused there and gave me a last, speculative look. Then she was gone.

And suddenly it was easy. Peredur agreed to pay us a hoard of silver. He showed us the hoard that had been hidden in a back room. There were coins, broken jewelry, battered cups, and three candleholders that had been taken from the church, and when I weighed the silver, using a balance fetched from the marketplace, I discovered there was three hundred and sixteen shillings' worth, which was not negligible. Asser divided it into two piles, one only half the size of the other. "We shall give you the smaller portion tonight," the monk said, "and the rest you will get when Dreyndynas is recovered."

"You think I am a fool?" I asked, knowing that after the fight it would be hard to get the rest of the silver.

"You take me for one?" he retorted, knowing that if he gave us all the silver then *Fyrdraca* would vanish in the dawn.

We agreed in the end that we would take the one-third now and that the other two-thirds would be carried to the battlefield so that it was easily accessible. Peredur had hoped I would leave that larger portion in his hall, and then I would have faced an uphill fight through his dung-spattered streets, and that was a fight I would have lost, and it was probably the prospect of such a battle that had stopped Callyn's men from attacking Peredur's hall. They hoped to starve him, or at least Asser believed that.

"Tell me about Iseult," I demanded of the monk when the bargaining was done.

He sneered at that. "I can read you like a missal," he said.

"Whatever a missal is," I said, pretending ignorance.

"A book of prayers," he said, "and you will need prayers if you touch her." He made the sign of the cross. "She is evil," he said vehemently.

"She's a queen, a young queen," I said, "so how can she be evil?"

"What do you know of the Britons?"

"That they stink like stoats," I said, "and thieve like jackdaws."

He gave me a sour look and, for a moment, I thought he would refuse to say more, but he swallowed his British pride. "We are Christians," he said, "and God be thanked for that great mercy, but among our people there are still some old superstitions. Pagan ways. Iseult is part of that."

"What part?"

He did not like talking about it, but he had raised the subject of Iseult's evil and so he reluctantly explained. "She was born in the springtime," he said, "eighteen years ago, and at her birth there was an eclipse of the sun, and the folk here are credulous fools and they believe a dark child born at the sun's death has power. They have made her into a"—he paused, not knowing the Danish word—"a *gwrach*," he said, a word that meant nothing to me. "*Dewines,*" he said irritably and, when I still showed incomprehension, he at last found a word. "A sorceress."

"A witch?"

"And Peredur married her. Made her his shadow queen. That is what kings did with such girls. They take them into their households so they may use their power."

"What power?"

"The skills the devil gives to shadow queens, of course," he said irritably. "Peredur believes she can see the future. But it is a skill she will retain only so long as she is a virgin."

I laughed at that. "If you disapprove of her, monk, then I would be doing you a favor if I raped her." He ignored that, or at least he made no reply other than to give me a harsh scowl. "Can she see the future?" I asked.

"She saw you victorious," he said, "and told the king he could trust you, so you tell me?"

"Then assuredly she can see the future," I said.

Brother Asser sneered at that answer. "They should have strangled her with her own birth cord," he snarled. "She is a pagan bitch, a devil's thing, evil."

There was a feast that night, a feast to celebrate our pact, and I hoped Iseult would be there, but she was not. Peredur's older wife was present, but she was a sullen, grubby creature with two weeping boils on her neck and she hardly spoke. Yet it was a surprisingly good feast. There was fish, beef, mutton, bread, ale, mead, and cheese, and while we ate Asser told me he had come from the kingdom of Dyfed, which lay north of the Sæfern Sea, and that his king, who had an impossible British name that sounded like a man coughing and spluttering, had sent him to Cornwalum to dissuade the British kings from supporting the Danes.

I was surprised by that, so surprised that I looked away from the girls serving the food. A harpist played at the hall's end and two of the girls swayed in time to the music as they walked. "You don't like Danes," I said.

"You are pagans," Asser said scornfully.

"So how come you speak the pagan tongue?" I asked.

"Because my abbot would have us send missionaries to the Danes."

"You should go," I said. "It would be a quick route to heaven for you."

He ignored that. "I learned Danish among many other tongues," he said loftily, "and I speak the language of the Saxons, too. And you, I think, were not born in Denmark?"

"How do you know?"

"Your voice," he said. "You are from Northumbria?"

"I am from the sea," I said.

He shrugged. "In Northumbria," he said severely, "the Danes have corrupted the Saxons so that they think of themselves as Danes." He was wrong, but I was scarcely in a position to correct him. "Worse," he went on, "they have extinguished the light of Christ."

"Is the light of Thor too bright for you?"

"The West Saxons are Christians," he said, "and it is our duty to support them, not because of a love for them, but because of our fellow love for Christ."

"You have met Alfred of Wessex?" I asked sourly.

"I look forward to meeting him," he said fervently, "for I hear he is a good Christian."

"I hear the same."

"And Christ rewards him," Asser went on.

"Rewards him?"

"Christ sent the storm that destroyed the Danish fleet," Asser said, "and Christ's angels destroyed Ubba. That is proof of God's power. If we fight against Alfred then we range ourselves against Christ, so we must not do it. That is my message to the kings of Cornwalum."

I was impressed that a British monk at the end of the land of Britain knew so much of what happened in Wessex, and I reckoned Alfred would have been pleased to hear Asser's nonsense, though of course Alfred had sent many messengers to the British. His messengers had all been priests or monks and they had preached the gospel of their god slaughtering the Danes, and Asser had evidently taken up their message enthusiastically. "So why are you fighting Callyn?" I asked.

"He would join the Danes," Asser said.

"And we're going to win," I said, "so Callyn is sensible."

Asser shook his head. "God will prevail."

"You hope," I said, fingering Thor's hammer. "But if you are wrong, monk, then we'll take Wessex and Callyn will share the spoils."

"Callyn will share nothing," Asser said spitefully, "because you will kill him tomorrow."

The Britons have never learned to love the Saxons. Indeed they hate us, and in those years when the last English kingdom was on the edge of destruction, they could have tipped the balance by joining Guthrum. Instead they held back their sword arms, and for that the Saxons can thank the church. Men like Asser had decided that the Danish heretics were a worse enemy than English Christians, and if I were a Briton I would resent that, because the Britons might have taken back much of their lost lands if they had allied themselves with the pagan Northmen. Religion makes strange bedfellows.

So does war, and Peredur offered Haesten and myself two of the serving girls to seal our bargain. I had sent Cenwulf back to *Fyrdraca* with a message for Leofric, warning him to be ready to fight in the morning, and I thought perhaps Haesten and I should retreat to the ship, but the serving girls were pretty and so we stayed, and I need not have worried for no one tried to kill us in the night, and no one even tried when Haesten and I carried the first third of the silver down to the water's edge where a small boat carried us to our ship. "There's twice as much as that waiting for us," I told Leofric.

He stirred the sack of silver with his foot. "And where were you last night?"

"In bed with a Briton."

"Earsling," he said. "So who are we fighting?"

"A pack of savages."

We left ten men as ship guards. If Peredur's men made a real effort to capture *Fyrdraca* then those ten would have had a hard fight, and probably a losing fight, but they had the three hostages who may or may not have been Peredur's sons, so that was a risk we had to take, and it seemed safe enough because Peredur had assembled his army on the eastern side of the town. I say army, though it was only forty men, and I brought thirty more, and my thirty were well armed and looked ferocious in their leather. Leofric, like me, wore mail, as did half a dozen of my crewmen, and I had my fine helmet with its faceplate so I, at least, looked like a lord of battles.

Peredur was in leather, and he had woven black horsetails into his hair and onto the twin forks of his beard so that the horsetails hung down wild and long and scary. His men were mostly armed with spears, though Peredur himself possessed a fine sword. Some of his men had shields and a few had helmets, and though I did not doubt their bravery I did not reckon them formidable. My crewmen were formidable. They had fought Danish ships off the Wessex coast and they had fought in the shield wall at Cynuit and I had no doubt that we could destroy whatever troops Callyn had placed in Dreyndynas.

It was afternoon before we climbed the hill. We should have gone in the morning, but some of Peredur's men were recovering from their night's drinking, and the women of his settlement kept pulling others away, not wanting them to die, and then Peredur and his advisers huddled and talked about how they should fight the battle, though what there was to talk about I did not know. Callyn's men were in the fort. We were outside it, so we had to assault the bastards. Nothing clever, just an attack, but they talked for a long time, and Father Mardoc said a prayer, or rather he shouted it, and then I refused to advance because the rest of the silver had not been fetched.

It came, carried in a chest by two men, and so at last, under the afternoon sun, we climbed the eastern hill. Some women followed us, shrieking their battle screams, which was a waste of breath because the enemy was still too far away to hear them.

"So what do we do?" Leofric asked me.

"Form a wedge," I guessed. "Our best men in the front rank and you and me in front of them, then kill the bastards."

He grimaced. "Have you ever assaulted one of the old people's forts?"

"Never."

"It can be hard," he warned me.

"If it's too hard," I said, "we'll just kill Peredur and his men and take their silver anyway."

Brother Asser, his neat black robes muddied about their skirts, hurried over to me. "Your men are Saxons!" he said accusingly.

"I hate monks," I snarled at him. "I hate them more than I hate priests. I like killing them. I like slitting their bellies. I like watching the bastards die. Now run off and die before I cut your throat."

He ran off to Peredur with his news that we were Saxons. The king stared at us morosely. He had thought he had recruited a crew of Danish Vikings, and now he discovered we were West Saxons and he was not happy, so I drew Serpent-Breath and banged her blade against my limewood shield. "You want to fight this battle or not?" I asked him through Asser.

Peredur decided he wanted to fight, or rather he wanted us to fight the battle for him, and so we slogged on up the hill, which had a couple of false crests so it was well into the afternoon before we emerged onto the long, shallow summit and could see Dreyndynas's green turf walls on the skyline. A banner flew there. It was a triangle of cloth, supported on its pole by a small cross-staff, and the banner showed a white horse prancing on a green field.

I stopped then. Peredur's banner was a wolf's tail hung from a pole. I carried none though, like most Saxons, mine would have been a rectangular flag. I only knew one people who flew triangular banners and I turned on Brother Asser as he sweated up the hill. "They're Danes," I accused him.

"So?" he demanded. "I thought you were a Dane, and all the world knows the Danes will fight anyone for silver, even other Danes. But are you frightened of them, Saxon?"

"Your mother didn't give birth to you," I told him, "but farted you out of her shriveled arsehole."

"You've taken Peredur's silver," Asser said, "so you must fight."

"Say one more word, monk," I said, "and I'll cut off your scrawny balls." I was gazing uphill, trying to estimate numbers. Everything had changed since I had seen the white horse banner because instead of fighting against half-armed British savages we would have to take on a crew of lethal Danes, but if I was surprised by that, then the Danes were equally surprised to see us. They were crowding Dreyn-dynas's wall, which was made of earth fronted with a ditch and topped with a thorn fence. It would be a hard wall to attack, I thought, especially if it was defended by Danes. I counted over forty men on the skyline and knew there would be others I could not see, and the numbers alone told me this assault would fail. We could attack, and we might well get as far as the thorn palisade, but I doubted we could hack our way through, and the Danes would kill a score of us as we tried, and we would be lucky to retreat down the hill without greater loss.

"We're in a cesspit," Leofric said to me.

"Up to our necks."

"So what do we do? Turn on them and take the money?"

I did not answer because the Danes had dragged a section of the thorn fence aside and three of them now jumped down from the ramparts and strolled toward us. They wanted to talk.

"Who the hell is that?" Leofric asked.

He was staring at the Danish leader. He was a huge man, big as Steapa Snotor, and dressed in a mail coat that had been polished with sand until it shone. His helmet, as highly polished as his mail, had a faceplate modeled as a boar's mask with a squat, broad snout, and from the helmet's crown there flew a white horsetail. He wore arm-rings over his mail, rings of silver and gold that proclaimed him to be a warrior chief, a sword Dane, a lord of war. He walked the hillside as if he owned it, and in truth he did own it because he possessed the fort.

Asser hurried to meet the Danes, going with Peredur and two of his courtiers. I went after them and found Asser trying to convert the Danes. He told them that God had brought us and we would slaughter them all and their best course was to surrender now and yield their heathen souls to God. "We shall baptize you," Asser said, "and there will be much rejoicing in heaven."

The Danish leader slowly pulled off his helmet and his face was almost as frightening as the boar-snouted mask. It was a broad face, hardened by sun and wind, with the blank, expressionless eyes of a killer. He was around thirty years old and had a tightly cropped beard and a scar running from the corner of his left eye down across his cheek. He gave the helmet to one of his men and, without saying a word, hauled up the skirt of his mail coat and began pissing on Asser's robe. The monk leaped back.

The Dane, still pissing, looked at me. "Who are you?"

"Uhtred Ragnarson. And you?"

"Svein of the White Horse." He said it defiantly, as though I would know his reputation, and for a heartbeat I said nothing. Was this the same Svein who was said to be gathering troops in Wales? Then what was he doing here?

"You're Svein of Ireland?" I asked.

"Svein of Denmark," he said. He let the mail coat drop and glared at Asser, who was threatening the Danes with heaven's vengeance. "If you want to live," he told Asser, "shut your filthy mouth." Asser shut his mouth. "Ragnarson." Svein looked back to me. "Earl Ragnar? Ragnar Ravnson? The Ragnar who served Ivar?"

"The same," I said.

"Then you are the Saxon son?"

"I am. And you?" I asked. "You're the Svein who has brought men from Ireland?"

"I have brought men from Ireland," he admitted.

"And gather forces in Wales?"

"I do what I do," he said vaguely. He looked at my men, judging how well they would fight. Then he looked me up and down, noting my mail and helmet, and noting especially my arm rings, and when the inspection was done he jerked his head to indicate that he and I should walk away a few paces and talk privately.

Asser objected, saying anything that was spoken should be heard by all, but I ignored him and followed Svein uphill. "You can't take this fort," Svein told me.

"True."

"So what do you do?"

"Go back to Peredur's settlement, of course."

He nodded. "And if I attack the settlement?"

"You'll take it," I said, "but you'll lose men. Maybe a dozen?"

"Which will mean a dozen fewer oarsmen," he said, thinking, and then he looked past Peredur to where two men carried the box. "Is that your battle price?'

"It is."

"Split it?" he suggested.

I hesitated a heartbeat. "And we'll split what's in the town?" I asked.

"Agreed," he said, then looked at Asser who was hissing urgently at Peredur. "He knows what we're doing," he said grimly, "so a necessary deception is about to happen." I was still trying to understand what he meant when he struck me in the face. He struck hard, and my hand went to Serpent-Breath and his two men ran to him, swords in hand.

"I'll come out of the fort and join you," Svein said to me softly. Then, louder, "You bastard piece of goat-dropping."

I spat at him as his two men pretended to drag him away. Then I stalked back to Asser. "We kill them all," I said savagely. "We kill them all!"

"What did he say to you?" Asser asked. He had feared, rightly, as it happened, that Svein and I had made our own alliance, but Svein's quick display had put doubts in the monk's mind, and I fed the doubts by raging like a madman, screaming at the retreating Svein that I would send his miserable soul to Hel, who was the goddess of the dead. "Are you going to fight?" Asser demanded.

"Of course we're going to fight!" I shouted at him. Then I crossed to Leofric. "We're on the same side as the Danes," I told him quietly. "We kill these Britons, capture their settlement, and split everything with the Danes. Tell the men, but tell them quietly."

Svein, true to his word, brought his men out of Dreyndynas. That should have warned Asser and Peredur of treachery, for no sensible man would abandon a fine defensive position like a thorn-topped earth wall to fight a battle on open ground, but they put it down to Danish arrogance. They assumed Svein believed he could destroy us all in open battle, and he made that assumption more likely by parading a score of his men on horseback, suggesting that he intended to tear our shield wall open with his swords and axes and then pursue the survivors with spear-armed cavalry. He made his own shield wall in front of the horsemen, and I made another shield wall on the left of Peredur's line, and once we were in the proper array we shouted insults at each other. Leofric was going down our line, whispering to the men, and I sent Cenwulf and two others to the rear with their own orders, and just then Asser ran across to us.

"Attack," the monk demanded, pointing at Svein.

"When we're ready," I said, for Leofric had not yet given every man his orders.

"Attack now!" Asser spat at me, and I almost gutted the bastard on the spot and would have saved myself a good deal of future trouble if I had, but I kept my patience and Asser went back to Peredur, where he began praying, both hands held high in the air, demanding that God send fire from heaven to consume the pagans.

"You trust Svein?" Leofric had come back to my side.

"I trust Svein," I said. Why? Only because he was a Dane and I liked the Danes. These days, of course, we are all agreed that they are the spawn of Satan, untrustworthy pagans, savages, and anything else we care to call them, but in truth the Danes are warriors and they like other warriors, and though it is true that Svein might have persuaded me to attack Peredur so that he could then attack us, I did not believe it. Besides, there was something I wanted in Peredur's hall and, to get her, I needed to change sides.

"*Fyrdraca!*" I shouted, and that was our signal, and we swung our shield wall around to the right and went at it.

It was, of course, an easy slaughter. Peredur's men had no belly for a fight. They had been hoping that we would take the brunt of the Danish assault and that they could then scavenge for plunder among Svein's wounded, but instead we turned on them, attacked them, and cut them down, and Svein came on their right, and Peredur's men fled. That was when Svein's horsemen kicked back their heels, leveled their spears, and charged.

It was not a fight, it was a massacre. Two of Peredur's men put up some resistance, but Leofric swatted their spears aside with his ax and they died screaming, and Peredur went down to my sword, and he put up no fight at all, but seemed resigned to his death that I gave him quickly enough. Cenwulf and his two companions did what I had ordered them to do, which was to intercept the chest of silver, and we rallied around them as Svein's riders chased down the fugitives. The only man to escape was Asser, the monk, which he managed by running north instead of west. Svein's horsemen were ranging down the hill, spearing Peredur's men in the backs, and Asser saw that only death lay that way and so, with surprising quickness, he changed direction and sprinted past my men, his skirts clutched up about his knees, and I shouted at the men on the right of the line to kill the bastard, but they simply looked at me and let him go. "I said kill him!" I snarled.

"He's a monk!" one of them answered. "You want me to go to hell?"

I watched Asser run slantwise into the valley and, in truth, I did not much care whether he lived or died. I thought Svein's horsemen would catch him, but perhaps they did not see him. They did catch Father Mardoc and one of them took off the priest's head with a single swing of his sword, which made some of my men cross themselves.

The horsemen made their killing, but Svein's other Danes made a shield wall that faced us, and in its center, beneath the white horse banner, was Svein himself in his boar-mask helmet. His shield had a white horse painted on its boards and his weapon was an ax, the largest war ax I had ever seen. My men shifted nervously. "Stand still!" I snarled at them.

"Up to our necks in it," Leofric said quietly.

Svein was staring at us and I could see the death light in his eyes. He was in a killing mood, and we were Saxons, and there was a knocking sound as his men hefted shields to make the wall, and so I tossed Serpent-Breath into the air. Tossed her high so that the big blade whirled about in the sun, and of course they were all wondering whether I would catch her or whether she would thump onto the grass.

I caught her, winked at Svein, and slid the blade into her scabbard. He laughed and the killing mood passed as he realized he could not afford the casualties he would inevitably take in fighting us. "Did you really think I was going to attack you?" he called across the springy turf.

"I was hoping you would attack me," I called back, "so I wouldn't have to split the plunder with you."

He dropped the ax and walked toward us, and I walked toward him and we embraced. Men on both sides lowered weapons. "Shall we take the bastard's miserable village?" Svein asked.

So we all went back down the hill, past the bodies of Peredur's men, and there was no one defending the thorn wall about the settlement so it was an easy matter to get inside, and a few men tried to protect their homes, but very few. Most of the folk fled to the beach, but there were not enough boats to take them away, and so Svein's Danes rounded them up and began sorting them into the useful and the dead. The useful were the young women and those who could be sold as slaves, the dead were the rest.

I took no part in that. Instead, with all of my men, I went straight to Peredur's hall. Some Danes, reckoning that was where the silver would be, were also climbing the hill, but I reached the hall first, pushed open the door, and saw Iseult waiting there.

I swear she was expecting me for her face showed no fear and no surprise. She was sitting in the king's throne, but stood as if welcoming me as I walked up the hall. Then she took the silver from her neck and her wrists and her ankles and held it mutely out as an offering and I took it all and tossed it to Leofric. "We divide it with Svein," I said.

"And her?" He sounded amused. "Do we share her, too?"

For answer I took the cloak from about Iseult's neck. Beneath it she wore a black dress. I still had Serpent-Breath drawn and I used the bloodied blade to slash at the cloak until I could tear a strip from its hem. Iseult watched me, her face showing nothing. When the strip was torn away I gave her back the cloak, then tied one end of the cloth strip about her neck and tied the other end to my belt. "She's mine," I said.

More Danes were coming into the hall and some stared wolfishly at Iseult, and then Svein arrived and snarled at his men to start digging up the hall floor to search for hidden coins or silver. He grinned when he saw Iseult's leash. "You can have her, Saxon," he said. "She's pretty, but I like them with more meat on the bone."

I kept Iseult with me as we feasted that night. There was a good deal of ale and mead in the settlement and so I ordered my men not to fight with the Danes, and Svein told his men not to fight with us, and on the whole we were obeyed, though inevitably some men quarreled over the captured women and one of the boys I had brought from my estate got a knife in his belly and died in the morning.

Svein was amused that we were a West Saxon ship. "Alfred sent you?" he asked me.

"No."

"He doesn't want to fight, does he?"

"He'll fight," I said, "except he thinks his god will do the fighting for him."

"Then he's an idiot," Svein said. "The gods don't do our bidding.

I wish they did." He sucked on a pork bone. "So what are you doing here?" he asked.

"Looking for money," I said, "the same as you."

"I'm looking for allies," he said.

"Allies?"

He was drunk enough to speak more freely than he had when we first met, and I realized this was indeed the Svein who was said to be gathering men in Wales. He admitted as much, but added that he did not have enough warriors. "Guthrum can lead two thousand men to battle, maybe more! I have to match that."

So he was a rival to Guthrum. I tucked that knowledge away. "You think the Cornishmen will fight with you?"

"They promised they would," he said, spitting out a shred of gristle. "That's why I came here. But the bastards lied. Callyn isn't a proper king, he's a village chief! I'm wasting my time here."

"Could the two of us beat Callyn?" I asked.

Svein thought about it, then nodded. "We could." He frowned suddenly, staring into the hall's shadows, and I saw he was looking at one of his men who had a girl on his lap. He evidently liked the girl for he slapped the table, pointed to her, beckoned, and the man reluctantly brought her. Svein sat her down, pulled her tunic open so he could see her breasts, then gave her his pot of ale. "I'll think about it," he told me.

"Or are you thinking of attacking me?" I asked.

He grinned. "You are Uhtred Ragnarson," he said, "and I heard about the fight on the river where you killed Ubba."

I evidently had more of a reputation among my enemies than I did among my so-called friends. Svein insisted I tell the tale of Ubba's death, which I did, and I told him the truth, which was that Ubba had slipped and fallen, and that had let me take his life.

"But men say you fought well," Svein said.

Iseult listened to all this. She did not speak our language, but her big eyes seemed to follow every word. When the feast was over I took her to the small rooms at the back of the hall and she used my makeshift leash to pull me into her wood-walled chamber. I made a bed from our cloaks. "When this is done," I told her in words she could not understand, "you'll have lost your power."

She touched a finger to my lips to silence me and she was a queen so I obeyed her.

In the morning we finished ravaging the town. Iseult showed me which houses might have something of value and generally she was right though the search meant demolishing the houses, for folk hide their small treasures in their thatch. So we scattered rats and mice as we hauled down the moldy straw and sifted through it, and afterward we dug under every hearth, or wherever else a man might bury silver, and we collected every scrap of metal, every cooking pot or fishhook, and the search took all day. That night we divided the hoard on the beach.

Svein had evidently thought about Callyn and, being sober by the time he did his thinking, he had decided that the king was too strong. "We can easily beat him," he said, "but we'll lose men."

A ship's crew can only endure so many losses. We had lost none in the fight against Peredur, but Callyn was a stronger king and he was bound to be suspicious of Svein, which meant that he would have his household troops ready and armed. "And he's got little enough to take," Svein said scornfully.

"He's paying you?"

"He's paying me," Svein said, "just as Peredur paid you."

"I split that with you," I said.

"Not the money he paid you before the fight," Svein said with a grin, "you didn't split that."

"What money?" I asked.

"So we're even," he said, and we had both done well enough out of Peredur's death, for Svein had slaves and we each now possessed over nine hundred shillings' worth of silver and metal, which was not a fortune, especially once it was divided among the men, but it was better than I had done so far on the voyage. I also had Iseult. She was no longer leashed to me, but she stayed beside me and I sensed that she was happy about that. She had taken a vicious pleasure in seeing her home destroyed and I decided she must have hated Peredur. He had feared her and she had hated him, and if it was true that she had been able to see the future, then she had seen me and given her husband bad advice to make that future come true.

"So where do you go now?" Svein asked. We were walking along the beach, past the huddled slaves who watched us with dark, resentful eyes.

"I have a mind," I said, "to go into the Sæfern Sea."

"There's nothing left there," he said scornfully.

"Nothing?"

"It's been scoured," he said, meaning that Danish and Norse ships had bled the coasts dry of any treasure. "All you'll find in the Sæfern Sea," he went on, "are our ships bringing men from Ireland."

"To attack Wessex?"

"No!" He grinned at me. "I've a mind to start trading with the Welsh kingdoms."

"And I have a mind," I said, "to take my ship to the moon and build a feasting hall there."

He laughed. "But speaking of Wessex," he said, "I hear they're building a church where you killed Ubba."

"I hear the same."

"A church with an altar of gold."

"I've heard that, too," I allowed. I hid my surprise that he knew of Odda the Younger's plans, but I should not have been surprised. A rumor of gold would spread like couch grass. "I've heard it," I said again, "but I don't believe it."

"Churches have money," he said thoughtfully, then frowned, "but that's a strange place to build a church."

"Strange? Why?"

"So close to the sea? An easy place to attack?"

"Or perhaps they want you to attack," I said, "and have men ready to defend it?"

"A lure, you mean?" He thought about that.

"And hasn't Guthrum given orders that the West Saxons aren't to be provoked?" I said.

"Guthrum can order what he wants," Svein said harshly, "but I am Svein of the White Horse and I don't take orders from Guthrum." He walked on, frowning as he threaded the fishing nets that men now dead had hung to dry. "Men say Alfred is not a fool."

"Nor is he."

"If he has put valuables beside the sea," he said, "he will not leave them unguarded." He was a warrior, but like the best warriors he was no madman. When folk speak of the Danes these days they have an idea that they were all savage pagans, unthinking in their terrible violence, but most were like Svein and feared losing men. That was always the great Danish fear, and the Danish weakness. Svein's ship was called the *White Horse* and had a crew of fifty-three men, and if a dozen of those men were to be killed or gravely wounded, then the *White Horse* would be fatally weakened. Once in a fight, of course, he was like all Danes, terrifying, but there was always a good deal of thinking before there was any fighting. He scratched at a louse, then gestured toward the slaves his men had taken. "Besides, I have these."

He meant he would not go to Cynuit. The slaves, once they were sold, would bring him silver and he must have reckoned Cynuit was not worth the casualties.

Svein needed my help next morning. His own ship was in Callyn's harbor and he asked me to take him and a score of his men to fetch it. We left the rest of his crew at Peredur's settlement. They guarded the slaves he would take away, and they also burned the place as we carried Svein east up the coast to Callyn's settlement. We waited a day there as Svein settled his accounts with Callyn, and we used the time to sell fleeces and tin to Callyn's traders and, though we received a poor enough price, it was better to travel with silver than with bulky cargo. The *Fyrdraca* was glittering with silver now and the crewmen, knowing they would receive their proper share, were happy. Haesten wanted to go with Svein, but I refused his request. "I saved your life," I told him, "and you have to serve me longer to pay for that." He accepted that and was pleased when I gave him a second arm ring as a reward for the men he had killed at Dreyndynas.

Svein's *White Horse* was smaller than *Fyrdraca*. Her prow had a carved horse's head and her stern a wolf's head, while at her masthead was a wind vane decorated with a white horse. I asked Svein about the horse and he laughed. "When I was sixteen," he said, "I wagered my father's stallion against our king's white horse. I had to beat the king's champion at wrestling and sword play. My father beat me for making the wager, but I won! So the white horse is lucky. I ride only white horses." And so his ship was the *White Horse* and I followed her back up the coast to where a thick plume of smoke marked where Peredur had ruled.

"Are we staying with him?" Leofric asked, puzzled that we were going back west rather than turning toward Defnascir.

"I have a mind to see where Britain ends," I said, and I had no wish to return to the Uisc and to Mildrith's misery.

Svein put the slaves into the belly of his boat. We spent one last night in the bay, under the thick smoke, and in the morning, as the rising sun flickered across the sea, we rowed away. As we passed the western headland, going into the wide ocean, I saw a man watching us from the cliff's top and I saw he was robed in black and, though he was a long way off, I thought I recognized Asser. Iseult saw him, too, and she hissed like a cat, made a fist, and threw it at him, opening her fingers at the last moment as if casting a spell at the monk.

Then I forgot him because *Fyrdraca* was back in the open sea and we were going to the place where the world ended.

And I had a shadow queen for company.

FOUR

I love the sea. I grew up beside it, though in my memories the seas off Bebbanburg are gray, usually sullen, and rarely sunlit. They are nothing like the great waters that roll from beyond the Isles of the Dead to thunder and shatter against the rocks at the west of Britain. The sea heaves there, as if the ocean gods flexed their muscles, and the white birds cry endlessly, and the wind rattles the spray against the cliffs. *Fyrdraca*, running before that bright wind, left a path in the sea and the steering oar fought me, pulsing with the life of the water and the flexing of the ship and the joy of the passage. Iseult stared at me, astonished by my happiness, but then I gave her the oar and watched her thin body heave against the sea's strength until she understood the power of the oar and could move the ship, and then she laughed. "I would live on the sea," I told her, though she did not understand me. I had given her an arm ring from Peredur's hoard and a silver toe ring and a necklace of monster's teeth, all sharp and long and white, strung on a silver wire.

I turned and watched Svein's *White Horse* cut through the water. Her bow would sometimes break from a wave so that the forepart of her hull, all green and dark with growth, would rear skyward with her horse's head snarling at the sun, and then she would crash down and the seas would explode white about her timbers. Her oars, like ours, were inboard and the oar holes plugged, and we both ran under sail. *Fyrdraca* was the faster ship, which was not because she was more cunningly built, but because her hull was longer.

There is such joy in a good ship, and a greater joy to have the ship's belly fat with other men's silver. It is the Viking joy, driving a dragon-headed hull through a wind-driven sea toward a future full of feasts and laughter. The Danes taught me that and I love them for it, pagan swine though they might be. At that moment, running before Svein's *White Horse*, I was as happy as a man could be, free of all the churchmen and laws and duties of Alfred's Wessex, but then I gave orders that the sail was to be lowered and a dozen men uncleated the lines and the big yard scraped down the mast. We had come to Britain's ending and I would turn about, and I waved to Svein as the *White Horse* swept past us. He waved back, watching the *Fyrdraca* wallow in the long ocean swells.

"Seen enough?" Leofric asked me.

I was staring at the end of Britain where the rocks endured the sea's assault. "Penwith," Iseult said, giving me the British name for the headland.

"You want to go home?" I asked Leofric.

He shrugged. The crew was turning the yard, lining it fore and aft so it could be stowed on its crutches while other men were binding the sail so it did not flap. The oars were being readied to take us eastward and the *White Horse* was getting smaller as it swept up into the Sæfern Sea.

I stared after Svein, envying him. "I need to be rich," I said to Leofric.

He laughed at that.

"I have a path to follow," I said, "and it goes north. North back to Bebbanburg. And Bebbanburg has never been captured, so I need many men to take it. Many good men and many sharp swords."

"We have silver," he said, gesturing into the boat's bilge.

"Not enough," I answered sourly. My enemies had money and Alfred claimed that I owed the church money, and the courts of

Defnascir would be chasing me for wergild. I could only go home if I had enough silver to pay off the church, to bribe the courts, and to attract men to my banner. I stared at the *White Horse* that was now little more than a sail above the wind-fretted sea and I felt the old temptation to go with the Danes. Wait till Ragnar was free and give him my sword arm, but then I would be fighting against Leofric and I would still need to make money, raise men, go north, and fight for my birthright. I touched Thor's hammer and prayed for a sign.

Iseult spat. That was not quite true. She said a word that sounded like someone clearing their throat, spitting and choking all at the same time, and she was pointing over the ship's side and I saw a strange fish arcing out of the water. The fish was as big as a deerhound and had a triangular fin. "Porpoise," Leofric said.

"*Llamhydydd*," Iseult said again, giving the fish its British name.

"They bring sailors luck," Leofric said.

I had never seen a porpoise before, but suddenly there were a dozen of the creatures. They were gray and their backs glistened in the sun and they were all going north.

"Put the sail back up," I told Leofric.

He stared at me. The crew was unlashing the oars and taking the plugs from the oar holes. "You want the sail up?" Leofric asked.

"We're going north." I had prayed for a sign and Thor had sent me the porpoise.

"There's nothing in the Sæfern Sea," Leofric said. "Svein told you that."

"Svein told me there was no plunder in the Sæfern Sea," I said, "because the Danes have taken it all, so that means the Danes have the plunder." I felt a surge of happiness so intense that I punched Leofric's shoulder and gave Iseult a hug. "And he told me that their ships are coming from Ireland."

"So?" Leofric rubbed his shoulder.

"Men from Ireland!" I told Leofric. "Danes coming from Ireland to attack Wessex. And if you brought a ship's crew from Ireland, what would you bring with you?"

"Everything you possess," Leofric said flatly.

"And they don't know we're here! They're sheep, and we are a fire-dragon."

He grinned. "You're right," he said.

"Of course I'm right! I'm a lord! I'm right and I'm going to be rich! We're all going to be rich! We shall eat off gold plates, piss down our enemies' throats, and make their wives into our whores." I was shouting this nonsense as I walked down the boat's center, casting off the sail's lashings. "We'll all be rich with silver shoes and golden bonnets. We'll be richer than kings! We'll wallow in silver, shower our whores with gold, and shit lumps of amber! Tie those oars up! Plug the holes, we're going north, we're going to be rich as bishops, every man of us!" The men were grinning, pleased because I was roaring my enthusiasm, and men like to be led.

They did have qualms about going north, for that would take us out of sight of land, and I had never been that far from the shore, and I was frightened, too, for Ragnar the Older had often told me tales of Norsemen who had been tempted out into the sea wastes, to sail ever farther westward, and he said there were lands out there, lands beyond the Isles of the Dead, lands where ghosts walked, but I am not sure if he told the truth. I am sure, though, that he told me that many of those ships never returned. They voyage into the dying sun and they go onward because they cannot bear to turn back and so they sail to where the lost ships die at the world's dark ending.

Yet the world did not end to the north. I knew that, though I was not certain what did lie northward. Dyfed was there, somewhere, and Ireland, and there were other places with barbarous names and savage people who lived like hungry dogs on the wild edges of the land, but there was also a waste of sea, a wilderness of empty waves, and so, once the sail was hoisted and the wind was thrusting the *Fyrdraca* northward, I leaned on the oar to take her somewhat to the east for fear we would otherwise be lost in the ocean's vastness.

"You know where you're going?" Leofric asked me.

"No."

"Do you care?"

I grinned at him for answer. The wind, which had been southerly, came more from the west, and the tide took us eastward, so that by the afternoon I could see land, and I thought it must be the land of the Britons on the north side of the Sæfern, but as we came closer I saw it was an island. I later discovered it was the place the Northmen call Lundi, because that is their word for the puffin, and the island's high cliffs were thick with the birds that shrieked at us when we came into a cove on the western side of the island that was an uncomfortable place to anchor for the night because the big seas rolled in. And so we dropped the sail, took out the oars, and rowed around the cliffs until we found shelter on the eastern side.

I went ashore with Iseult and we dug some puffin burrows to find eggs, though all were hatched, so we contented ourselves with killing a pair of goats for the evening meal. There was no one living on the island, though there had been because there were the remains of a small church and a field of graves. The Danes had burned everything, pulled down the church, and dug up the graves in search of gold. We climbed to a high place and I searched the evening sea for ships, but saw none, though I wondered if I could see land to the south. It was hard to be sure for the southern horizon was thick with dark cloud, but a darker strip within the cloud could have been hills and I assumed I was looking at Cornwalum or the western part of Wessex. Iseult sang to herself.

I watched her. She was gutting one of the dead goats, doing it clumsily for she was not accustomed to such work. She was thin, so thin that she looked like the ælfcynn, the elf-kind, but she was happy. In time I would learn just how much she had hated Peredur. He had valued her and made her a queen, but he had also kept her a prisoner in his hall so that he alone could profit from her powers. Folk would pay Peredur to hear Iseult's prophecies and one of the reasons Callyn had fought his neighbor was to take Iseult for himself. Shadow queens were valued among the Britons for they were part of the old mysteries, the powers that had brooded over the land before the monks arrived, and Iseult was one of the last shadow queens. She had been born in the sun's darkness, but now she was free and I was to find she had a soul as wild as a falcon. Mildrith, poor Mildrith, wanted order and routine. She wanted the hall swept, the clothes clean, the cows milked, the sun to rise, the sun to set, and for nothing to change, but Iseult was different. She was strange, shadow-born, and full of mystery. Nothing she said to me those first days made any sense, for we had no language in common, but on the island, as the sun set and I took the knife to finish cutting the entrails from the goat, she plucked twigs and wove a small cage. She showed me the cage, broke it, and then, with her long white fingers, mimed a bird flying free. She pointed to herself, tossed the twig scraps away, and laughed.

Next morning, still ashore, I saw boats. There were two of them and they were sailing to the west of the island, going northward. They were small craft, probably traders from Cornwalum, and they were running before the southwest wind toward the hidden shore where I assumed Svein had taken the *White Horse*.

We followed the two small ships. By the time we had waded out to *Fyrdraca*, raised her anchor, and rowed her from the lee of the island, both boats were almost out of sight, but once our sail was hoisted we began to overhaul them. They must have been terrified to see a dragon ship shoot out from behind the island, but I lowered the sail a little to slow us down and so followed them for much of the day until, at last, a blue-gray line showed at the sea's edge. Land. We hoisted the sail fully and seethed past the two small, tubby boats, and so, for the first time, I came to the shore of Wales. The Britons had another name for it, but we simply called it Wales, which means "for-eigners," and much later I worked out that we must have made that landfall in Dyfed, which is the name of the churchman who con-verted the Britons of Wales to Christianity and had the westernmost kingdom of the Welsh named for him.

We found a deep inlet for shelter. Rocks guarded the entrance, but once inside we were safe from wind and sea. We turned the ship so that her bows faced the open sea, and the cove was so narrow that our stern scraped stone as we slewed *Fyrdraca* about, and then we slept on board, men and their women sprawled under the rowers' benches. There were a dozen women aboard, all captured from Pere-dur's tribe, and one of them managed to escape that night, presum-ably sliding over the side and swimming to shore. It was not Iseult. She and I slept in the small black space beneath the steering platform, a hole screened by a cloak, and Leofric woke me there in the dawn, worried that the missing woman would raise the country against us. I shrugged. "We won't be here long."

But we stayed in the cove all day. I wanted to ambush ships com-ing around the coast and we saw two, but they traveled together and I could not attack more than one ship at a time. Both ships were under sail, riding the southwest wind, and both were Danish, or per-haps Norse, and both were laden with warriors. They must have come from Ireland, or perhaps from the east coast of Northumbria, and doubtless they traveled to join Svein, lured by the prospect of captur-ing good West Saxon land. "Burgweard should have the whole fleet up here," I said. "He could tear through these bastards."

Two horsemen came to look at us in the afternoon. One had a glinting chain about his neck, suggesting he was of high rank, but neither man came down to the shingle beach. They watched from the head of the small valley that fell to the cove and after a while they went away. The sun was low now, but it was summer, so the days were long. "If they bring men—" Leofric said as the two horsemen rode away, but he did not finish the thought.

I looked up at the high bluffs either side of the cove. Men could rain rocks down from those heights and the *Fyrdraca* would be crushed like an egg. "We could put sentries up there," I suggested, but just then Eadric, who led the men who occupied the forward steorbord benches, shouted that there was a ship in sight. I ran forward and there she was.

The perfect prey.

She was a large ship, not so big as *Fyrdraca*, but large all the same, and she was riding low in the water for she was so heavily laden. Indeed, she carried so many people that her crew had not dared raise the sail for, though the wind was not heavy, it would have bent her leeward side dangerously close to the water. So she was being rowed and now she was close inshore, evidently looking for a place where she could spend the night and her crew had plainly been tempted by our cove and only now realized that we already filled it. I could see a man in her bows pointing farther up the coast and meanwhile my men were arming themselves, and I shouted at Haesten to take the steering oar. He knew what to do and I was confident he would do it well, even though it might mean the death of fellow Danes. We cut the lines that had tethered us to the shore as Leofric brought me my mail coat, helmet, and shield. I dressed for battle as the oars were shipped, then pulled on my helmet so that suddenly the edges of my vision were darkened by its faceplate.

"Go!" I shouted, and the oars bit and the *Fyrdraca* surged out. Some of the oar blades struck rock as we pulled, but none broke, and I was staring at the ship ahead, so close now, and her prow was a snarling wolf, and I could see men and women staring at us, not believing what they saw. They thought they saw a Danish ship, one of their own, yet we were armed and we were coming for them. A man shouted a warning and they scrambled for their weapons, and Leofric yelled at our men to put their hearts into the oars, and the long shafts bent under the strain as *Fyrdraca* leaped across the small waves and I yelled at the men to leave the oars, to come to the bows, and Cenwulf and the twelve men he commanded were already there as our big bows slammed through the enemy oars, snapping them.

Haesten had done well. I had told him to steer for the forward part of the ship, where her freeboard was low, and our bows rode up across her strakes, plunging her low in the water, and we staggered with the impact, but then I jumped down into the wolf-headed ship's belly. Cenwulf and his men were behind me, and there we began the killing.

The enemy ship was so loaded with men that they probably out-numbered us, but they were bone weary from a long day's rowing, they had not expected an attack, and we were hungry for wealth. We had done this before and the crew was well trained, and they chopped their way down the boat, swords and axes swinging, and the sea was slopping over the side so that we waded through water as we clambered over the rowers' benches. The water about our feet grew red. Some of our victims jumped overboard and clung to shattered oars in an attempt to escape us. One man, big-bearded and wild-eyed, came at us with a great sword and Eadric drove a spear into his chest and Leofric struck the man's head with his ax, struck again, and blood sprayed up to the sail that was furled fore and aft on its long yard. The man sank to his knees and Eadric ground the spear deeper so that blood spilled down to the water. I half fell as a wave tilted the half-swamped ship. A man screamed and lunged a spear at me. I took it on my shield, knocked it aside, and rammed Serpent-Breath at his face. He half fell, trying to escape the lunge, and I knocked him over the side with my shield's heavy boss. I sensed movement to my right and swung Serpent-Breath like a reaping scythe and struck a woman in the head. She went down like a felled calf, a sword in her hand. I kicked the sword away and stamped on the woman's belly. A child screamed and I shoved her aside, lunged at a man in a leather jerkin, raised my shield to block his ax blow, and then spitted him on Serpent-Breath. The sword went deep into his belly, so deep that the blade stuck and I had to stand on him to tug it free. Cenwulf went past me, his snarling face covered in blood, sword swinging. The water was up to my knees, and then I staggered and almost fell as the whole ship lurched and I realized we had drifted ashore and struck rocks. Two horses were tethered in the ship's belly and the beasts screamed at the smell of blood. One broke its tether and jumped overboard, swimming white-eyed toward the open sea.

"Kill them! Kill them!" I heard myself shouting. It was the only way to take a ship, to empty her of fighting men, but she was now emptying herself as the survivors jumped onto the rocks and clambered away through the sucking backwash of blood-touched water. A half dozen men had been left aboard *Fyrdraca* and they were fending her off the rocks with oars. A blade stabbed the back of my right ankle and I turned to see a wounded man trying to hamstring me with a short knife. I stabbed down again and again, butchering him in the weltering water, and I think he was the last man to die on board, though a few Danes were still clinging to the ship's side and those we cut away.

The *Fyrdraca* was seaward of the doomed ship now, and I shouted at the men aboard to bring her close. She heaved up and down, much higher than the half-sunk ship, and we threw our plunder up and over the side. There were sacks, boxes, and barrels. Many were heavy, and some clinked with coin. We stripped the enemy dead of their valuables, taking six coats of mail and a dozen helmets and we found another three coats of mail in the flooded bilge. I took eight arm rings off dead men. We tossed weapons aboard *Fyrdraca*, then cut away the captured ship's rigging. I loosed the remaining horse that stood shivering as the water rose. We took the ship's yard and sail, and all the time her survivors watched from the shore where some had found a precarious refuge above the sea-washed rocks. I went to the space beneath her sleeping platform and found a great war-helm there, a beautiful thing with a decorated faceplate and a wolf's head molded in silver on the crown, and I tossed my old helmet onto *Fyrdraca* and donned the new one, and then passed out sacks of coin. Beneath the sacks was what I thought must be a small shield wrapped in black cloth and I half thought of leaving it where it was, then threw it into *Fyrdraca* anyway. We were rich.

"Who are you?" a man shouted from onshore.

"Uhtred," I called back.

He spat at me and I laughed. Our men were climbing back on board *Fyrdraca* now. Some were retrieving oars from the water, and Leofric was pushing *Fyrdraca* away, fearful that she would be caught on the rocks. "Get on board!" he shouted at me, and I saw I was the last man, and so I took hold of *Fyrdraca*'s stern, put a foot on an oar, and heaved myself over her side. "Row!" Leofric shouted, and so we pulled away from the wreck.

Two young women had been thrown up with the plunder and I

found them weeping by *Fyrdraca*'s mast. One spoke no language that I
recognized and later we discovered she was from Ireland, but the other
was Danish and, as soon as I squatted beside her, she lashed out at me
and spat in my face. I slapped her back, and that made her lash out
again. She was a tall girl, strong, with a tangled mass of fair hair and
bright blue eyes. She tried to claw her fingers through the eyeholes of
my new helmet and I had to slap her again, which made my men
laugh. Some were shouting at her to keep fighting me, but instead she
suddenly burst into tears and leaned back against the mast root. I took
off the helmet and asked her name, and her only answer was to wail
that she wanted to die, but when I said she was free to throw herself
off the ship she did not move. Her name was Freyja, she was fifteen
years old, and her father had been the owner of the ship we had sunk.
He had been the big man with the sword, and his name had been Ivar
and he had held land at Dyflin, wherever that was, and Freyja began to
weep again when she looked at my new helmet, which had belonged
to her father. "He died without cutting his nails," she said accusingly,
as if I was responsible for that ill luck, and it was bad fortune indeed
because now the grim things of the underworld would use Ivar's nails
to build the ship that would bring chaos at the world's end.

"Where were you going?" I asked her.

To Svein, of course. Ivar had been unhappy in Dyflin, which was
in Ireland and had more Norsemen than Danes and also possessed
savagely unfriendly native tribes, and he had been lured by the
prospect of land in Wessex and so he had abandoned his Irish
steading, put all his goods and wealth aboard his ships, and sailed
eastward.

"Ships?" I asked her.

"There were three when we left," Freyja said, "but we lost the others in the night."

I guessed they were the two ships we had seen earlier, but the
gods had been good to me for Freyja confirmed that her father had
put his most valuable possessions into his own ship, and that was the
one we had captured, and we had struck lucky for there were barrels

of coins and boxes of silver. There was amber, jet, and ivory. There were weapons and armor. We made a rough count as the *Fyrdraca* wallowed offshore and we could scarce believe our fortune. One box contained small lumps of gold, roughly shaped as bricks, but best of all was the wrapped bundle that I had thought was a small shield, but, when we unwrapped the cloth, proved to be a great silver plate on which was modeled a crucifixion. All about the death scene, ringing the plate's heavy rim, were saints. Twelve of them. I assumed they were the apostles and that the plate had been the treasure of some Irish church or monastery before Ivar had captured it. I showed the plate to my men. "This," I said reverently, "is not part of the plunder. This must go back to the church."

Leofric caught my eye, but did not laugh.

"It goes back to the church," I said again, and some of my men, the more pious ones, muttered that I was doing the right thing. I wrapped the plate and put it under the steering platform.

"How much is your debt to the church?" Leofric asked me.

"You have a mind like a goat's arsehole," I told him.

He laughed, then looked past me. "Now what do we do?" he asked.

I thought he was asking what we should do with the rest of our charmed lives, but instead he was gazing at the shore where, in the evening light, I could see armed men lining the cliff top. The Britons of Dyfed had come for us, but too late. Yet their presence meant we could not go back into our cove, and so I ordered the oars to be manned and for the ship to row eastward. The Britons followed us along the shore. The woman who had escaped in the night must have told them we were Saxons and they must have been praying we would seek refuge on land so they could kill us. Few ships stayed at sea overnight, not unless they were forced to, but I dared not seek shelter and so I turned south and rowed away from the shore, while in the west the sun leaked red fire through rifts in the cloud so that the whole sky glowed as if a god had bled across the heavens.

"What will you do with the girl?" Leofric asked me.

"Freyja?"

"Is that her name? You want her?"

"No," I said.

"I do."

"She'll eat you alive," I warned him. She was probably a head taller than Leofric.

"I like them like that," he said.

"All yours," I said, and such is life. One day Freyja was the pampered daughter of an earl and the next she was a slave.

I gave the coats of mail to those who deserved them. We had lost two men, and another three were badly injured, but that was a light cost. We had, after all, killed twenty or thirty Danes and the survivors were ashore where the Britons might or might not treat them well. Best of all, we had become rich and that knowledge was a consolation as night fell.

Hoder is the god of the night and I prayed to him. I threw my old helmet overboard as a gift to him, because all of us were scared of the dark that swallowed us, and it was a complete dark because clouds had come from the west to smother the sky. No moon, no stars. For a time there was the gleam of firelight on the northern shore, but that vanished and we were blind. The wind rose, the seas heaved us, and we brought the oars inboard and let the air and water carry us for we could neither see nor steer. I stayed on deck, peering into the dark, and Iseult stayed with me, under my cloak, and I remembered the look of delight on her face when we had gone into battle.

Dawn was gray and the sea was white-streaked gray and the wind was cold, and there was no land in sight, but two white birds flew over us and I took them for a sign and rowed in the direction they had gone, and late that day, in a bitter sea and cold rain, we saw land and it was the Isle of Puffins again where we found shelter in the cove and made fires ashore.

"When the Danes know what we've done—" Leofric said.

"They'll look for us." I finished the sentence for him.

"Lots of them will look for us."

"Then it's time to go home," I said.

The gods had been good to us and, next dawn, in a calming sea, we rowed south to the land and followed the coast toward the west. We would go around the wild headlands where the porpoises swam, turn east, and so find home.

Much later I discovered what Svein had done after we parted company and, because what he did affected my life and made the enmity between me and Alfred worse, I shall tell it here.

I suspect that the thought of a gold altar at Cynuit had gnawed into his heart, for he carried the dream back to Glwysing where his men gathered. Glwysing was another kingdom of the Britons in the south of Wales, a place where there were good harbors and where the king welcomed the Danes, for their presence prevented Guthrum's men from raiding across the Mercian border.

Svein ordered a second ship and its crew to accompany him and together they attacked Cynuit. They came in the dawn, hidden by a mist, and I can imagine their beast-headed ships appearing in the early grayness like monsters from nightmare. They went upriver, oars splashing, then grounded the boats and the crews streamed ashore, men in mail and helmets, spear Danes, sword Danes, and they found the half-built church and monastery.

Odda the Younger was making the place, but he knew it was too close to the sea and so he had decided to make it a fortified building. The church's tower was to be of stone, and high enough for men to keep a watch from its summit, and the priests and monks were to be surrounded by a palisade and a flooded ditch, but when Svein came ashore none of the work was finished and so was indefensible, and besides, there were scarce forty troops there and those all died or fled within minutes of the Danes landing. The Danes then burned what work had been done and cut down the high wooden cross that customarily marked a monastery and had been the first thing the builders had made.

The builders were monks, many of them novices, and Svein

herded them together and demanded they show him where the valuables were hidden and promised them mercy if they told the truth.
Which they did. There was not much of value, certainly no altar of
gold, but supplies and timber needed to be purchased and so the
monks had a chest of silver pennies, which was reward enough for
the Danes, who then pulled down the half-constructed church tower,
wrecked the unfinished palisade, and slaughtered some cattle. Then
Svein asked the monks where Ubba was buried and was met by sullen
silence, and the swords were drawn again and the question asked a
second time, and the monks were forced to confess that the church
was being built directly over the dead chieftain's grave. That grave
had been an earth mound, but the monks had dug it up and thrown
the body into the river, and when the Danes heard that story the
mercy fled from their souls.

The monks were made to wade in the river until some bones were
found, and those bones were placed on a funeral pyre made from the
timbers of the half-constructed buildings. It was, by all accounts, a
huge pyre, and when it was lit, and when the bones were at the heart
of a furnace blaze, the monks were thrown onto the flames. While
their bodies burned the Danes selected two girls, captured from the
soldiers' shelters, raped them, and then strangled them, sending their
souls to be company for Ubba in Valhalla. We heard all this from two
children who survived by hiding in a nettle patch, and some folk from
the nearby town who were dragged to see the end of the funeral pyre.
"Svein of the White Horse did this," they were told, and made to
repeat the words. It was a Danish custom to leave some witnesses to
their horror, so that the tales would spread fear and make cowards of
other folk who might be attacked, and sure enough the story of the
burned monks and murdered girls went through Wessex like a high
wind through dry grass. It became exaggerated, as such tales do. The
number of dead monks went from sixteen to sixty, the raped girls from
two to twenty, and the stolen silver from a chest of pennies to a hoard
worthy of the gods. Alfred sent a message to Guthrum, demanding to
know why he should not slaughter the hostages he held, and

Guthrum sent him a present of gold, two captured gospel books, and a groveling letter in which he claimed that the two ships had not been from his forces, but were pirates from beyond the sea. Alfred believed him and so the hostages lived and the peace prevailed, but Alfred commanded that a curse should be pronounced on Svein in every church of Wessex. The Danish chieftain was to be damned through all eternity, his men were to burn in the fires of hell, and his children and his children's children were all to bear the mark of Cain. I asked a priest what the mark was, and he explained that Cain was the son of Adam and Eve and the first murderer, but he did not know what mark he had carried. He thought God would recognize it.

So Svein's two ships sailed away, leaving a pillar of smoke on the Wessex shore, and I knew none of it. In time I would know all of it, but for now I was going home.

We went slowly, sheltering each night, retracing our steps that took us past the blackened hillside where Peredur's settlement had stood, and on we went, under a summer's sun and rain, until we had returned to the Uisc.

The *Heahengel* was afloat now and her mast was stepped, which meant Leofric could take her and the *Eftwyrd*, for the *Fyrdraca* was no more, back to Hamtun. We divided the plunder first and, though Leofric and I took the greater share, every man went away wealthy. I was left with Haesten and Iseult, and I took them up to Oxton where Mildrith wept with relief because she had thought I might be dead. I told her we had been patrolling the coast, which was true enough, and that we had captured a Danish ship laden with wealth, and I spilled the coins and gold bricks onto the floor and gave her a bracelet of amber and a necklace of jet, and the gifts distracted her from Iseult who watched her with wide, dark eyes, and if Mildrith saw the British girl's jewelry she said nothing.

We had come back in time for the harvest, though it was poor for there had been much rain that summer. There was a black growth on the rye, which meant it could not even be fed to the animals, though the straw was good enough to thatch the hall I built. I have always enjoyed building. I made the hall from clay, gravel, and straw, all packed together to make thick walls. Oak beams straddled the walls, and oak rafters held a high, long roof that looked golden when the thatch was first combed into place. The walls were painted with powdered lime in water, and one of the local men poured oxblood into the mix so that the walls were the color of a summer sky at sunset. The hall's great door faced east toward the Uisc and I paid a man from Exanceaster to carve the doorposts and lintels with writhing wolves, for the banner of Bebbanburg, my banner, is a wolf's head. Mildrith wanted the carving to show saints, but she got wolves. I paid the builders well, and when other men heard that I had silver they came looking for employment, and though they were there to build my hall I took only those who had experience fighting. I equipped them with spades, axes, adzes, weapons, and shields.

"You are making an army," Mildrith accused me. Her relief at my homecoming had soured quickly when it was apparent that I was no more a Christian than when I had left her.

"Seventeen men? An army?"

"We are at peace," she said. She believed that because the priests preached it, and the priests only said what they were told to say by the bishops, and the bishops took their orders from Alfred. A traveling priest sought shelter with us one night and he insisted that the war with the Danes was over.

"We still have Danes on the border," I said.

"God has calmed their hearts," the priest insisted, and told me that God had killed the Lothbrok brothers, Ubba, Ivar, and Halfdan, and that the rest of the Danes were so shocked by the deaths that they no longer dared to fight against Christians. "It is true, lord," the priest said earnestly. "I heard it preached in Cippanhamm, and the king was there and he praised God for the truth of it. We are to beat our swords into ard points and our spear blades into reaping hooks."

I laughed at the thought of melting Serpent-Breath into a tool to plow Oxton's fields, but then I did not believe the priest's nonsense.

The Danes were biding their time, that was all, yet it did seem peaceful as the summer slid imperceptibly into autumn. No enemies crossed the frontier of Wessex and no ships harried our coasts. We threshed the corn, netted partridges, hunted deer on the hill, staked nets in the river, and practiced with our weapons. The women spun thread, gathered nuts, and picked mushrooms and blackberries. There were apples and pears, for this was the time of plenty, the time when the livestock was fattened before the winter slaughter. We ate like kings and, when my hall was finished, I gave a feast and Mildrith saw the ox head over the door and knew it was an offering to Thor, but said nothing.

Mildrith hated Iseult, which was hardly surprising, for I had told Mildrith that Iseult was a queen of the Britons and that I held her for the ransom that the Britons would offer. I knew no such ransom would ever come, and the story went some way to explaining Iseult's presence, but Mildrith resented that the British girl was given her own house. "She is a queen," I said.

"You take her hunting," Mildrith said resentfully.

I did more than that, but Mildrith chose to be blind to much of it. Mildrith wanted little more than her church, her baby, and an unvarying routine. She had charge of the women who milked the cows, churned the butter, spun wool, and collected honey, and she took immense pride that those things were done well. If a neighbor visited there would be a flurry of panic as the hall was cleaned, and she worried much about those neighbors' opinions. She wanted me to pay Oswald's wergild. It did not matter to Mildrith that the man had been caught thieving, because to pay the wergild would make peace in the valley of the Uisc. She even wanted me to visit Odda the Younger. "You could be friends," she pleaded.

"With that snake?"

"And Wirken says you have not paid the tithe."

Wirken was the priest in Exanmynster, and I hated him. "He eats and drinks the tithe," I snarled. The tithe was the payment all landholders were supposed to make to the church, and by rights I should have sent Wirken part of my harvest, but I had not. Yet the priest was often at Oxton, coming when he thought I was hunting, and he ate my food and drank my ale and was growing fat on them.

"He comes to pray with us," Mildrith said.

"He comes to eat," I said.

"And he says the bishop will take the land if we don't pay the debt."

"The debt will be paid," I said.

"When? We have the money!" She gestured at the new hall. "When?" she insisted.

"When I want to," I snarled. I did not tell her when, or how, because if I had then Wirken the priest would know, and the bishop would know. It was not enough to pay the debt. Mildrith's father had foolishly donated part of our land's future produce to the church, and I wanted that burden taken away so the debt would not go on through eternity, and to do that I needed to surprise the bishop, and so I kept Mildrith ignorant, and inevitably those arguments would end with her tears. I was bored with her and she knew it. I found her beating Iseult's maid one day. The girl was a Saxon I had given to Iseult as a servant, but she also worked in the dairy and Mildrith was beating her because some cheeses had not been turned. I dragged Mildrith away, and that, of course, provoked another argument and Mildrith proved not to be so blind after all for she accused me of trying to whelp bastards on Iseult, which was true enough, but I reminded her that her own father had sired enough bastards, half a dozen of whom now worked for us. "You leave Iseult and her maid alone," I said, causing more tears. They were not happy days.

It was the time when Iseult learned to speak English, or at least the Northumbrian version of English for she learned it mostly from me. "You're my mon," she said. I was Mildrith's man and Iseult's mon. She said she had been born again on the day I came into Peredur's hall. "I had dreamed of you," she said, "tall and golden-haired."

"Now you don't dream?" I asked, knowing that her powers of scrying came from dreams.

"I do still dream," she said earnestly. "My brother speaks to me."

"Your brother?" I asked, surprised.

"I was born a twin," she told me, "and my brother came first and then, as I was born, he died. He went to the shadow world and he speaks to me of what he sees there."

"What does he see?"

"He sees your king."

"Alfred," I said sourly. "Is that good or bad?"

"I don't know. The dreams are shadowy."

She was no Christian. Instead she believed that every place and every thing had its own god or goddess: a nymph for a stream, a dryad for a wood, a spirit for a tree, a god for the fire and another for the sea. The Christian god, like Thor or Odin, was just one more deity among this unseen throng of powers, and her dreams, she said, were like eavesdropping on the gods. One day, as she rode beside me on the hills above the empty sea, she suddenly said that Alfred would give me power.

"He hates me," I told her. "He'll give me nothing."

"He will give you power," she said flatly. I stared at her and she gazed to where the clouds met the waves. Her black hair was unbound and the sea wind stirred it. "My brother told me," she said. "Alfred will give you power and you will take back your northern home and your woman will be a creature of gold."

"My woman?"

She looked at me and there was sadness in her face. "There," she said, "now you know," and she kicked back her heels and made the horse run along the ridgetop, her hair streaming, her eyes wet with tears. I wanted to know more, but she said she had told me what she had dreamed and I must be content.

At summer's end we drove the swine into the forests to feed on the fallen beech nuts and acorns. I bought bags of salt because the killing time was coming and the meat of our pigs and cattle would have to be salted into barrels to feed us through the winter. Some of that food would come from the men who rented land at the edge of the estate, and I visited them all so they would know I expected payment of wheat, barley, and livestock, and to show them what would happen if they tried to cheat me I bought a dozen good swords from a smith in Exanceaster. I gave the swords to my men, and in the shortening days we practiced with them. Mildrith might not believe war was coming, but I did not think God had changed Danish hearts.

The late autumn brought heavy rain and the shire reeve to Oxton. The reeve was called Harald and he was charged with keeping the peace of Defnascir, and he came on horseback and with him were six other horsemen, all in mail coats and helmets, and all with swords or spears. I waited for him in the hall, making him dismount and come into the smoky shadows. He came cautiously, expecting an ambush. Then his eyes became accustomed to the gloom and he saw me standing by the central hearth. "You are summoned to the shire court," he told me.

His men had followed Harald into the hall. "You bring swords into my house?" I asked.

Harald looked around the hall and he saw my men armed with their spears and axes. I had seen the horsemen approaching and summoned my men and ordered them to arm themselves.

Harald had the reputation of being a decent man, sensible and fair, and he knew how weapons in a hall could lead to slaughter. "You will wait outside," he told his men, and I gestured for my men to put their weapons down. "You are summoned . . ." Harald began again.

"I heard you," I said.

"There is a debt to be paid," he said, "and a man's death to make good."

I said nothing. One of my hounds growled softly and I put a hand into its fur to silence it.

"The court will meet on All Saints' Day," Harald said, "at the cathedral."

"I shall be there," I said.

He took off his helmet to reveal a balding pate fringed with brown hair. He was at least ten years older than I, a big man, with two fingers missing from his shield hand. He limped slightly as he walked toward me. I calmed the hounds, waited.

"I was at Cynuit," he said to me, speaking softly.

"So was I," I said, "though men pretend I was not."

"I know what you did," he said.

"So do I."

He ignored my surliness. He was showing me sympathy, though I was too proud to show I appreciated it. "The ealdorman has sent men," he warned me, "to take this place once judgment is given."

There was a gasp behind me and I realized Mildrith had come into the hall. Harald bowed to her.

"The hall will be taken?" Mildrith asked.

"If the debt is not paid," Harald said, "the land will be given to the church." He stared up at the newly hewn rafters as if wondering why I would build a hall on land doomed to be given to God.

Mildrith came to stand beside me. She was plainly distressed by Harald's summons, but she made a great effort to compose herself. "I am sorry," she said, "about your wife."

A flicker of pain crossed Harald's face as he made the sign of the cross. "She was sick a long time, lady. It was merciful of God, I think, to take her."

I had not known he was a widower, nor did I care much. "She was a good woman," Mildrith said.

"She was," Harald said.

"And I pray for her."

"I thank you for that," Harald said.

"As I pray for Odda the Elder," Mildrith went on.

"God be praised, he lives." Harald made the sign of the cross again. "But he is feeble and in pain." He touched his scalp showing where Odda the Elder had been wounded.

"So who is the judge?" I asked harshly, interrupting the two.

"The bishop," Harald said.

"Not the ealdorman?"

"He is at Cippanhamm."

Mildrith insisted on giving Harald and his men ale and food. She and Harald talked a long time, sharing news of neighbors and family. They were both from Defnascir and I was not, and so I knew few of the folk they talked about, but I pricked up my ears when Harald said that Odda the Younger was marrying a girl from Mercia. "She's in exile here," he said, "with her family."

"Well born?" Mildrith asked.

"Exceedingly," Harald said.

"I wish them much joy," Mildrith said with evident sincerity. She was happy that day, warmed by Harald's company, though when he had gone she chided me for being churlish. "Harald is a good man," she insisted, "a kind man. He would have given you advice. He would have helped you!"

I ignored her, but two days later I went into Exanceaster with Iseult and all my men. Including Haesten I now had eighteen warriors and I had armed them, given them shields and leather coats, and I led them through the market that always accompanied the court's sittings. There were stilt walkers and jugglers, a man who ate fire, and a dancing bear. There were singers, harpists, storytellers, beggars, and pens of sheep, goats, cattle, pigs, geese, ducks, and hens. There were fine cheeses, smoked fish, bladders of lard, pots of honey, trays of apples, and baskets of pears. Iseult, who had not been to Exanceaster before, was amazed at the size of the city, and the life of it, and the seething closeness of its houses, and I saw folk make the sign of the cross when they saw her for they had heard of the shadow queen held at Oxton and they knew her for a foreigner and a pagan.

Beggars crowded at the bishop's gate. There was a crippled woman with a blind child, men who had lost arms or legs in the wars, a score of them, and I threw them some pence. Then, because I was on horseback, I ducked under the archway of the courtyard beside the cathedral where a dozen chained felons were awaiting their fate. A group of young monks, nervous of the chained men, were plaiting beehives, while a score of armed men were clustered around three fires. They eyed my followers suspiciously as a young priest, his hands flapping, hurried across the puddles. "Weapons are not to be brought into the precinct!" he told me sternly.

"They've got weapons," I said, nodding at the men warming themselves by the flames.

"They are the reeve's men."

"Then the sooner you deal with my business," I said, "the sooner my weapons will be gone."

He looked up at me, his face anxious. "Your business?"

"Is with the bishop."

"The bishop is at prayer," the priest said reprovingly, as though I should have known that. "And he cannot see every man who comes here. You can talk to me."

I smiled and raised my voice a little. "In Cippanhamm, two years ago," I said, "your bishop was friends with Eanflæd. She has red hair and works her trade out of the Corncrake tavern. Her trade is whoring."

The priest's hands were flapping again in an attempt to persuade me to lower my voice.

"I've been with Eanflæd," I said, "and she told me about the bishop. She said—"

The monks had stopped making beehives and were listening, but the priest cut me off by half shouting. "The bishop might have a moment free."

"Then tell him I'm here," I said pleasantly.

"You are Uhtred of Oxton?" he asked.

"No," I said. "I am Lord Uhtred of Bebbanburg."

"Yes, lord."

"Sometimes known as Uhtredærwe," I added mischievously. Uhtred the Wicked.

"Yes, lord," the priest said again and hurried away.

The bishop was called Alewold and he was really the Bishop of Cridianton, but that place had not been thought as safe as Exanceaster and so for years the Bishops of Cridianton had lived in the larger town, which, as Guthrum had shown, was not the wisest decision.

Guthrum's Danes had pillaged the cathedral and the bishop's house, which was still scantily furnished, and I discovered Alewold, sitting behind a table that looked as if it had once belonged to a butcher, for its hefty top was scored with knife cuts and stained with old blood. He looked at me indignantly. "You should not be here," he said snidely.

"Why not?"

"You have business before the court tomorrow."

"Tomorrow," I said, "you sit as a judge. Today you are a bishop."

He acknowledged that with a small nod. He was an elderly man with a heavy jowled face and a reputation as a severe judge. He had been with Alfred in Scireburnan when the Danes arrived in Exanceaster, which is why he was still alive, and, like all the bishops in Wessex, he was a fervent supporter of the king, and I had no doubt that Alfred's dislike of me was known to Alewold, which meant I could expect little clemency when the court sat.

"I am busy," Alewold said, gesturing at the parchments on the stained table. Two clerks shared the table and a half dozen resentful priests had gathered behind the bishop's chair.

"My wife," I said, "inherited a debt to the church."

Alewold looked at Iseult who alone had come into the house with me. She looked beautiful, proud, and wealthy. There was silver at her throat and in her hair, and her cloak was fastened with two brooches, one of jet and the other of amber. "Your wife?" the bishop asked snidely.

"I would discharge the debt," I said, ignoring his question, and I tipped a bag onto his butcher's table and the big silver plate we had taken from Ivar slid out. The silver made a satisfying noise as it thumped down and suddenly, in that small dark room ill lit by three rush lights and a small, wood-barred window, it seemed as if the sun had come out. The heavy silver glowed and Alewold just stared at it.

There are good priests. Beocca is one and Willibald another, but I have discovered in my long life that most churchmen preach the merits of poverty while they lust after wealth. They love money and the church attracts money like a candle brings moths. I knew Alewold was a greedy man, as greedy for wealth as he was for the delights of a red-haired whore in Cippanhamm, and he could not take his eyes from that plate. He reached out and caressed the thick rim as if he scarce believed what he was seeing, and then he pulled the plate toward him and examined the twelve apostles. "A pyx," he said reverently.

"A plate," I said casually.

One of the other priests leaned over a clerk's shoulder. "Irish work," he said.

"It looks Irish," Alewold agreed, then looked suspiciously at me. "You are returning it to the church?"

"Returning it?" I asked innocently.

"The plate was plainly stolen," Alewold said, "and you do well, Uhtred, to bring it back."

"I had the plate made for you," I said.

He turned the plate over, which took some effort for it was heavy, and once it was inverted he pointed to the scratches in the silver. "It is old," he said.

"I had it made in Ireland," I said grandly, "and doubtless it was handled roughly by the men who brought it across the sea."

He knew I was lying. I did not care. "There are silversmiths in Wessex who could have made you a pyx," one of the priests snapped.

"I thought you might want it," I said, then leaned forward and pulled the plate out of the bishop's hands, "but if you prefer West Saxon work," I went on, "then I can—"

"Give it back!" Alewold said and, when I made no move to obey, his voice became pleading. "It is a beautiful thing." He could see it in his church, or perhaps in his hall, and he wanted it. There was silence as he stared at it. If he had known that the plate existed, if I had told Mildrith of it, then he would have had a response ready, but as it was he was overwhelmed by desire for the heavy silver. A maid brought in a flagon and he waved her out of the room. She was, I noted, red-haired. "You had the plate made," Alewold said skeptically.

"In Dyflin," I said.

"Is that where you went in the king's ship?" the priest who had snapped at me asked.

"We patrolled the coast," I said, "nothing more."

"The value of the plate—" Alewold began, then stopped.

"Is far and above the debt Mildrith inherited," I said. That was probably not true, but it was close to the amount, and I could see Alewold did not care. I was going to get what I wanted.

The debt was discharged. I insisted on having that written down, and written three times, and I surprised them by being able to read and so discovering that the first scrap of parchment made no mention of the church yielding their rights to the future produce of my estate, but that was corrected and I let the bishop keep one copy while I took two. "You will not be arraigned for debt," the bishop said as he pressed his seal into the wax of the last copy, "but there is still the matter of Oswald's wergild."

"I rely on your good and wise judgment, bishop," I said, and I opened the purse hanging at my waist and took out a small lump of gold, making sure he could see there was more gold inside as I placed the small lump on the plate. "Oswald was a thief."

"His family will make oaths that he was not," the priest said.

"And I will bring men who will swear he was," I said. A trial relied heavily on oaths, but both sides would bring as many liars as they could muster, and judgment usually went to the better liars or, if both sides were equally convincing, to the side who had the sympathy of the onlookers. It was better, though, to have the sympathy of the judge. Oswald's family would have many supporters around Exanceaster, but gold is much the best argument in a law court.

And so it proved. To Mildrith's astonishment the debt was paid and Oswald's family denied two hundred shillings of wergild. I did not even bother to go to the court, relying on the persuasive power of gold, and sure enough the bishop peremptorily dismissed the demand for wergild, saying it was well known that Oswald had been a thief, and so I won. That did not make me any more popular. To the folk who lived in the Uisc's valley I was a Northumbrian interloper and, worse, it was known I was a pagan, but none dared confront me for I went nowhere beyond the estate without my men and my men went nowhere without their swords.

The harvest was in the storehouses. Now was the time for the Danes to come, when they could be sure to find food for their armies, but neither Guthrum nor Svein crossed the frontier. The winter came instead and we slaughtered the livestock, salted the meat, scraped hides, and made calves' foot jelly. I listened for the sound of church bells ringing at an unusual time, for that would have been a sign that the Danes had attacked, but the bells did not ring.

Mildrith prayed that the peace would continue and I, being young and bored, prayed it would not. She prayed to the Christian god and I took Iseult to the high woods and made a sacrifice to Hoder, Odin, and Thor and the gods were listening, for in the dark beneath the gallows tree, where the three spinners make our lives, a red thread was woven into my life. Fate is everything, and just after Yule the spinners brought a royal messenger to Oxton and he, in turn, brought me a summons. It seemed possible that Iseult's dream was true, and that Alfred would give me power for I was ordered to Cippanhamm to see the king. I was summoned to the witan.

FIVE

Mildrith was excited by the summons. The witan gave the king advice and her father had never been wealthy or important enough to receive such a summons, and she was overjoyed that the king wanted my presence. The witanegemot, as the meeting was called, was always held on the Feast of Saint Stephen, the day after Christmas, but my summons required me to be there on the twelfth day of Christmas and that gave Mildrith time to wash clothes for me. They had to be boiled and scrubbed and dried and brushed, and three women did the work and it took three days before Mildrith was satisfied that I would not disgrace her by appearing at Cippanhamm looking like a vagabond. She was not summoned, nor did she expect to accompany me, but she made a point of telling all our neighbors that I was to give counsel to the king. "You mustn't wear that," she told me, pointing to my Thor's hammer amulet.

"I always wear it," I said.

"Then hide it," she said, "and don't be belligerent!"

"Belligerent?"

"Listen to what others say," she said. "Be humble. And remember to congratulate Odda the Younger."

"For what?"

"He's to be married. Tell him I pray for them both." She was happy again, sure that by paying the church its debt I had regained Alfred's favor and her good mood was not even spoiled when I announced I would take Iseult with me. She bridled slightly at the news, then said that it was only right that Iseult should be taken to Alfred. "If she is a queen," Mildrith said, "then she belongs in Alfred's court. This isn't a fit place for her." She insisted on taking silver coins to the church in Exanceaster where she donated the money to the poor and gave thanks that I had been restored to Alfred's favor. She also thanked God for the good health of our son, Uhtred. I saw little of him, for he was still a baby and I have never had much patience for babies, but the women of Oxton constantly assured me that he was a lusty, strong boy.

We allowed two days for the journey. I took Haesten and six men as an escort for though the shire reeve's men patrolled the roads, there were plenty of wild places where outlaws preyed on travelers. We were in mail coats or leather tunics, with swords, spears, axes, and shields. We all rode. Iseult had a small black mare I had bought for her, and I had also given her an otter-skin cloak, and when we passed through villages folk would stare at her for she rode like a man, her black hair bound up with a silver chain. They would kneel to her, as well as to me, and call out for alms. She did not take her maid for I remembered how crowded every tavern and house had been in Exanceaster when the witan met, and I persuaded Iseult that we would be hard-pressed to find accommodation for ourselves, let alone a maid.

"What does the king want of you?" she asked as we rode up the Uisc valley. Rainwater puddled in the long furrows, gleaming in the winter sunshine, while the woods were glossy with holly leaves and bright with the berries of rowan, thorn, elder, and yew.

"Aren't you supposed to tell me that?" I asked her.

She smiled. "Seeing the future," she said, "is like traveling a strange road. Usually you cannot see far ahead, and when you can it is only a glimpse. And my brother doesn't give me dreams about everything."

"Mildrith thinks the king has forgiven me," I said.

"Has he?"

I shrugged. "Perhaps." I hoped so, not because I wanted Alfred's forgiveness, but because I wanted to be given command of the fleet again. I wanted to be with Leofric. I wanted the wind in my face and the sea rain on my cheek. "It's odd, though," I went on, "that he didn't want me there for the whole witanegemot."

"Maybe," Iseult suggested, "they discussed religious things at first?"

"He wouldn't want me there for that," I said.

"So that's it," she said. "They talk about their god, but at the end they will talk of the Danes, and that is why he summoned you. He knows he needs you."

"Or perhaps he just wants me there for the feast," I suggested.

"The feast?"

"The Twelfth Night feast," I explained, and that seemed to me the likeliest explanation; that Alfred had decided to forgive me and, to show he now approved of me, would let me attend the winter feast. I secretly hoped that was true, and it was a strange hope. I had been ready to kill Alfred only a few months before, yet now, though I still hated him, I wanted his approval. Such is ambition. If I could not rise with Ragnar then I would make my reputation with Alfred.

"Your road, Uhtred," Iseult went on, "is like a bright blade across a dark moor. I see it clearly."

"And the woman of gold?"

She said nothing to that.

"Is it you?" I asked.

"The sun dimmed when I was born," she said, "so I am a woman of darkness and of silver, not of gold."

"So who is she?"

"Someone far away, Uhtred, far away," and she would say no more. Perhaps she knew no more, or perhaps she was guessing.

We reached Cippanhamm late on the eleventh day of Yule. There was still frost on the furrows and the sun was a gross red ball poised low above the tangling black branches as we came to the town's western gate. The city was full, but I was known in the Corncrake tavern where the redheaded whore called Eanflæd worked and she found us shelter in a half-collapsed cattle byre where a score of hounds had been kenneled. The hounds, she said, belonged to Huppa, Ealdorman of Thornsæta, but she reckoned the animals could survive a night or two in the yard. "Huppa may not think so," she said, "but he can rot in hell."

"He doesn't pay?" I asked her.

She spat for answer, then looked at me curiously. "I hear Leofric's here."

"He is?" I said, heartened by the news.

"I haven't seen him," she said, "but someone said he was here. In the royal hall. Maybe Burgweard brought him?" Burgweard was the new fleet commander, the one who wanted his ships to sail two by two in imitation of Christ's disciples. "Leofric had better not be here," Eanflæd finished.

"Why not?"

"Because he hasn't come to see me," she said indignantly, "that's why!" She was five or six years older than I with a broad face, a high forehead, and springy hair. She was popular, so much so that she had a good deal of freedom in the tavern that owed its profits more to her abilities than to the quality of the ale. I knew she was friendly with Leofric, but I suspected from her tone that she wanted to be more than friends. "Who's she?" she asked, jerking her head at Iseult.

"A queen," I said.

"That's another name for it, I suppose. How's your wife?"

"Back in Defnascir."

"You're like all the rest, aren't you?" She shivered. "If you're cold tonight bring the hounds back in to warm you. I'm off to work."

We were cold, but I slept well enough and, next morning, the twelfth after Christmas, I left my six men at the Corncrake and took Iseult and Haesten to the king's buildings that lay behind their own palisade to the south of the town where the river curled about the walls. A man expected to attend the witanegemot with retainers, though not usually with a Dane and a Briton, but Iseult wanted to see

Alfred and I wanted to please her. Besides, there was the great feast that evening and, though I warned her that Alfred's feasts were poor things, Iseult still wanted to be there. Haesten, with his mail coat and sword, was there to protect her for I suspected she might not be allowed into the hall while the witanegemot debated and so might have to wait until evening for her chance to glimpse Alfred.

The gatekeeper demanded that we surrender our weapons, a thing I did with a bad grace, but no man, except the king's own household troops, could go armed in Alfred's presence. The day's talking had already begun, the gatekeeper told us, and so we hurried past the stables and past the big new royal chapel with its twin towers. A group of priests was huddled by the main door of the great hall and I recognized Beocca, my father's old priest, among them. I smiled in greeting, but his face, as he came toward us, was drawn and pale. "You're late," he said sharply.

"You're not pleased to see me?" I asked sarcastically.

He looked up at me. Beocca, despite his squint, red hair, and palsied left hand, had grown into a stern authority. He was now a royal chaplain, confessor, and a confidant to the king, and the responsibilities had carved deep lines on his face. "I prayed," he said, "never to see this day." He made the sign of the cross. "Who's that?" he asked, staring at Iseult.

"A queen of the Britons," I said.

"She's what?"

"A queen. She's with me. She wants to see Alfred."

I don't know whether he believed me, but he seemed not to care. Instead he was distracted, worried, and, because he lived in a strange world of kingly privilege and obsessive piety, I assumed his misery had been caused by some petty theological dispute. He had been Bebbanburg's mass priest when I was a child and, after my father's death, he had fled Northumberland because he could not abide living among the pagan Danes. He had found refuge in Alfred's court where he had become a friend of the king. He was also a friend to me, a man who had preserved the parchments that proved my claim to the lordship of Bebbanburg, but on that twelfth day of Yule he was anything but pleased to see me. He plucked my arm, drawing me toward the door. "We must go in," he said, "and may God in his mercy protect you."

"Protect me?"

"God is merciful," Beocca said, "and you must pray for that mercy," and then the guards opened the door and we walked into the great hall. No one stopped Iseult and, indeed, there were a score of other women watching the proceedings from the edge of the hall.

There were also more than a hundred men there, though only forty or fifty comprised the witanegemot, and those thegns and senior churchmen were on chairs and benches set in a half circle in front of the dais where Alfred sat with two priests and with Ælswith, his wife, who was pregnant. Behind them, draped with a red cloth, was an altar on which stood thick candles and a heavy silver cross while all about the walls were platforms where, in normal times, folk slept or ate to be out of the fierce drafts. This day, though, the platforms were crammed with the followers of the thegns and noblemen of the witan and among them, of course, were a lot of priests and monks for Alfred's court was more like a monastery than a royal hall. Beocca gestured that Iseult and Haesten should join those spectators, then he drew me toward the half circle of privileged advisers.

No one noticed my arrival. It was dark in the hall, for little of the wintry sunshine penetrated the small high windows. Braziers tried to give some warmth, but failed, succeeding only in thickening the smoke in the high rafters. There was a large central hearth, but the fire had been taken away to make room for the witanegemot's circle of stools, chairs, and benches. A tall man in a blue cloak was on his feet as I approached. He was talking of the necessity of repairing bridges, and how local thegns were skimping the duty, and he suggested that the king appoint an official to survey the kingdom's roads. Another man interrupted to complain that such an appointment would encroach on the privileges of the shire ealdormen, and that started a chorus of voices, some for the proposal, most against, and two priests, seated at a small table beside Alfred's dais, tried to write down all the comments. I recognized Wulfhere, the Ealdorman of Wiltunscir, who yawned prodigiously. Close to him was Alewold, the Bishop of Exanceaster, who was swathed in furs. Still no one noticed me. Beocca had held me back, as if waiting for a lull in the proceedings before finding me a seat. Two servants brought in baskets of logs to feed the braziers, and it was then that Ælswith saw me and she leaned across and whispered in Alfred's ear. He had been paying close attention to the discussion, but now looked past his council to stare at me.

And a silence fell on that great hall. There had been a murmur of voices when men saw the king being distracted from the argument about bridges and they had all turned to look at me and then there was the silence that was broken by a priest's sneeze and a sudden odd scramble as the men closest to me, those sitting beside the cold stones of the hearth, moved to one side. They were not making way for me, but avoiding me.

Ælswith was smiling and I knew I was in trouble then. My hand instinctively went to my left side, but of course I had no sword so could not touch her hilt for luck. "We shall talk of bridges later," Alfred said. He stood. He wore a bronze circlet as a crown and had a fur-trimmed blue robe, matching the gown worn by his wife.

"What is happening?" I asked Beocca.

"You will be silent!" It was Odda the Younger who spoke. He was dressed in his war glory, in shining mail covered by a black cloak, in high boots and with a red-leather sword belt from which hung his weapons, for Odda, as commander of the king's troops, was permitted to go armed into the royal hall. I looked into his eyes and saw triumph there, the same triumph that was on the Lady Ælswith's pinched face, and I knew I had not been brought to receive the king's favor, but summoned to face my enemies.

I was right. A priest was called from the dark gaggle beside the door. He was a young man with a pouchy, scowling face. He moved briskly, as if the day did not have enough hours to complete his work. He bowed to the king, then took a parchment from the table where the two clerks sat and came to stand in the center of the witan's circle.

"There is an urgent matter," Alfred said, "which, with the witan's permission, we shall deal with now." No one there was likely to disagree so a low murmur offered approval of interrupting the more mundane discussions. Alfred nodded. "Father Erkenwald will read the charges," the king said, and took his throne again.

Charges? I was confused like a boar trapped between hounds and spears, and I seemed incapable of movement so I just stood there as Father Erkenwald unrolled the parchment and cleared his throat. "Uhtred of Oxton," he said, speaking in a high and precise voice, "you are this day charged with the crime of taking a king's ship without our king's consent, and with taking that ship to the country of Cornwalum and there making war against the Britons, again without our king's consent, and this we can prove by oaths." There was a small murmur in the hall, a murmur that was stilled when Alfred raised a thin hand. "You are further charged," Erkenwald went on, "with making an alliance with the pagan called Svein, and with his help you murdered Christian folk in Cornwalum, despite those folk living in peace with our king, and this also we can prove by oaths." He paused, and now there was complete silence in the hall. "And you are charged"—Erkenwald's voice was lower now, as though he could scarce believe what he was reading—"with joining the pagan Svein in an attack on our blessed king's realm by committing vile murder and impious church-robbery at Cynuit." This time there was no murmur, but a loud outburst of indignation and Alfred made no move to check it, so Erkenwald had to raise his voice to finish the indictment. "And this also," he was shouting now, and men hushed to listen, "we shall prove by oaths." He lowered the parchment, gave me a look of pure loathing, then walked back to the edge of the dais.

"He's lying," I snarled.

"You will have a chance to speak," a fierce-looking churchman sitting beside Alfred said. He was in monk's robes, but over them he wore a priest's half cape richly embroidered with crosses. He had a full head of white hair and a deep, stern voice.

"Who's that?" I asked Beocca.

"The most holy Æthelred," Beocca said softly and, seeing I did not recognize the name, "Archbishop of Contwaraburg, of course."

The archbishop leaned over to speak with Erkenwald. Ælswith was staring at me. She had never liked me, and now she was watching my destruction and taking a great pleasure from it. Alfred, meanwhile, was studying the roof beams as though he had never noticed them before, and I realized he intended to take no part in this trial, for trial it was. He would let other men prove my guilt, but doubtless he would pronounce sentence, and not just on me, it seemed, because the archbishop scowled. "Is the second prisoner here?"

"He is held in the stables," Odda the Younger said.

"He should be here," the archbishop said indignantly. "A man has a right to hear his accusers."

"What other man?" I demanded.

It was Leofric, who was brought into the hall in chains, and there was no outcry against him because men perceived him as my follower. The crime was mine, Leofric had been snared by it, and now he would suffer for it, but he plainly had the sympathy of the men in the hall as he was brought to stand beside me. They knew him, he was of Wessex, while I was a Northumbrian interloper. He gave me a rueful glance as the guards led him to my side. "Up to our arses in it," he muttered.

"Quiet!" Beocca hissed.

"Trust me," I said.

"Trust you?" Leofric asked bitterly.

But I had glanced at Iseult and she had given me the smallest shake of her head, an indication, I reckoned, that she had seen the outcome of this day and it was good. "Trust me," I said again.

"The prisoners will be silent," the archbishop said.

"Up to our royal arses," Leofric said quietly.

The archbishop gestured at Father Erkenwald. "You have oath-takers?" he asked.

"I do, lord."

"Then let us hear the first."

Erkenwald gestured to another priest who was standing by the door leading to the passage at the back of the hall. The door was opened and a slight figure in a dark cloak entered. I could not see his face for he wore a hood. He hurried to the front of the dais and there bowed low to the king and went on his knees to the archbishop who held out a hand so that his heavy jeweled ring could be kissed. Only then did the man stand, push back his hood, and turn to face me.

It was the Ass. Asser, the Welsh monk. He stared at me as yet another priest brought him a gospel book on which he laid a thin hand. "I make oath," he said in accented English, still staring at me, "that what I say is truth, and God so help me in that endeavor and condemn me to the eternal fires of hell if I dissemble." He bent and kissed the gospel book with the tenderness of a man caressing a lover.

"Bastard," I muttered.

Asser was a good oath-taker. He spoke clearly, describing how I had come to Cornwalum in a ship that bore a beast head on its prow and another on its stern. He told how I had agreed to help King Peredur, who was being attacked by a neighbor assisted by the pagan Svein, and how I had betrayed Peredur by allying myself with the Dane. "Together," Asser said, "they made great slaughter, and I myself saw a holy priest put to death."

"You ran like a chicken," I said to him. "You couldn't see a thing."

Asser turned to the king and bowed. "I did run, lord king. I am a brother monk, not a warrior, and when Uhtred turned that hillside red with Christian blood I did take flight. I am not proud of that, lord king, and I have earnestly sought God's forgiveness for my cowardice."

Alfred smiled and the archbishop waved away Asser's remarks as if they were nothing. "And when you left the slaughter," Erkenwald asked, "what then?"

"I watched from a hilltop," Asser said, "and I saw Uhtred of Oxton leave that place in the company of the pagan ship. Two ships sailing westward."

"They sailed westward?" Erkenwald asked.

"To the west," Asser confirmed.

Erkenwald glanced at me. There was silence in the hall as men leaned forward to catch each damning word. "And what lay to the west?" Erkenwald asked.

"I cannot say," Asser said. "But if they did not go to the end of the world then I assume they turned about Cornwalum to go into the Sæfern Sea."

"And you know no more?" Erkenwald asked.

"I know I helped bury the dead," Asser said, "and I said prayers for their souls, and I saw the smoldering embers of the burned church, but what Uhtred did when he left the place of slaughter I do not know. I only know he went westward."

Alfred was pointedly taking no part in the proceedings, but he plainly liked Asser for, when the Welshman's testimony was done, he beckoned him to the dais and rewarded him with a coin and a moment of private conversation. The witan talked among themselves, sometimes glancing at me with the curiosity we give to doomed men. Lady Ælswith, suddenly so gracious, smiled on Asser.

"You have anything to say?" Erkenwald demanded of me when Asser had been dismissed.

"I shall wait," I said, "till all your lies are told."

The truth, of course, was that Asser had told the truth, and told it plainly, clearly, and persuasively. The king's councillors had been impressed, just as they were impressed by Erkenwald's second oath-taker.

It was Steapa Snotor, the warrior who was never far from Odda the Younger's side. His back was straight, his shoulders square, and his feral face with its stretched skin was grim. He glanced at me, bowed to the king, then laid a huge hand on the gospel book and let Erkenwald lead him through the oath, and he swore to tell the truth on pain of hell's eternal agony, and then he lied. He lied calmly in a flat, toneless voice. He said he had been in charge of the soldiers who guarded the place at Cynuit where the new church was being built, and how two ships had come in the dawn and how warriors streamed from the ships, and how he had fought against them and killed six of them, but there were too many, far too many, and he had been forced to retreat, but he had seen the attackers slaughter the priests and he had heard the pagan leader shout his name as a boast. "Svein, he was called."

"And Svein brought two ships?"

Steapa paused and frowned, as though he had trouble counting to two, then nodded. "He had two ships."

"He led both?"

"Svein led one of the ships," Steapa said. Then he pointed a finger at me. "And he led the other."

The audience seemed to growl and the noise was so threatening that Alfred slapped the arm of his chair and finally stood to restore quiet. Steapa seemed unmoved. He stood, solid as an oak, and though he had not told his tale as convincingly as Brother Asser, there was something very damning in his testimony. It was so matter-of-fact, so unemotionally told, so straightforward, and none of it was true.

"Uhtred led the second ship," Erkenwald said, "but did Uhtred join in the killing?"

"Join it?" Steapa asked. "He led it." He snarled those words and the men in the hall growled their anger.

Erkenwald turned to the king. "Lord king," he said, "he must die."

"And his land and property must be forfeited!" Bishop Alewold shouted in such excitement that a whirl of his spittle landed and hissed in the nearest brazier. "Forfeited to the church!"

The men in the hall thumped their feet on the ground to show their approbation. Ælswith nodded vigorously, but the archbishop clapped his hands for silence. "He has not spoken," he reminded Erkenwald, then nodded at me. "Say your piece," he ordered curtly.

"Beg for mercy," Beocca advised me quietly.

When you are up to your arse in shit there is only one thing to do. Attack, and so I admitted I had been at Cynuit, and that admission provoked some gasps in the hall. "But I was not there this summer," I went on. "I was there in the spring at which time I killed Ubba Lothbrokson, and there are men in this hall who saw me do it! Yet Odda the Younger claimed the credit. He took Ubba's banner, which I laid low, and he took it to his king and he claimed to have killed Ubba. Now, lest I spread the truth, which is that he is a coward and a liar, he would have me murdered by lies." I pointed to Steapa. "His lies."

Steapa spat to show his scorn. Odda the Younger was looking furious, but he said nothing and some men noted it. To be called a coward and a liar is to be invited to do battle, but Odda stayed still as a stump.

"You cannot prove what you say," Erkenwald said.

"I can prove I killed Ubba," I said.

"We are not here to discuss such things," Erkenwald said loftily, "but to determine whether you broke the king's peace by an impious attack on Cynuit."

"Then summon my crewmen," I demanded. "Bring them here, put them on oath, and ask what they did in the summer." I waited, and Erkenwald said nothing. He glanced at the king as if seeking help, but Alfred's eyes were momentarily closed. "Or are you in so much of a hurry to kill me," I went on, "that you dare not wait to hear the truth?"

"I have Steapa's sworn testimony," Erkenwald said as if that made any other evidence unnecessary. He was flustered.

"And you can have my oath," I said, "and Leofric's oath, and the oath of a crewman who is here." I turned and beckoned Haesten who looked frightened at being summoned, but at Iseult's urging came to stand beside me. "Put him on oath," I demanded of Erkenwald.

Erkenwald did not know what to do, but some men in the witan called out that I had the right to summon oath-takers and the newcomer must be heard, and so a priest brought the gospel book to Haesten. I waved the priest away. "He will swear on this," I said, and took out Thor's amulet.

"He's not a Christian?" Erkenwald demanded in astonishment.

"He is a Dane," I said.

"How can we trust the word of a Dane?" Erkenwald demanded.

"But our lord king does," I retorted. "He trusts the word of Guthrum to keep the peace, so why should this Dane not be trusted?"

That provoked some smiles. Many in the witan thought Alfred far too trusting of Guthrum and I felt the sympathy in the hall move to my side, but then the archbishop intervened to declare that the oath of a pagan was of no value. "None whatsoever," he snapped. "He must stand down."

"Then put Leofric under oath," I demanded, "and then bring our crew here and listen to their testimony."

"And you will all lie with one tongue," Erkenwald said, "and what happened at Cynuit is not the only matter on which you are accused. Do you deny that you sailed in the king's ship? That you went to Cornwalum and there betrayed Peredur and killed his Christian people? Do you deny that Brother Asser told the truth?"

"But what if Peredur's queen were to tell you that Asser lies?" I asked. "What if she were to tell you that he lies like a hound at the hearth?" Erkenwald stared at me. They all stared at me and I turned and gestured at Iseult who stepped forward, tall and delicate, the silver glinting at her neck and wrists. "Peredur's queen," I announced, "whom I demand that you hear under oath, and thus hear how her husband was planning to join the Danes in an assault on Wessex."

That was rank nonsense, of course, but it was the best I could invent at that moment, and Iseult, I knew, would swear to its truth. Quite why Svein would fight Peredur if the Briton planned to support him was a dangerously loose plank in the argument, but it did not really matter for I had confused the proceedings so much that no one was sure what to do. Erkenwald was speechless. Men stood to look at Iseult, who looked calmly back at them, and the king and the archbishop bent their heads together. Ælswith, one hand clapped to her pregnant belly, hissed advice at them. None of them wanted to summon Iseult for fear of what she would say, and Alfred, I suspect, knew that the trial, which had already become mired in lies, could only get worse.

"You're good, earsling," Leofric muttered, "you're very good."

Odda the Younger looked at the king, then at his fellow members of the witan, and he must have known I was slithering out of his snare for he pulled Steapa to his side. He spoke to him urgently. The king was frowning, the archbishop looked perplexed, Ælswith's blotched face showed fury, while Erkenwald seemed helpless. Then Steapa rescued them. "I do not lie!" he shouted.

He seemed uncertain what to say next, but he had the hall's attention. The king gestured to him, as if inviting him to continue, and Odda the Younger whispered in the big man's ear.

"He says I lie," Steapa said, pointing at me, "and I say I do not, and my sword says I do not." He stopped abruptly, having made what was probably the longest speech of his life, but it was enough. Feet drummed on the floor and men shouted that Steapa was right, which he was not, but he had reduced the whole tangled morass of lies and accusations to a trial by combat and they all liked that. The archbishop still looked troubled, but Alfred gestured for silence.

He looked at me. "Well?" he asked. "Steapa says his sword will support his truth. Does yours?"

I could have said no. I could have insisted on letting Iseult speak and then allowing the witan to advise the king which side had spoken the greater truth, but I was ever rash, ever impetuous, and the invitation to fight cut through the whole entanglement. If I fought and won, then Leofric and I were innocent of every charge.

I did not even think about losing. I just looked at Steapa. "My sword," I told him, "says I tell the truth, and that you are a stinking bag of wind, a liar from hell, a cheat and a perjurer who deserves death."

"Up to our arses again," Leofric said.

Men cheered. They liked a fight to the death, which was much better entertainment than listening to Alfred's harpist chant the psalms. Alfred hesitated, and I saw Ælswith look from me to Steapa, and she must have thought him the greater warrior for she leaned forward, touched Alfred's elbow, and whispered urgently.

And the king nodded. "Granted," he said. He sounded weary, as if he was dispirited by the lies and the insults. "You will fight tomorrow. Swords and shields, nothing else." He held up a hand to stop the cheering. "My lord Wulfhere?"

"Sire?" Wulfhere struggled to his feet.

"You will arrange the fight. And may God grant victory to the truth." Alfred stood, pulled his robe about him, and left.

And Steapa, for the first time since I had seen him, smiled.

"You're a damned fool," Leofric told me. He had been released from his chains and allowed to spend the evening with me. Haesten was there, as was Iseult and my men who had been brought from the town. We were lodged in the king's compound, in a cattle byre that stank of dung, but I did not notice the smell. It was Twelfth Night so there was the great feast in the king's hall, but we were left out in the cold, watched there by two of the royal guards. "Steapa's good," Leofric warned me.

"I'm good."

"He's better," Leofric said bluntly. "He'll slaughter you."

"He won't," Iseult said calmly.

"Damn it, he's good!" Leofric insisted, and I believed him.

"It's that goddamned monk's fault," I said bitterly. "He went bleating to Alfred, didn't he?" In truth Asser had been sent by the king of Dyfed to assure the West Saxons that Dyfed was not planning war, but Asser had taken the opportunity of his embassy to recount the tale of the *Eftwyrd* and from that it was a small jump to conclude that we had stayed with Svein while he attacked Cynuit. Alfred had no proof of our guilt, but Odda the Younger had seen a chance to destroy me and so persuaded Steapa to lie.

"Now Steapa will kill you," Leofric grumbled, "whatever she says." Iseult did not bother to answer him. She was using handfuls of grubby straw to clean my mail coat. The armor had been fetched from the Corncrake tavern and given to me, but I would have to wait till morning to get my weapons, which meant they would not be newly sharpened. Steapa, because he served Odda the Younger, was one of the

king's bodyguards so he would have all night to put an edge on his
sword. The royal kitchens had sent us food, though I had no appetite.
"Just take it slow in the morning," Leofric told me.

"Slow?"

"You fight in a rage," he said, "and Steapa's always calm."

"So better to get in a rage," I said.

"That's what he wants. He'll fend you off and fend you off and
wait till you're tired. Then he'll finish you off. It's how he fights."

Harald told us the same thing. Harald was the shire reeve of Defnascir, the widower who had summoned me to the court in
Exanceaster, but he had also fought alongside us at Cynuit and that
makes a bond, and sometime in the dark he splashed through the
rain and mud and came into the light of the small fire that lit the cattle shed without warming it. He stopped in the doorway and gazed at
me reproachfully. "Were you with Svein at Cynuit?" he asked.

"No," I said.

"I didn't think so." Harald came into the byre and sat by the fire.
The two royal guards were at the door and he ignored them, and that
was interesting. All of them served Odda, and the young ealdorman
would not be pleased to hear that Harald had come to us, yet plainly
Harald trusted the two guards not to tell, which suggested that there
was unhappiness in Odda's ranks. Harald put a pot of ale on the floor.
"Steapa's sitting at the king's table," he said.

"So he's eating badly," I said.

Harald nodded, but did not smile. "It's not much of a feast," he
admitted. He stared into the fire for a moment, then looked at me.
"How's Mildrith?"

"Well."

"She is a dear girl," he said, then glanced at Iseult's dark beauty
before staring into the fire again. "There will be a church service at
dawn," he said, "and after that you and Steapa will fight."

"Where?"

"In a field on the other side of the river," he said, then pushed the
pot of ale toward me. "He's left-handed."

I could not remember fighting against a man who held his sword in his left hand, but nor could I see a disadvantage in it. We would both have our shields facing the other man's shield instead of his weapon, but that would be a problem to both of us. I shrugged.

"He's used to it," Harald explained, "and you're not. And he wears mail down to here"—he touched his calf—"and he has an iron strip on his left boot."

"Because that's his vulnerable foot?"

"He plants it forward," Harald said, "inviting attack, then chops at your sword arm."

"So he's a hard man to kill," I said mildly.

"No one's done it yet," Harald said gloomily.

"You don't like him?"

He did not answer at first, but drank ale, then passed the pot to Leofric. "I like the old man," he said, meaning Odda the Elder. "He's foul tempered, but he's fair enough. But the son?" He shook his head sadly. "I think the son is untested. Steapa? I don't dislike him, but he's like a hound. He only knows how to kill."

I stared into the feeble fire, looking for a sign from the gods in the small flames, but none came, or none that I saw. "He must be worried though," Leofric said.

"Steapa?" Harald asked. "Why should he be worried?"

"Uhtred killed Ubba."

Harald shook his head. "Steapa doesn't think enough to be worried. He just knows he'll kill Uhtred tomorrow."

I thought back to the fight with Ubba. He had been a great warrior, with a reputation that glowed wherever Norsemen sailed, and I had killed him, but the truth was that he had put a foot into the spilled guts of a dying man and slipped. His leg had shot sideways, he had lost his balance, and I had managed to cut the tendons in his arm. I touched the hammer amulet and thought that the gods had sent me a sign after all. "An iron strip in his boot?" I asked.

Harald nodded. "He doesn't care how much you attack him. He knows you're coming from his left and he'll block most of your attacks with his sword. Big sword, heavy thing. But some blows will get by and he won't care. You'll waste them on iron. Heavy mail, helmet, boot, doesn't matter. It'll be like hitting an oak tree, and after a while you'll make a mistake. He'll be bruised and you'll be dead."

He was right, I thought. Striking an armored man with a sword rarely achieved much except to make a bruise because the edge would be stopped by mail or helmet. Mail cannot be chopped open by a sword, which was why so many men carried axes into battle, but the rules of trial by combat said the fight had to be with swords. A sword lunge would pierce mail, but Steapa was not going to make himself an easy target for a lunge. "Is he quick?" I asked.

"Quick enough," Harald said, then shrugged. "Not as quick as you," he added grudgingly, "but he isn't slow."

"What does the money say?" Leofric asked, though he surely knew the answer.

"No one's wagering a penny on Uhtred," Harald said.

"You should," I retorted.

He smiled at that, but I knew he would not take the advice. "The big money," he said, "is what Odda will give Steapa when he kills you. A hundred shillings."

"Uhtred's not worth it," Leofric said with rough humor.

"Why does he want me dead so badly?" I wondered aloud. It could not be Mildrith, I thought, and the argument over who had killed Ubba was long in the past, yet still Odda the Younger conspired against me.

Harald paused a long time before answering. He had his bald head bowed and I thought he was in prayer, but then he looked up. "You threaten him," he said quietly.

"I haven't even seen him for months," I protested, "so how do I threaten him?"

Harald paused again, choosing his words carefully. "The king is frequently ill," he said after the pause, "and who can say how long he will live? And if, God forbid, he should die soon, then the witan will not choose his infant son to be king. They'll choose a nobleman with a reputation made on the battlefield. They'll choose a man who can stand up to the Danes."

"Odda?" I laughed at the thought of Odda as king.

"Who else?" Harald asked. "But if you were to stand before the witan and swear an oath to the truth about the battle where Ubba died, they might not choose him. So you threaten him, and he fears you because of that."

"So now he's paying Steapa to chop you to bits," Leofric added gloomily.

Harald left. He was a decent man, honest and hardworking, and he had taken a risk by coming to see me, and I had been poor company for I did not appreciate the gesture he made. It was plain he thought I must die in the morning, and he had done his best to prepare me for the fight, but despite Iseult's confident prediction that I would live, I did not sleep well. I was worried, and it was cold. The rain turned to sleet in the night and the wind whipped into the byre. By dawn the wind and sleet had stopped and instead there was a mist shrouding the buildings and icy water dripping from the mossy thatch. I made a poor breakfast of damp bread and it was while I was eating that Father Beocca came and said Alfred wished to speak with me.

I was sour. "You mean he wants to pray with me?"

"He wants to speak with you," Beocca insisted and, when I did not move, he stamped his lamed foot. "It is not a request, Uhtred. It is a royal order!"

I put on my mail, not because it was time to arm for the fight, but because its leather lining offered some warmth on a cold morning. The mail was not very clean, despite Iseult's efforts. Most men wore their hair short, but I liked the Danish way of leaving it long and so I tied it behind with a lace and Iseult plucked the straw scraps from it. "We must hurry," Beocca said, and I followed him through the mud past the great hall and the newly built church to some smaller buildings made of timber that had still not weathered gray. Alfred's father had used Cippanhamm as a hunting lodge, but Alfred was expanding it. The church had been his first new building, and he had built that even before he repaired and extended the palisade, and that was an indication of his priorities. Even now, when the nobility of Wessex was gathered just a day's march from the Danes, there seemed to be more churchmen than soldiers in the place, and that was another indication of how Alfred thought to protect his realm. "The king is gracious," Beocca hissed at me as we went through a door, "so be humble."

Beocca knocked on another door, did not wait for an answer, but pushed it open and indicated I should step inside. He did not follow me, but closed the door, leaving me in a gloomy half darkness.

A pair of beeswax candles flickered on an altar and by their light I saw two men kneeling in front of the plain wooden cross that stood between the candles. The men had their backs to me, but I recognized Alfred by his fur-trimmed blue cloak. The second man was a monk. They were both praying silently and I waited. The room was small, evidently a private chapel, and its only furniture was the draped altar and a kneeling stool on which was a closed book.

"In the name of the father," Alfred broke the silence.

"And of the son," the monk said, and he spoke English with an accent and I recognized the voice of the Ass.

"And of the holy ghost," Alfred concluded, "amen."

"Amen," Asser echoed, and both men stood, their faces suffused with the joy of devout Christians who have said their prayers well, and Alfred blinked as though he were surprised to see me, though he must have heard Beocca's knocking and the sound of the door opening and closing.

"I trust you slept well, Uhtred?" he said.

"I trust you did, lord."

"The pains kept me awake," Alfred said, touching his belly. Then he went to one side of the room and hauled open a big pair of wooden shutters, flooding the chapel with a wan, misty light. The window looked onto a courtyard and I was aware of men out there. The king shivered, for it was freezing in the chapel. "It is Saint Cedd's feast day," he told me.

I said nothing.

"You have heard of Saint Cedd?" he asked me and, when my silence betrayed ignorance, he smiled indulgently. "He was an East Anglian, am I not right, brother?"

"The most blessed Cedd was indeed an East Angle, lord," Asser confirmed.

"And his mission was in Lundene," Alfred went on, "but he concluded his days at Lindisfarena. You must know that house, Uhtred?"

"I know it, lord," I said. The island was a short ride from Bebbanburg and not so long before I had ridden to its monastery with Earl Ragnar and watched the monks die beneath Danish swords. "I know it well," I added.

"So Cedd is famous in your homeland?"

"I've not heard of him, lord."

"I think of him as a symbol," Alfred said, "a man who was born in East Anglia, did his life's work in Mercia, and died in Northumbria." He brought his long, pale hands together so that the fingers embraced. "The Saxons of England, Uhtred, gathered together before God."

"And united in joyful prayer with the Britons," Asser added piously.

"I beseech Almighty God for that happy outcome," the king said, smiling at me, and by now I recognized what he was saying. He stood there, looking so humble, with no crown, no great necklace, no arm rings, nothing but a small garnet brooch holding the cloak at his neck, and he spoke of a happy outcome, but what he was really seeing was the Saxon people gathered under one king. A king of Wessex. Alfred's piety hid a monstrous ambition.

"We must learn from the saints," Alfred told me. "Their lives are a guide to the darkness that surrounds us, and Saint Cedd's holy example teaches that we must be united, so I am loath to shed Saxon blood on Saint Cedd's feast day."

"There need be no bloodshed, lord," I said.

"I am pleased to hear it," Alfred interjected.

"If the charges against me are retracted."

The smile went from his face and he walked to the window and stared into the misty courtyard and I looked where he looked and saw that a small display was being mounted for my benefit. Steapa was being armored. Two men were dropping a massive mail coat over his wide shoulders, while a third stood by with an outsized shield and a monstrous sword.

"I talked with Steapa last night," the king said, turning from the window, "and he told me there was a mist when Svein attacked at Cynuit. A morning mist like this one." He waved at the whiteness sifting into the chapel.

"I wouldn't know, lord," I said.

"So it is possible," the king went on, "that Steapa was mistaken when he thought he saw you." I almost smiled. The king knew Steapa had lied, though he would not say as much. "Father Willibald also spoke to the crew of the *Eftwyrd*," the king went on, "and not one of them confirmed Steapa's tale."

The crew was still in Hamtun, so Willibald's report must have come from there and that meant the king had known I was innocent of the slaughter at Cynuit even before I was charged. "So I was falsely arraigned?" I said harshly.

"You were accused," the king corrected me, "and accusations must be proven or refuted."

"Or withdrawn."

"I can withdraw the charges," Alfred agreed. Steapa, outside the window, was making sure his mail coat was seated comfortably by swinging his great sword. And it was great. It was huge, a hammer of a blade. Then the king half closed the shutter, hiding Steapa. "I can withdraw the accusation about Cynuit," he said, "but I do not think Brother Asser lied to us."

"I have a queen," I said, "who says he does."

"A shadow queen," Asser hissed, "a pagan! A sorceress!" He looked at Alfred. "She is evil, lord," he said, "a witch! *Maleficos non patieris vivere!*"

"Thou shalt not permit a witch to live," Alfred translated for my benefit. "That is God's commandment, Uhtred, from the holy scriptures."

"Your answer to the truth," I sneered, "is to threaten a woman with death?"

Alfred flinched at that. "Brother Asser is a good Christian," he said vehemently, "and he tells the truth. You went to war without my orders. You used my ship, my men, and you behaved treacherously! You are the liar, Uhtred, and you are the cheat!" He spoke angrily, but managed to control his anger. "It is my belief," he went on, "that you have paid your debt to the church with goods stolen from other good Christians."

"Not true," I said harshly. I had paid the debt with goods stolen from a Dane.

"So resume the debt," the king said, "and we shall have no death on this blessed day of Saint Cedd."

I was being offered life. Alfred waited for my response, smiling. He was sure I would accept his offer because to him it seemed reasonable. He had no love for warriors, weapons, and killing. Fate decreed that he must spend his reign fighting, but it was not to his taste. He wanted to civilize Wessex, to give it piety and order, and two men fighting to the death on a winter's morning was not his idea of a well-run kingdom.

But I hated Alfred. I hated him for humiliating me at Exanceaster when he had made me wear a penitent's robe and crawl on my knees. Nor did I think of him as my king. He was a West Saxon and I was a Northumbrian, and I reckoned so long as he was king then Wessex had small chance of surviving. He believed God would protect him from the Danes, while I believed they had to be defeated by swords. I also had an idea how to defeat Steapa, just an idea, and I had no wish to take on a debt I had already paid, and I was young and I was foolish and I was arrogant and I was never able to resist a stupid impulse. "Everything I have said is the truth," I lied, "and I would defend that truth with my sword."

Alfred flinched from my tone. "Are you saying Brother Asser lied?" he demanded.

"He twists truth," I said, "like a woman wrings a hen's neck."

The king pulled the shutter open, showing me the mighty Steapa in his gleaming war glory. "You really want to die?" he asked me.

"I want to fight for the truth, lord king," I said stubbornly.

"Then you are a fool," Alfred said, his anger showing again. "You are a liar, a fool, and a sinner." He strode past me, pulled open the door, and shouted at a servant to tell Ealdorman Wulfhere that the fight was to take place after all. "Go," he added to me, "and may your soul receive its just reward."

Wulfhere had been charged with arranging the fight, but there was a delay because the ealdorman had disappeared. The town was searched, the royal buildings were searched, but there was no sign of him until a stable slave nervously reported that Wulfhere and his men had ridden away from Cippanhamm before dawn. No one knew why, though some surmised that Wulfhere wanted no part in a trial by combat, which made little sense to me for the ealdorman had never struck me as a squeamish man. Ealdorman Huppa of Thornsæta was appointed to replace him, and so it was close to midday when my swords were brought to me and we were escorted down to the meadow that lay across the bridge that led from the town's eastern gate. A huge crowd had gathered on the river's far bank. There were cripples, beggars, jugglers, women selling pies, dozens of priests, excited children, and, of course, the assembled warriors of the West Saxon nobility, all of them in Cippanhamm for the meeting of the witan, and all eager to see Steapa Snotor show off his renowned skill.

"You're a damned fool," Leofric said to me.

"Because I insisted on fighting?"

"You could have walked away."

"And men would have called me a coward," I said. And that, too, was the truth, that a man cannot step back from a fight and stay a man. We make much in this life if we are able. We make children and wealth and amass land and build halls and assemble armies and give great feasts, but only one thing survives us. Reputation. I could not walk away.

Alfred did not come to the fight. Instead, with the pregnant

Ælswith and their two children, and escorted by a score of guards and as many priests and courtiers, he had ridden westward. He was accompanying Brother Asser on the start of the monk's return journey to Dyfed, and the king was making a point that he preferred the company of the British churchman to watching two of his warriors fight like snarling hounds. But no one else in Wessex wanted to miss the battle. They were eager for it, but Huppa wanted everything to be orderly and so he insisted that the crowd push back from the damp ground beside the river to give us space. Eventually the folk were massed on a green bank overlooking the trampled grass and Huppa went to Steapa to enquire if he was ready.

He was ready. His mail shone in the weak sunlight. His helmet was glistening. His shield was a huge thing, bossed and rimmed with iron, a shield that must have weighed as much as a sack of grain and was a weapon in itself if he managed to hit me with it, but his chief weapon was his great sword that was longer and heavier than any I had seen.

Huppa, trailed by two guards, came to me. His feet squelched in the grass and I thought that the ground would prove treacherous. "Uhtred of Oxton," he said, "are you ready?"

"My name," I said, "is Uhtred of Bebbanburg."

"Are you ready?" he demanded, ignoring my correction.

"No," I said.

A murmur went through the folk nearest to me, and the murmur spread, and after a few heartbeats the whole crowd was jeering me. They thought me a coward, and that thought was reinforced when I dropped my shield and sword and made Leofric help as I stripped off the heavy coat of mail. Odda the Younger, standing beside his champion, was laughing. "What are you doing?" Leofric asked me.

"I hope you put money on me," I said.

"Of course I didn't."

"Are you refusing to fight?" Huppa asked me.

"No," I said, and when I was stripped of my armor I took Serpent-Breath back from Leofric. Just Serpent-Breath. No helmet, no shield, just my good sword. Now I was unburdened. The ground was heavy, Steapa was armored, but I was light and I was fast and I was ready.

"I'm ready," I told Huppa.

He went to the meadow's center, raised an arm, dropped it, and the crowd cheered.

I kissed the hammer around my neck, trusted my soul to the great god Thor, and walked forward.

Steapa came steadily toward me, shield up, sword held out to his left. There was no trace of concern in his eyes. He was a workman at his trade and I wondered how many men he had killed, and he must have thought my death would be easy for I had no protection, not even a shield. And so we walked toward each other until, a dozen paces from him, I ran. I ran at him, feinted right toward his sword, and then broke hard to my left, still running, going past him now, and I was aware of the huge blade swinging fast after me as he turned, but then I was behind him, he was still turning, and I dropped to my knees, ducked, heard the blade go over my head, and I was up again, lunging.

The sword pierced his mail, drew blood from just behind his left shoulder, but he was quicker than I had expected and had already checked that first great swing and was bringing the sword back and his turn pulled Serpent-Breath free. I had scratched him.

I danced back two paces. I went left again and he charged me, hoping to crush me with the weight of his shield, but I ran back to the right, fending off the sword with Serpent-Breath and the crack of the blades was like the bell of doomsday, and I lunged again, this time aiming at his waist, but he stepped back quickly. I kept going to the right, my arm jarred by the clash of the swords. I went fast, making him turn, and I feinted a lunge, brought him forward, and went back to the left. The ground was boggy. I feared slipping, but speed was my weapon. I had to keep him turning, keep him swinging into empty air, and snatch what chances I could to use Serpent-Breath's point. Bleed him enough, I thought, and he would tire, but he guessed my tactics and started making short rushes to frustrate me, and each rush would be accompanied by the hiss of that huge sword. He wanted to make me parry and hoped he could break Serpent-Breath when the blades met. I feared the same. She was well made, but even the best sword can break.

He forced me back, trying to crowd me against the spectators on the bank so he could hack me to pieces in front of them. I let him drive me, then dodged to my right where my left foot slipped and I went down on that knee and the crowd, close behind me now, took in a great breath and a woman screamed because Steapa's huge sword was swinging like an ax onto my neck, only I had not slipped, merely pretended to, and I pushed off with my right foot, came out from under the blow and around his right flank, and he thrust the shield out, catching my shoulder with the rim, and I knew I would have a bruise there, but I also had a heartbeat of opportunity and I darted Serpent-Breath forward and her point punctured his mail again to scrape against the ribs of his back and he roared as he turned, wrenching my blade free of his mail, but I was already going backward.

I stopped ten paces away. He stopped, too, and watched me. There was a slight puzzlement on his big face now. There was still no worry there, just puzzlement. He pushed his left foot forward, as Harald had warned me, and he was hoping I would attack it and he would rely on the hidden iron strip in the boot to protect him while he thumped and hacked and bludgeoned me to death. I smiled at him and threw Serpent-Breath from my right hand into my left and held her there, and that was a new puzzle for him. Some men could fight with either hand, and perhaps I was one of them? He drew his foot back. "Why do they call you Steapa Snotor?" I asked. "You're not clever. You've got the brains of an addled egg."

I was trying to enrage him and hoped that anger would make him careless, but my insult bounced off him. Instead of rushing me in fury he came slowly, watching the sword in my left hand, and the men on the hill called for him to kill me and I suddenly ran at him, broke right, and he swung at me a little late, thinking I was going to go left at the last moment and I swept Serpent-Breath back and she caught his sword arm and I could feel her blade scraping through the rings of his mail, but she did not slice through them and then I was away from him and put her back into my right hand, turned, charged him, and swerved away at the last moment so that his massive swing missed me by a yard.

He was still puzzled. This was like a bull-baiting and he was the bull, and his problem was to get me in a place where he could use his greater strength and weight. I was the dog, and my job was to lure him, tease him, and bite him until he weakened. He had thought I would come with mail and shield and we would batter each other for a few moments until my strength faded and he could drive me to the ground with massive blows and chop me to scraps with the big sword, but so far his blade had not touched me. But nor had I weakened him. My two cuts had drawn blood, but they were mere scratches. So now he came forward again, hoping to herd me back to the river. A woman screamed from the top of the bank, and I assumed she was trying to encourage him, and the screaming grew louder and I just went back faster, making Steapa lumber forward, but I had slipped away to his right and was coming back at him, making him turn, and then he suddenly stopped and stared past me and his shield went down and his sword dropped, too, and all I had to do was lunge. He was there for the killing. I could thrust Serpent-Breath into his chest or throat, or ram her into his belly, but I did none of those things. Steapa was no fool at fighting and I guessed he was luring me and I did not take the bait. If I lunged, I thought, he would crush me between his shield and sword. He wanted me to think him defenseless so that I could come into range of his weapons, but instead I stopped and spread my arms, inviting him to attack me as he was inviting me to attack him.

But he ignored me. He just stared past my shoulder. And the woman's screaming was shrill now and there were men shouting, and Leofric was yelling my name, and the spectators were no longer watching us, but running in panic.

So I turned my back on Steapa and looked toward the town on its hill that was cradled by the river's bend.

And I saw that Cippanhamm was burning. Smoke was darkening the winter sky and the horizon was filled with men, mounted men, men with swords and axes and shields and spears and banners, and more horsemen were coming from the eastern gate to thunder across the bridge.

Because all Alfred's prayers had failed and the Danes had come to Wessex.

SIX

Steapa recovered his wits before I did. He stared openmouthed at the Danes crossing the bridge and then just ran toward his master, Odda the Younger, who was shouting for his horses. The Danes were spreading out from the bridge, galloping across the meadow with drawn swords and leveled spears. Smoke poured into the low wintry clouds from the burning town. Some of the king's buildings were alight. A riderless horse, stirrups flapping, galloped across the grass, then Leofric grabbed my elbow and pulled me northward beside the river. Most of the folk had gone south and the Danes had followed them, so north seemed to offer more safety. Iseult had my mail coat and I took it from her, leaving her to carry Wasp-Sting, and behind us the screaming rose as the Danes chopped into the panicked mass. Folk scattered. Escaping horsemen thumped past us, the hooves throwing up spadefuls of damp earth and grass with every step. I saw Odda the Younger swerve away with three other horsemen. Harald, the shire reeve, was one of them, but I could not see Steapa and for a moment I feared the big man was looking for me. Then I forgot him as a band of Danes turned north in pursuit of Odda. "Where are our horses?" I shouted at Leofric, who looked bemused and I remembered he had not traveled to Cippanhamm with me. The beasts were probably still in the yard behind the Corncrake tavern, which meant they were lost.

There was a fallen willow in a stand of leafless alders by the river and we paused there for breath, hidden by the willow's trunk. I pulled on the mail coat, buckled on my swords, and took my helmet and shield from Leofric. "Where's Haesten?" I asked.

"He ran," Leofric said curtly. So had the rest of my men. They had joined the panic and were gone southward. Leofric pointed northward. "Trouble," he said curtly. There was a score of Danes riding down our bank of the river, blocking our escape, but they were still some distance away, while the men pursuing Odda had vanished, so Leofric led us across the water meadow to a tangle of thorns, alders, nettles, and ivy. At its center was an old wattle hut, perhaps a herdsman's shelter, and though the hut had half collapsed, it offered a better hiding place than the willow and so the three of us plunged into the nettles and crouched behind the rotting timbers.

A bell was ringing in the town. It sounded like the slow tolling that announced a funeral. It stopped abruptly, started again, and then finally ended. A horn sounded. A dozen horsemen galloped close to our hiding place and all had black cloaks and black painted shields, the marks of Guthrum's warriors.

Guthrum. Guthrum the Unlucky. He called himself King of East Anglia, but he wanted to be King of Wessex and this was his third attempt to take the country and this time, I thought, his luck had turned. While Alfred had been celebrating the twelfth night of Yule, and while the witan met to discuss the maintenance of bridges and the punishment of malefactors, Guthrum had marched. The army of the Danes was in Wessex, Cippanhamm had fallen, and the great men of Alfred's kingdom had been surprised, scattered, or slaughtered. The horn sounded again and the dozen black-cloaked horsemen turned and rode toward the sound.

"We should have known the Danes were coming," I said angrily.

"You always said they would," Leofric said.

"Didn't Alfred have spies at Gleawecestre?"

"He had priests praying here instead," Leofric said bitterly, "and he trusted Guthrum's truce."

I touched my hammer amulet. I had taken it from a boy in Eofer-wic. I had been a boy myself then, newly captured by the Danes, and my opponent had fought me in a whirl of fists and feet and I had hammered him down into the riverbank and taken his amulet. I still have it. I touch it often, reminding Thor that I live, but that day I touched it because I thought of Ragnar. The hostages would be dead, and was that why Wulfhere had ridden away at dawn? But how could he have known the Danes were coming? If Wulfhere had known, then Alfred would have known and the West Saxon forces would have been ready. None of it made sense, except that Guthrum had again attacked during a truce and the last time he had broken a truce he had showed that he was willing to sacrifice the hostages held to prevent just such an attack. It seemed certain he had done it again and so Ragnar would be dead and my world was diminished.

So many dead. There were corpses in the meadow between our hiding place and the river, and still the slaughter went on. Some of the Saxons had run back toward the town, discovered the bridge was guarded, and tried to escape northward and we watched them being ridden down by the Danes. Three men tried to resist, standing in a tight group with swords ready, but a Dane gave a great whoop and charged them with his horse, and his spear went through one man's mail, crushing his chest, and the other two were thrown aside by the horse's weight and immediately more Danes closed on them, swords and axes rose, and the horsemen spurred on. A girl screamed and ran in terrified circles until a Dane, long hair flying, leaned from his saddle and pulled her dress up over her head so she was blind and half naked. She staggered in the damp grass and a half dozen Danes laughed at her, then one slapped her bare rump with his sword and another dragged her southward, her screams muffled by the entangling dress. Iseult was shivering and I put a mail-clad arm around her shoulders.

I could have joined the Danes in the meadow. I spoke their lan-guage and, with my long hair and my arm rings, I looked like a Dane. But Haesten was somewhere in Cippanhamm and he might betray me, and Guthrum had no great love for me, and even if I survived then it would go hard with Leofric and Iseult. These Danes were in a rampant mood, flushed by their easy success and if a dozen decided they wanted Iseult then they would take her whether they thought I was a Dane or not. They were hunting in packs and so it was best to stay hidden until the frenzy had passed. Across the river, at the top of the low hill on which Cippanhamm was built, I could see the town's largest church burning. The thatched roof was whirling into the sky in great ribbons of flame and plumes of spark-riddled smoke.

"What in God's name were you doing back there?" Leofric asked me.

"Back there?" His question confused me.

"Dancing around Steapa like a gnat! He could have endured that all day!"

"I wounded him," I said, "twice."

"Wounded him? Sweet Christ, he's hurt himself worse when he was shaving!"

"Doesn't matter now, does it?" I said. I guessed Steapa was dead by now. Or perhaps he had escaped. I did not know. None of us knew what was happening except that the Danes had come. And Mildrith? My son? They were far away, and presumably they would receive warning of the Danish attack, but I had no doubt that the Danes would keep going deep into Wessex and there was nothing I could do to protect Oxton. I had no horse, no men, and no chance of reaching the south coast before Guthrum's mounted soldiers.

I watched a Dane ride past with a girl across his saddle. "What happened to that Danish girl you took home," I asked Leofric, "the one we captured off Wales?"

"She's still in Hamtun," he said, "and now that I'm not there she's probably in someone else's bed."

"Probably? Certainly."

"Then the bastard's welcome to her," he said. "She cries a lot."

"Mildrith does that," I said and then, after a pause, "Eanflæd was angry with you."

"Eanflæd? Angry with me! Why?"

"Because you didn't go to see her."

"How could I? I was in chains." He looked satisfied that the whore had asked after him. "Eanflæd doesn't cry, does she?"

"Not that I've seen."

"Good girl that. I reckon she'd like Hamtun."

If Hamtun still existed. Had a Danish fleet come from Lundene? Was Svein attacking across the Sæfern Sea? I knew nothing except that Wessex was suffering chaos and defeat. It began to rain again, a thin winter's rain, cold and stinging. Iseult crouched lower and I sheltered her with my shield. Most of the folk who had gathered to watch the fight by the river had fled south and only a handful had come our way, which meant there were fewer Danes near our hiding place, and those that were in the northern river meadows were now gathering their spoils. They stripped corpses of weapons, belts, mail, clothes, anything of value. A few Saxon men had survived, but they were being led away with the children and younger women to be sold as slaves. The old were killed. A wounded man was crawling on hands and knees and a dozen Danes tormented him like cats playing with an injured sparrow, nicking him with swords and spears, bleeding him to a slow death. Haesten was one of the tormentors. "I always liked Haesten," I said sadly.

"He's a Dane," Leofric said scornfully.

"I still liked him."

"You kept him alive," Leofric said, "and now he's gone back to his own. You should have killed him." I watched as Haesten kicked the wounded man who called out in agony, begging to be killed, but the group of young men went on jabbing him, laughing, and the first ravens came. I have often wondered if ravens smell blood, for the sky can be clear of them all day, but when a man dies they come from nowhere on their shining black wings. Perhaps Odin sends them, for the ravens are his birds, and now they flapped down to start feasting on eyes and lips, the first course of every raven feast. The dogs and foxes would soon follow.

"The end of Wessex," Leofric said sadly.

"The end of England," I said.

"What do we do?" Iseult asked.

There was no answer from me. Ragnar must be dead, which meant I had no refuge among the Danes, and Alfred was probably dead or else a fugitive, and my duty now was to my son. He was only a baby, but he was my son and he carried my name. Bebbanburg would be his if I could take it back, and if I could not take it back then it would be his duty to recapture the stronghold, and so the name Uhtred of Bebbanburg would go on till the last weltering chaos of the dying world.

"We must get to Hamtun," Leofric said, "and find the crew."

Except the Danes would surely be there already. Or else on their way. They knew where the power of Wessex lay, where the great lords had their halls, where the soldiers gathered, and Guthrum would be sending men to burn and kill and so disarm the Saxons' last kingdom.

"We need food," I said, "food and warmth."

"Light a fire here," Leofric grumbled, "and we're dead."

So we waited. The small rain turned to sleet. Haesten and his new companions, now that their victim was dead, wandered away, leaving the meadow empty but for the corpses and their attendant ravens. And still we waited, but Iseult, who was as thin as Alfred, was shivering uncontrollably and so, in the late afternoon, I took off my helmet and unbound my hair so it hung loose.

"What are you doing?" Leofric asked.

"For the moment," I said, "we're Danes. Just keep your mouth shut."

I led them toward the town. I would have preferred to wait until dark, but Iseult was too cold to wait longer, and I just hoped the Danes had calmed down. I might look like a Dane, but it was still dangerous. Haesten might see me, and if he told others how I had ambushed the Danish ship off Dyfed, then I could expect nothing but a slow death. So we went nervously, stepping past bloodied bodies along the riverside path. The ravens protested as we approached, flapped indignantly into the winter willows, and returned to their feast when we had passed. There were more corpses piled by the bridge where the young folk captured for slavery were being made to dig a grave. The Danes guarding them were drunk and none challenged us as we went across the wooden span and under the gate arch that was still hung with holly and ivy in celebration of Christmas.

The fires were dying now, damped by rain or else extinguished by the Danes who were ransacking houses and churches. I stayed in the narrowest alleys, edging past a smithy, a hide-dealer's shop, and a place where pots had been sold. Our boots crunched through the pottery shards. A young Dane was vomiting in the alley's entrance and he told me that Guthrum was in the royal compound where there would be a feast that night. He straightened up, gasping for breath, but was sober enough to offer me a bag of coins for Iseult. There were women screaming or sobbing in houses and their noise was making Leofric angry, but I told him to stay quiet. Two of us could not free Cippanhamm, and if the world had been turned upside down and it had been a West Saxon army capturing a Danish town, it would have sounded no different. "Alfred wouldn't allow it," Leofric said sullenly.

"You'd do it anyway," I said. "You've done it."

I wanted news, but none of the Danes in the street made any sense. They had come from Gleawecestre, leaving long before dawn, they had captured Cippanhamm, and now they wanted to enjoy whatever the town offered. The big church had burned, but men were raking through the smoking embers looking for silver. For lack of anywhere else to go we climbed the hill to the Corncrake tavern where we always drank and found Eanflæd, the redheaded whore, being held on a table by two young Danes while three others, none of them more than seventeen or eighteen, took turns raping her. Another dozen Danes were drinking peaceably enough, taking scant notice of the rape.

"You want her," one of the young men said, "you'll have to wait."

"I want her now," I said.

"Then you can jump in the shit pit," he said. He was drunk. He had a wispy beard and insolent eyes. "You can jump in the shit pit," he said again, evidently liking the insult, then pointed to Iseult, "and I'll have her while you drown." I hit him, breaking his nose and spattering his face with blood, and while he gasped I kicked him hard between the legs. He went down, whimpering, and I hit a second man in the belly while Leofric loosed all his day's frustration in a savage attack on another. The two who had been holding Eanflæd turned on us and one of them squealed when Eanflæd grabbed his hair and hooked sharp fingernails into his eyes. Leofric's opponent was on the floor and he stamped on the boy's throat and I head-slapped my boy until I had him by the door, then I thumped another in the ribs, rescued Eanflæd's victim, and broke his jaw, then went back to the lad who had threatened to rape Iseult. I ripped a silver loop from his ear, took off his one arm ring, and stole his pouch that clinked with coins. I dropped the silver into Eanflæd's lap, then kicked the groaning man between the legs, did it again, and hauled him out into the street.

"Go jump in a shit pit," I told him, then slammed the door. The other Danes, still drinking on the tavern's far side, had watched the fight with amusement and now gave us ironic applause.

"Bastards," Eanflæd said, evidently talking of the men we had driven away. "I'm sore as hell. What are you two doing here?"

"They think we're Danes," I said.

"We need food," Leofric said.

"They've had most of it," Eanflæd said, jerking her head at the seated Danes, "but there might be something left in the back." She tied her girdle. "Edwulf's dead." Edwulf had owned the tavern. "And thanks for helping me, you spavined bastards!" She shouted this at the Danes, who did not understand her and just laughed at her. Then she went toward the back room to find us food, but one of the men held out a hand to stop her.

"Where are you going?" he asked her in Danish.

"She's going past you," I called.

"I want ale," he said, "and you? Who are you?"

"I'm the man who's going to cut your throat if you stop her fetching food," I said.

"Quiet, quiet!" an older man said, then frowned at me. "Don't I know you?"

"I was with Guthrum at Readingum," I said, "and at Werham."

"That must be it. He's done better this time, eh?"

"He's done better," I agreed.

The man pointed at Iseult. "Yours?"

"Not for sale."

"Just asking, friend, just asking."

Eanflæd brought us stale bread, cold pork, wrinkled apples, and a rock-hard cheese in which red worms writhed. The older man carried a pot of ale to our table, evidently as a peace offering, and he sat and talked with me and I learned a little more of what was happening. Guthrum had brought close to three thousand men to attack Cippanhamm. Guthrum himself was now in Alfred's hall and half his men would stay in Cippanhamm as a garrison while the rest planned to ride either south or east in the morning. "Keep the bastards on the run, eh?" the man said, then frowned at Leofric. "He doesn't say much."

"He's dumb," I said.

"I knew a man who had a dumb wife. He was ever so happy." He looked jealously at my arm rings. "So who do you serve?"

"Svein of the White Horse."

"Svein? He wasn't at Readingum. Or at Werham."

"He was in Dyflin," I said, "but I was with Ragnar the Older then."

"Ah, Ragnar! Poor bastard."

"I suppose his son's dead now?" I asked.

"What else?" the man said. "Hostages, poor bastards." He thought for a heartbeat, then frowned again. "What's Svein doing here? I thought he was coming by ship?"

"He is," I said. "We're just here to talk to Guthrum."

"Svein sends a dumb man to talk to Guthrum?"

"He sent me to talk," I said, "and sent him," I jerked a thumb at the glowering Leofric, "to kill people who ask too many questions."

"All right, all right!" The man held up a hand to ward off my belligerence.

We slept in the stable loft, warmed by straw, and we left before dawn, and at that moment fifty West Saxons could have retaken Cippanhamm for the Danes were drunk, sleeping, and oblivious to the world. Leofric stole a sword, ax, and shield from a man snoring in the tavern. Then we walked unchallenged out of the western gate. In a field outside we found over a hundred horses, guarded by two men sleeping in a thatched hut, and we could have taken all the beasts, but we had no saddles or bridles and so, reluctantly, I knew we must walk. There were four of us now, because Eanflæd had decided to come with us. She had swathed Iseult in two big cloaks, but the British girl was still shivering.

We walked west and south along a road that twisted through small hills. We were heading for Baðum, and from there I could strike south toward Defnascir and my son, but it was clear the Danes were already ahead of us. Some must have ridden this way the previous day, for in the first village we reached there were no cocks crowing, no sound at all, and what I had taken for a morning mist was smoke from burned cottages. Heavier smoke showed ahead, suggesting the Danes might already have reached Baðum, a town they knew well for they had negotiated one of their truces there. Then, that afternoon, a horde of mounted Danes appeared on the road behind us and we were driven west into the hills to find a hiding place.

We wandered for a week. We found shelter in hovels. Some were deserted while others still had frightened folk, but every short winter's day was smeared with smoke as the Danes ravaged Wessex. One day we discovered a cow, trapped in its byre in an otherwise deserted homestead. The cow was with calf and bellowing with hunger, and that night we feasted on fresh meat. Next day we could not move for it was bitterly cold and a slanting rain slashed on an east wind and the trees thrashed as if in agony and the building that gave us shelter leaked and the fire choked us and Iseult just sat, eyes wide and empty, staring into the small flames.

"You want to go back to Cornwalum?" I asked her.

She seemed surprised I had spoken. It took her a few heartbeats to gather her thoughts. Then she shrugged. "What is there for me?"

"Home," Eanflæd said.

"Uhtred is home for me."

"Uhtred is married," Eanflæd said harshly.

Iseult ignored that. "Uhtred will lead men," she said, rocking back and forth, "hundreds of men. A bright horde. I want to see that."

"He'll lead you into temptation, that's all he'll do," Eanflæd said. "Go home, girl, say your prayers, and hope the Danes don't come."

We kept trying to go southward and we made some small progress every day, but the bitter days were short and the Danes seemed to be everywhere. Even when we traveled across countryside far from any track or path, there would be a patrol of Danes in the distance, and to avoid them we were constantly driven west. To our east was the Roman road that ran from Baðum and eventually to Exanceaster, the main thoroughfare in this part of Wessex, and I supposed the Danes were using it and sending patrols out to either side of the road, and it was those patrols that drove us ever nearer the Sæfern Sea, but there could be no safety there, for Svein would surely have come from Wales.

I also supposed that Wessex had finally fallen. We met a few folk, fugitives from their villages and hiding in the woods, but none had any news, only rumor. No one had seen any West Saxon soldiers, no one had heard about Alfred, they only saw Danes and the ever-present smoke. From time to time we would come across a ravaged village or a burned church. We would see ragged ravens flapping black and follow them to find rotting bodies. We were lost and any hope I had of reaching Oxton was long gone, and I assumed Mildrith had fled west into the hills as the folk around the Uisc always did when the Danes came. I hoped she was alive, I hoped my son lived, but what future he had was as dark as the long winter nights.

"Maybe we should make our peace," I suggested to Leofric one night. We were in a shepherd's hut, crouched around a small fire that filled the low turf-roofed building with smoke. We had roasted a dozen mutton ribs cut from a sheep's half-eaten corpse. We were all filthy, damp, and cold. "Maybe we should find the Danes," I said, "and swear allegiance."

"And be made slaves?" Leofric answered bitterly.

"We'll be warriors," I said.

"Fighting for a Dane?" He poked the fire, throwing up a new burst of smoke. "They can't have taken all Wessex," he protested.

"Why not?"

"It's too big. There have to be some men fighting back. We just have to find them."

I thought back to the long-ago arguments in Lundene. Back then I had been a child with the Danes, and their leaders had argued that the best way to take Wessex was to attack its western heartland and there break its power. Others had wanted to start the assault by taking the old kingdom of Kent, the weakest part of Wessex and the part that contained the great shrine of Contwaraburg, but the boldest argument had won. They had attacked in the west and that first assault had failed, but now Guthrum had succeeded. Yet how far had he succeeded? Was Kent still Saxon? Defnascir?

"And what happens to Mildrith if you join the Danes?" Leofric asked.

"She'll have hidden," I spoke dully and there was a silence, but I saw Eanflæd was offended and I hoped she would hold her tongue.

She did not. "Do you care?" she challenged me.

"I care," I said.

Eanflæd scorned that answer. "Grown dull, has she?"

"Of course he cares." Leofric tried to be a peacemaker.

"She's a wife," Eanflæd retorted, still looking at me. "Men tire of wives," she went on and Iseult listened, her big dark eyes going from me to Eanflæd.

"What do you know of wives?" I asked.

"I was married," Eanflæd said.

"You were?" Leofric asked, surprised.

"I was married for three years," Eanflæd said, "to a man who was in Wulfhere's guard. He gave me two children, then died in the battle that killed King Æthelred."

"Two children?" Iseult asked.

"They died," Eanflæd said harshly. "That's what children do. They die."

"You were happy with him," Leofric asked, "your husband?"

"For about three days," she said, "and in the next three years I learned that men are bastards."

"All of them?" Leofric asked.

"Most." She smiled at Leofric, then touched his knee. "Not you."

"And me?" I asked.

"You?" She looked at me for a heartbeat. "I wouldn't trust you as far as I could spit," she said, and there was real venom in her voice, leaving Leofric embarrassed and me surprised. There comes a moment in life when we see ourselves as others see us. I suppose that is part of growing up, and it is not always comfortable. Eanflæd, at that moment, regretted speaking so harshly, for she tried to soften it. "I don't know you," she said, "except you're Leofric's friend."

"Uhtred is generous," Iseult said loyally.

"Men are usually generous when they want something," Eanflæd retorted.

"I want Bebbanburg," I said.

"Whatever that is," Eanflæd said, "and to get it you'd do any-thing. Anything."

There was silence. I saw a snowflake show at the half-covered door. It fluttered into the firelight and melted. "Alfred's a good man." Leofric broke the awkward silence.

"He tries to be good," Eanflæd said.

"Only tries?" I asked sarcastically.

"He's like you," she said. "He'd kill to get what he wants, but there is a difference. He has a conscience."

"He's frightened of the priests, you mean."

"He's frightened of God. And we should all be that. Because one day we'll answer to God."

"Not me," I said.

Eanflæd sneered at that, but Leofric changed the conversation by saying it was snowing, and after a while we slept. Iseult clung to me in her sleep and she whimpered and twitched as I lay awake, half dreaming, thinking of her words that I would lead a bright horde. It seemed an unlikely prophecy, indeed I reckoned her powers must have gone with her virginity, and then I slept, too, waking to a world made white. The twigs and branches were edged with snow, but it was already melting, dripping into a misty dawn. When I went outside I found a tiny dead wren just beyond the door and I feared it was a grim omen.

Leofric emerged from the hut, blinking at the dawn's brilliance. "Don't mind Eanflæd," he said.

"I don't."

"Her world's come to an end."

"Then we must remake it," I said.

"Does that mean you won't join the Danes?"

"I'm a Saxon," I said.

Leofric half smiled at that. He undid his breeches and had a piss. "If your friend Ragnar was alive," he asked, watching the steam rise from his urine, "would you still be a Saxon?"

"He's dead, isn't he?" I said bleakly. "Sacrificed to Guthrum's ambition."

"So now you're a Saxon?"

"I'm a Saxon," I said again, sounding more certain than I felt, for I did not know what the future held. How can we? Perhaps Iseult had told the truth and Alfred would give me power and I would lead a shining horde and have a woman of gold, but I was beginning to doubt Iseult's powers. Alfred might already be dead and his kingdom was doomed, and all I knew at that moment was that the land stretched away south to a snow-covered ridgeline, and there it ended in a strange empty brightness. The skyline looked like the world's rim, poised above an abyss of pearly light. "We'll keep going south," I said. There was nothing else to do except walk toward the brightness.

We did. We followed a sheep track to the ridgetop and there I saw that the hills fell steeply away, dropping to the vast marshes of the sea. We had come to the great swamp, and the brightness I had seen was the winter light reflecting from the long meres and winding creeks.

"What now?" Leofric asked, and I had no answer. So we sat under the berries of a wind-bent yew and stared at the immensity of bog, water, grass, and reeds. This was the vast swamp that stretched inland from the Sæfern, and if I was to reach Defnascir I either had to go around it or try to cross it. If we went around it then we would have to go to the Roman road, and that was where the Danes were, but if we tried to cross the swamp we would face other dangers. I had heard a thousand stories of men being lost in its wet tangles. It was said there were spirits there, spirits that showed at night as flickering lights, and there were paths that led only to quicksand or to drowning pools, but there were also villages in the swamp, places where folk trapped fish and eels. The people of the swamp were protected by the spirits and by the sudden surges in the tide that could drown a road in an eyeblink. Now, as the last snows melted from the reed banks, the swamp looked like a great stretch of waterlogged land, its streams and meres swollen by the winter rains, but when the tide rose it would resemble an inland sea dotted with islands. We could see one of those islands not so far off, and there was a cluster of huts on that speck of higher ground; that would be a place to find food and warmth if we could ever reach it. Eventually we might cross the whole swamp, finding a way from island to island, but it would take far longer than a day, and we would have to find refuge at every high tide. I gazed at the long, cold stretches of water, almost black beneath the leaden clouds that came from the sea, and my spirits sank for I did not know where we were going, or why, or what the future held.

It seemed to get colder as we sat, and then a light snow began drifting from the dark clouds. Just a few flakes, but enough to convince me that we had to find shelter soon. Smoke was rising from the nearest swamp village, evidence that some folk still lived there. There would be food in their hovels and a meager warmth. "We have to get to that island," I said, pointing.

But the others were staring westward to where a flock of pigeons had burst from the trees at the foot of the slope. The birds rose and flew in circles. "Someone's there," Leofric said.

We waited. The pigeons settled in the trees higher up the hill. "Maybe it's a boar?" I suggested.

"Pigeons won't fly from a boar," Leofric said. "Boars don't startle pigeons, any more than stags do. There are folk there."

The thought of boars and stags made me wonder what had happened to my hounds. Had Mildrith abandoned them? I had not even told her where I had hidden the remains of the plunder we had taken off the coast of Wales. I had dug a hole in a corner of my new hall and buried the gold and silver down by the post stone, but it was not the cleverest hiding place and if there were Danes in Oxton then they were bound to delve into the edges of the hall floor, especially if a probing spear found a place where the earth had been disturbed. A flight of ducks flew overhead. The snow was falling harder, blurring the long view across the swamp.

"Priests," Leofric said.

There were a half dozen men off to the west. They were robed in black and had come from the trees to walk along the swamp's margin, plainly seeking a path into its tangled vastness, but there was no obvious track to the small village on its tiny island and so the priests came nearer to us, skirting the ridge's foot. One of them was carrying a long staff and, even at a distance, I could see a glint at its head and I suspected it was a bishop's staff, the kind with a heavy silver cross. Another three carried heavy sacks. "You think there's food in those bundles?" Leofric asked wistfully.

"They're priests," I said savagely. "They'll be carrying silver."

"Or books," Eanflæd suggested. "Priests like books."

"It could be food," Leofric said, though not very convincingly.

A group of three women and two children now appeared. One of the women appeared to be wearing a swathing cloak of silver fur, while another carried the smaller child. The women and children were not far behind the priests, who waited for them, and then they all walked eastward until they were beneath us and there they discovered some kind of path twisting into the marshes. Five of the priests led the women into the swamp while the sixth man, evidently younger than the others, hurried back westward. "Where's he going?" Leofric asked.

Another skein of ducks flighted low overhead, skimming down the slope to the long meres of the swamp. Nets, I thought. There must be nets in the swamp villages and we could trap fish and wildfowl. We could eat well for a few days. Eels, duck, fish, geese. If there were enough nets we could even trap deer by driving them into the tangling meshes.

"They're not going anywhere," Leofric said scornfully, nodding at the priests who had stranded themselves a hundred paces out in the swamp. The path was deceptive. It had offered an apparent route to the village, but then petered out amid a patch of reeds where the priests huddled. They did not want to come back and did not want to go forward, and so they stayed where they were, lost and cold and despairing. They looked as though they were arguing.

"We must help them," Eanflæd said and, when I said nothing, she protested that one of the women was holding a baby. "We have to help them!" she insisted.

I was about to retort that the last thing we needed was more hungry mouths to feed, but her harsh words in the night had persuaded me that I had to do something to show her I was not as treacherous as she evidently believed, so I stood, hefted my shield, and started down the hill. The others followed, but before we were even halfway down I heard shouts from the west. The lone priest who had gone that way was now with four soldiers and they turned as horsemen came from the trees. There were six horsemen, then eight more appeared, then another ten, and I realized a whole column of mounted soldiers was streaming from the dead winter trees. They had black shields and black cloaks, so they had to be Guthrum's men. One of the priests stranded in the swamp ran back along the path and I saw he had a sword and was going to help his companions.

It was a brave thing for the lone priest to do, but quite useless. The four soldiers and the single priest were surrounded now. They were standing back-to-back and the Danish horsemen were all around them, hacking down, and then two of the horsemen saw the priest with his sword and spurred toward him. "Those two are ours," I said to Leofric.

That was stupid. The four men were doomed, as was the priest if we did not intervene, but there were only two of us and, even if we killed the two horsemen, we would still face overwhelming odds, but I was driven by Eanflæd's scorn and I was tired of skulking through the winter countryside and I was angry. So I ran down the hill, careless of the noise I made as I crashed through brittle undergrowth. The lone priest had his back to the swamp now and the horsemen were charging at him as Leofric and I burst from the trees and came at them from their left side.

I hit the nearest horse's flank with my heavy shield. There was a scream from the horse and an explosion of wet soil, grass, snow, and hooves as man and beast went down sideways. I was also on the ground, knocked there by the impact, but I recovered first and found the rider tangled with his stirrups, one leg trapped under the struggling horse, and I chopped Serpent-Breath down hard. I cut into his throat, stamped on his face, chopped again, slipped in his blood, then left him and went to help Leofric who was fending off the second man who was still on horseback. The Dane's sword thumped on Leofric's shield, then he had to turn his horse to face me and Leofric's ax took the horse in the face and the beast reared, the rider slid backward, and I met his spine with Serpent-Breath's tip. Two down. The priest with the sword, not a half dozen paces away, had not moved. He was just staring at us. "Get back into the marsh!" I shouted at him. "Go! Go!" Iseult and Eanflæd were with us now and they seized the priest and hurried him toward the path. It might lead nowhere, but it was better to face the remaining Danes there than on the firm ground at the hill's foot.

And those black-cloaked Danes were coming. They had slaughtered the handful of soldiers, seen their two men killed, and now came for vengeance. "Come on!" I snarled at Leofric and, taking the wounded horse by the reins, I ran onto the small twisting path.

"A horse won't help you here," Leofric said.

The horse was nervous. Its face was wounded and the path was slippery, but I dragged it along the track until we were close to the small patch of land where the refugees huddled, and by now the Danes were also on the path, following us. They had dismounted. They could only come two abreast and, in places, only one man could use the track and in one of those places I stopped the horse and exchanged Serpent-Breath for Leofric's ax. The horse looked at me with a big brown eye. "This is for Odin," I said, and I swung the ax into its neck, chopping down through mane and hide, and a woman screamed behind me as the blood spurted bright and high in the dull day. The horse whinnied and tried to rear, and I swung again; this time the beast went down, thrashing hooves, blood, and water splashing. Snow turned red as I axed it a third time, finally stilling it, and now the dying beast was an obstacle athwart the track and the Danes would have to fight across its corpse. I took Serpent-Breath back.

"We'll kill them one by one," I told Leofric.

"For how long?" He nodded westward and I saw more Danes coming, a whole ship's crew of mounted Danes streaming along the swamp's edge. Fifty men? Maybe more, but even so they could only use the path in ones or twos and they would have to fight over the dead horse into Serpent-Breath and Leofric's ax. He had lost his own ax, taken from him when he was brought to Cippanhamm, but he seemed to like his stolen weapon. He made the sign of the cross, touched the blade, then hefted his shield as the Danes came.

Two young men came first. They were wild and savage, wanting to make a reputation, but the first to come was stopped by Leofric's ax banging into his shield and I swept Serpent-Breath beneath the shield to slice his ankle and he fell, cursing, to tangle his companion, and Leofric wrenched the wide-bladed ax free and slashed it down again. The second man stumbled on the horse and Serpent-Breath took him under the chin, above his leather coat, and the blood ran down her blade in a sudden flood and now there were two Danish corpses added to the horse-flesh barricade. I was taunting the other Danes, calling them corpse worms, telling them I had known children who could fight better. Another man came, screaming in rage as he leaped over the horse, and he was checked by Leofric's shield and Serpent-Breath met his sword with a dull crack and his blade broke, and two more men were trying to get past the horse, struggling in water up to their knees and I rammed Serpent-Breath into the belly of the first, pushing her through his leather armor, left him to die, and swung right at the man trying to get through the water. Serpent-Breath's tip flicked across his face to spray blood into the thickening snowfall. I went forward, feet sinking, lunged again, and he could not move in the mire and Serpent-Breath took his gullet. I was screaming with joy because the battle calm had come, the same blessed stillness I had felt at Cynuit. It is a joy, that feeling, and the only other joy to compare is that of being with a woman.

It is as though life slows. The enemy moves as if he is wading in mud, but I was kingfisher fast. There is rage, but it is a controlled rage, and there is joy, the joy that the poets celebrate when they speak of battle, and a certainty that death is not in that day's fate. My head was full of singing, a keening note, high and shrill, death's anthem. All I wanted was for more Danes to come to Serpent-Breath and it seemed to me that she took on her own life in those moments. To think was to act. A man came across the horse's flank; I thought to slice at his ankle, knew he would drop his shield and so open his upper body to an attack, and before the thought was even coherent it was done and Serpent-Breath had taken one of his eyes. She had gone down and up, was already moving to the right to counter another man trying to get around the horse, and I let him get past the stallion's bloodied head, then scornfully drove him down into the water and there I stood on him, holding his head under my boot as he drowned. I screamed at the Danes, told them I was Valhalla's gate-keeper, that they had been weaned on coward's milk, and that I wanted them to come to my blade. I begged them to come, but five men were dead around the horse and the others were now wary.

I stood on the dead horse and spread my arms. I held the shield high to my left and the sword to my right, and my mail coat was spattered with blood, and the snow fell about my wolf-crested helmet, and all I knew was the young man's joy of slaughter. "I killed Ubba Lothbrokson!" I shouted at them. "I killed him! So come and join him! Taste his death! My sword wants you!"

"Boats," Leofric said. I did not hear him. The man I thought I had drowned was still alive and he suddenly reared from the marsh, choking and vomiting water, and I jumped down off the horse and put my foot on his head again.

"Let him live!" a voice shouted behind me. "I want a prisoner!"

The man tried to fight my foot, but Serpent-Breath put him down. He struggled again and I broke his spine with Serpent-Breath and he was still.

"I said I wanted a prisoner," the voice behind protested.

"Come and die!" I shouted at the Danes.

"Boats," Leofric said again, and I glanced behind and saw three punts coming through the marsh. They were long flat boats, propelled by men with poles, and they grounded on the other side of the huddled refugees who hurried aboard. The Danes, knowing Leofric and I had to retreat if we were to gain the safety of the boats, readied for a charge and I smiled at them, inviting them.

"One boat left," Leofric said. "Room for us. You'll have to run like hell."

"I'll stay here," I shouted, but in Danish. "I'm enjoying myself."

Then there was a stir on the path as a man came to the front rank of the Danes and the others edged aside to give him room. He was in chain mail and had a silvered helmet with a raven's wing at its crown, but as he came closer he took the helmet off and I saw the gold-tipped bone in his hair. It was Guthrum himself. The bone was one of his mother's ribs and he wore it out of love for her memory. He stared at me, his gaunt face sad, and then looked down at the men we had killed. "I shall hunt you like a dog, Uhtred Ragnarson," he said, "and I shall kill you like a dog."

"My name," I said, "is Uhtred Uhtredson."

"We have to run," Leofric hissed at me.

The snow whirled above the swamp, thick enough now so that I could hardly see the ridgetop from where we had glimpsed the pigeons circling. "You are a dead man, Uhtred," Guthrum said.

"I never met your mother," I called to him, "but I would have liked to meet her."

His face took on the reverent look that any mention of his mother always provoked. He seemed to regret that he had spoken so harshly to me for he made a conciliatory gesture. "She was a great woman," he said.

I smiled at him. At that moment, looking back, I could have changed sides so easily and Guthrum would have welcomed me if I had just given his mother a compliment, but I was a belligerent young man and the battle joy was on me. "I would have spat in her ugly face," I told Guthrum, "and now I piss on your mother's soul, and tell you that the beasts of Niflheim are humping her rancid bones."

He screamed with rage and they all charged, some splashing through the shallows, all desperate to reach me and avenge the terrible insult, but Leofric and I were running like hunted boars, and we charged through the reeds and into the water and hurled ourselves onto the last punt. The first two were gone, but the third had waited for us and, as we sprawled on its damp boards, the man with the pole pushed hard and the craft slid away into the black water. The Danes tried to follow, but we were going surprisingly fast, gliding through the snowfall, and Guthrum was shouting at me and a spear was thrown, but the marshman poled again and the spear plunged harmlessly into the mud.

"I shall find you!" Guthrum shouted.

"Why should I care?" I called back. "Your men only know how to die!" I raised Serpent-Breath and kissed her sticky blade. "And your mother was a whore to dwarves!"

"You should have let that one man live," a voice said behind me, "because I wanted to question him." The punt only contained the one passenger besides Leofric and myself, and that one man was the priest who had carried a sword and now he was sitting in the punt's flat bow, frowning at me. "There was no need to kill that man," he said sternly and I looked at him with such fury that he recoiled. Damn all priests, I thought. I had saved the bastard's life and all he did was reprove me, and then I saw that he was no priest at all.

It was Alfred.

The punt slid over the swamp, sometimes gliding across black water, sometimes rustling through grass or reeds. The man poling it was a bent, dark-skinned creature with a massive beard, otter-skin clothes, and a toothless mouth. Guthrum's Danes were far behind now, carrying their dead back to firmer ground. "I need to know what they plan to do," Alfred complained to me. "The prisoner could have told us."

He spoke more respectfully. Looking back I realized I had frightened him, for the front of my mail coat was sheeted in blood and there was more blood on my face and helmet.

"They plan to finish Wessex," I said curtly. "You don't need a prisoner to tell you that."

"Lord," he said.

I stared at him.

"I am a king!" he insisted. "You address a king with respect."

"A king of what?" I asked.

"You're not hurt, lord?" Leofric asked Alfred.

"No, thank God. No." He looked at the sword he carried. "Thank God." I saw he was not wearing priest's robes, but a swathing black cloak. His long face was very pale. "Thank you, Leofric," he said, then looked up at me and seemed to shudder. We were catching up with the other two punts and I saw that Ælswith, pregnant and swathed in a silver fox-fur cloak, was in one. Iseult and Eanflæd were also in that punt whereas the priests were crowded onto the other, and I saw that Bishop Alewold of Exanceaster was one of them.

"What happened, lord?" Leofric asked.

Alfred sighed. He was shivering now, but he told his story. He had ridden from Cippanhamm with his family, his bodyguard, and a score of churchmen to accompany the monk Asser on the first part of his journey. "We had a service of thanksgiving," he said, "in the church at Soppan Byrg. It's a new church," he added earnestly to Leofric, "and very fine. We sang psalms, said prayers, and Brother Asser went on his way rejoicing." He made the sign of the cross. "I pray he's safe."

"I hope the lying bastard's dead," I snarled.

Alfred ignored that. After the church service they had all gone to a nearby monastery for a meal, and it was while they were there that the Danes had come. The royal group had fled, finding shelter in nearby woods while the monastery burned. After that they had tried to ride east into the heart of Wessex but, like us, they had constantly been headed off by patrolling Danes. One night, sheltering in a farm, they had been surprised by Danish troops who had killed some of Alfred's guards and captured all his horses and ever since they had been wandering, as lost as us, until they came to the swamp. "God knows what will happen now," Alfred said.

"We fight," I said. He just looked at me and I shrugged. "We fight," I said again.

Alfred stared across the swamp. "Find a ship," he said, but so softly that I hardly heard him. "Find a ship and go to Frankia." He pulled the cloak tighter around his thin body. The snow was thickening as it fell, though it melted as soon as it met the dark water. The Danes had vanished, lost in the snow behind. "That was Guthrum?" Alfred asked me.

"That was Guthrum," I said. "And he knew it was you he pursued?"

"I suppose so."

"What else would draw Guthrum here?" I asked. "He wants you dead. Or captured."

Yet, for the moment, we were safe. The island village had a score of damp hovels thatched with reeds and a few storehouses raised on stilts. The buildings were the color of mud, the street was mud, the goats and the people were mud-covered, but the place, poor as it was, could provide food, shelter, and a meager warmth. The men of the village had seen the refugees and, after a discussion, decided to rescue them. I suspect they wanted to pillage us rather than save our lives, but Leofric and I looked formidable and, once the villagers understood that their king was their guest, they did their clumsy best for him and his family. One of them, in a dialect I could scarcely understand, wanted to know the king's name. He had never heard of Alfred. He knew about the Danes, but said their ships had never reached the village, or any of the other settlements in the swamp. He told us the villagers lived off deer, goats, fish, eels, and wildfowl, and they had plenty of food, though fuel was scarce.

Ælswith was pregnant with her third child, while her first two were in the care of nurses.

There was Edward, Alfred's heir, who was three years old and sick. He coughed, and Ælswith worried about him, though Bishop Alewold insisted it was just a winter's cold. Then there was Edward's elder sister, Æthelflaed, who was now six and had a bright head of golden curls, a beguiling smile, and clever eyes. Alfred adored her, and in those first days in the swamp, she was his one ray of light and hope. One night, as we sat by a small, dying fire and Æthelflaed slept with her golden head in her father's lap, he asked me about my son.

"I don't know where he is," I said. There were only the two of us, everyone else was sleeping, and I was sitting by the door staring across the frost-bleached marsh that lay black and silver under a half-moon.

"You want to go and find him?" he asked earnestly.

"You truly want me to do that?" I asked. He looked puzzled. "These folk are giving you shelter," I explained, "but they'd as soon cut your throats. They won't do that while I'm here."

He was about to protest, then understood I probably spoke the truth. He stroked his daughter's hair. Edward coughed. He was in his mother's hut. The coughing had become worse, much worse, and we all suspected it was the whooping cough that killed small children. Alfred flinched at the sound. "Did you fight Steapa?" he asked.

"We fought," I said curtly. "The Danes came, and we never fin-ished. He was bleeding, I was not."

"He was bleeding?"

"Ask Leofric. He was there."

He was silent a long time, then, softly, "I am still king."

Of a swamp, I thought, and said nothing.

"And it is customary to call a king 'lord,' " he went on.

I just stared at his thin, pale face that was lit by the dying fire. He looked so solemn, but also frightened, as if he was making a huge effort to hold on to the shreds of his dignity. Alfred never lacked for bravery, but he was not a warrior and he did not much like the com-pany of warriors. In his eyes I was a brute: dangerous, uninteresting, but suddenly indispensable. He knew I was not going to call him lord, so he did not insist. "What do you notice about this place?" he asked.

"It's wet," I said.

"What else?"

I looked for the trap in the question and found none. "It can only be reached by punts," I said, "and the Danes don't have punts. But when they do have punts it'll need more than Leofric and me to fight them off."

"It doesn't have a church," he said.

"I knew I liked it," I retorted.

He ignored that. "We know so little of our own kingdom," he said in wonderment. "I thought there were churches everywhere." He closed his eyes for a few heartbeats, then looked at me plaintively. "What should I do?"

That morning I had told him to fight, but I could see no fight in him now, just despair. "You can go south," I said, thinking that was what he wanted to hear, "go south across the sea."

"To be another exiled Saxon king," he said bitterly.

"We hide here," I said, "and when we think the Danes aren't watching, we go to the south coast and find a ship."

"How do we hide?" he asked. "They know we're here. And they're on both sides of the swamp." The marshman had told us that a Danish fleet had landed at Cynuit, which lay at the swamp's western edge. That fleet, I assumed, was led by Svein and he would surely be wondering how to find Alfred. The king, I reckoned, was doomed, and his family, too. If Æthelflaed was lucky she would be raised by a Danish family, as I had been, but more probably they would all be killed so that no Saxon could ever again claim the crown of Wessex. "And the Danes will be watching the south coast," Alfred went on.

"They will," I agreed.

He looked out at the marsh where the night wind rippled the waters, shaking the long reflection of a winter moon. "The Danes can't have taken all Wessex," he said, then flinched because Edward was coughing so painfully.

"Probably not," I agreed.

"If we could find men," he said, then fell silent.

"What would we do with men?" I asked.

"Attack the fleet," he said, pointing west. "Get rid of Svein, if it is Svein at Cynuit, then hold the hills of Defnascir. Gain one victory and more men will come. We get stronger and one day we can face Guthrum."

I thought about it. He had spoken dully, as if he did not really believe in the words he had said, but I thought they made a perverse kind of sense. There were men in Wessex, men who were leaderless, but they were men who wanted a leader, men who would fight, and perhaps we could secure the swamp, then defeat Svein, then capture Defnascir, and so, piece by piece, take back Wessex. Then I thought about it more closely and reckoned it was a dream. The Danes had won. We were fugitives.

Alfred was stroking his daughter's golden hair. "The Danes will hunt us here, won't they?"

"Yes."

"Can you defend us?"

"Just me and Leofric?"

"You're a warrior, aren't you? Men tell me it was really you who defeated Ubba."

"You knew I killed Ubba?" I asked.

"Can you defend us?"

I would not be deflected. "Did you know I won your victory at Cynuit?" I demanded.

"Yes," he said simply.

"And my reward was to crawl to your altar? To be humiliated?" My anger made my voice too loud and Æthelflaed opened her eyes and stared at me.

"I have made mistakes," Alfred said, "and when this is all over, and when God returns Wessex to the West Saxons, I shall do the same. I shall put on the penitent's robe and submit myself to God."

I wanted to kill the pious bastard then, but Æthelflaed was watching me with her big eyes. She had not moved, so her father did not know she was awake, but I did, so instead of giving my anger a loose rein I cut it off abruptly. "You'll find that penitence helps," I said.

He brightened at that. "It helped you?" he asked.

"It gave me anger," I said, "and it taught me to hate. And anger is good. Hatred is good."

"You don't mean that," he said.

I half drew Serpent-Breath and little Æthelflaed's eyes grew wider. "This kills," I said, letting the sword slide back into its fleece-lined scabbard, "but anger and hate are what gives it the strength to kill. Go into battle without anger and hate and you'll be dead. You need all the blades, anger, and hate you can muster if we're to survive."

"But can you do it?" he asked. "Can you defend us here? Long enough to evade the Danes while we decide what to do?"

"Yes," I said. I had no idea whether I spoke the truth, indeed I doubted that I did, but I had a warrior's pride so gave a warrior's answer. Æthelflaed had not taken her eyes from me. She was only six, but I swear she understood all that we talked about.

"So I give you charge of that task," Alfred said. "Here and now I appoint you as the defender of my family. Do you accept that responsibility?"

I was an arrogant brute. Still am. He was challenging me, of course, and he knew what he was doing even if I did not. I just bridled. "Of course I accept it," I said, "yes."

"Yes what?" he asked.

I hesitated, but he had flattered me, given me a warrior's responsibility, and so I gave him what he wanted and what I had been determined not to give to him. "Yes, lord," I said.

He held out his hand. I knew he wanted more now. I had never meant to grant him this wish, but I had called him "lord" and so I knelt to him and, across Æthelflaed's body, I took his hand in both mine.

"Say it," he demanded, and he put the crucifix that hung about his neck between our hands.

"I swear to be your man," I said, looking into his pale eyes, "until your family is safe."

He hesitated. I had given him the oath, but I had qualified it. I had let him know that I would not remain his man forever, but he accepted my terms. He should have kissed me on both cheeks, but that would have disturbed Æthelflaed and so he raised my right hand and kissed the knuckles, then kissed the crucifix. "Thank you," he said.

The truth, of course, was that Alfred was finished, but, with the perversity and arrogance of foolish youth, I had just given him my oath and promised to fight for him.

And all, I think, because a six-year-old stared at me. And she had hair of gold.

SEVEN

The kingdom of Wessex was now a swamp and, for a few days, it possessed a king, a bishop, four priests, two soldiers, the king's pregnant wife, two nurses, a whore, two children, one of whom was sick, and Iseult.

Three of the four priests left the swamp first. Alfred was suffering, struck by the fever and belly pains that so often afflicted him, and he seemed incapable of rousing himself to any decision, so I gathered the three youngest priests, told them they were useless mouths we could not afford to feed, and ordered them to leave the swamp and discover what was happening on dry ground. "Find soldiers," I told them, "and say the king wants them to come here." Two of the priests begged to be spared the mission, claiming they were scholars incapable of surviving the winter or of confronting the Danes or of enduring discomfort or of doing any real work, and Alewold, the Bishop of Exanceaster, supported them, saying that their joint prayers were needed to keep the king healthy and safe, so I reminded the bishop that Eanflæd was present.

"Eanflæd?" He blinked at me as though he had never heard the name.

"The whore," I said, "from Cippanhamm." He still looked ignorant. "Cippanhamm," I went on, "where you and she rutted in the Corncrake tavern and she says . . ."

"The priests will travel," he said hastily.

"Of course they will," I said, "but they'll leave their silver here."

"Silver?"

The priests had been carrying Alewold's hoard, which included the great pyx I had given him to settle Mildrith's debts. That hoard was my next weapon. I took it all and displayed it to the marshmen. There would be silver, I said, for the food they gave us and the fuel they brought us and the punts they provided and the news they told us, news of the Danes on the swamp's far side. I wanted the marshmen on our side, and the sight of the silver encouraged them, but Bishop Alewold immediately ran to Alfred and complained that I had stolen from the church. The king was too low in spirits to care, so Ælswith, his wife, entered the fray. She was a Mercian and Alfred had married her to tighten the bonds between Wessex and Mercia, though that did little good for us now because the Danes ruled Mercia. There were plenty of Mercians who would fight for a West Saxon king, but none would risk their lives for a king reduced to a soggy realm in a tidal swamp. "You will return the pyx!" Ælswith ordered me. She looked ragged, her greasy hair tangled, her belly swollen, and her clothes filthy. "Give it back now. This instant!"

I looked at Iseult. "Should I?"

"No," Iseult said.

"She has no say here!" Ælswith shrieked.

"But she's a queen," I said, "and you're not." That was one cause of Ælswith's bitterness, that the West Saxons never called the king's wife a queen. She wanted to be Queen Ælswith and had to be content with less. She tried to snatch back the pyx, but I tossed it on the ground and, when she reached for it, I swung Leofric's ax. The blade chewed into the big plate, mangling the silver crucifixion, and Ælswith squealed in alarm and backed away as I hacked again. It took several blows, but I finally reduced the heavy plate into shreds of mangled silver that I tossed onto the coins I had taken from the priests. "Silver for your help!" I told the marshmen.

Ælswith spat at me, then went back to her son. Edward was three years old and it was evident now that he was dying. Alewold had claimed it was a mere winter's cold, but it was plainly worse, much worse. Every night we would listen to the coughing, an extraordinary hollow racking sound from such a small child, and all of us lay awake, dreading the next bout, flinching from the desperate, rasping sound, and when the coughing fits ended we feared they would not start again. Every silence was like the coming of death, yet somehow the small boy lived, clinging on through those cold wet days in the swamp. Bishop Alewold and the women tried all they knew. A gospel book was laid on his chest and the bishop prayed. A concoction of herbs, chicken dung, and ash was pasted on his chest and the bishop prayed. Alfred traveled nowhere without his precious relics, and the toe ring of Mary Magdalene was rubbed on the child's chest and the bishop prayed, but Edward just became weaker and thinner. A woman of the swamp, who had a reputation as a healer, tried to sweat the cough from him, and when that did not work she attempted to freeze it from him, and when that did not work she tied a live fish to his chest and commanded the cough and the fever to flee to the fish, and the fish certainly died, but the boy went on coughing and the bishop prayed and Alfred, as thin as his sick son, was in despair. He knew the Danes would search for him, but so long as the child was ill he dared not move, and he certainly could not contemplate the long walk south to the coast where he might find a ship to carry him and his family into exile.

He was resigned to that fate now. He had dared to hope he might recover his kingdom, but the cold reality was more persuasive. The Danes held Wessex and Alfred was king of nothing, and his son was dying. "It is a retribution," he said. It was the night after the three priests had left and Alfred unburdened his soul to me and Bishop Alewold. We were outside, watching the moon silver the marsh mists, and there were tears on Alfred's face. He was not really talking to either of us, only to himself.

"God would not take a son to punish the father," Alewold said.

"God sacrificed his own son," Alfred said bleakly, "and he commanded Abraham to kill Isaac."

"He spared Isaac," the bishop said.

"But he is not sparing Edward," Alfred said, and flinched as the awful coughing sounded from the hut. He put his head in his hands, covering his eyes.

"Retribution for what?" I asked, and the bishop hissed in reprimand for such an indelicate question.

"Æthelwold," Alfred said bleakly. Æthelwold was his nephew, the drunken, resentful son of the old king.

"Æthelwold could never have been king," Alewold said. "He is a fool!"

"If I name him king now," Alfred said, ignoring what the bishop had said, "perhaps God will spare Edward?"

The coughing ended. The boy was crying now, a gasping, grating, pitiful crying, and Alfred covered his ears with his hands.

"Give him to Iseult," I said.

"A pagan!" Alewold warned Alfred. "An adulteress!" I could see Alfred was tempted by my suggestion, but Alewold was having the better of the argument. "If God will not cure Edward," the bishop said, "do you think he will let a witch succeed?"

"She's no witch," I said.

"Tomorrow," Alewold said, ignoring me, "is Saint Agnes's Eve. A holy day, lord, a day of miracles! We shall pray to Saint Agnes and she will surely unleash God's power on the boy." He raised his hands to the dark sky. "Tomorrow, lord, we shall summon the strength of the angels, we shall call heaven's aid to your son, and the blessed Agnes will drive the evil sickness from young Edward."

Alfred said nothing, just stared at the swamp's pools that were edged with a thin skim of ice that seemed to glow in the wan moonlight.

"I have known the blessed Agnes perform miracles!" the bishop pressed the king. "There was a child in Exanceaster who could not walk, but the saint gave him strength and now he runs!"

"Truly?" Alfred asked.

"With my own eyes," the bishop said, "I witnessed the miracle."

Alfred was reassured. "Tomorrow then," he said.

I did not stay to see the power of God unleashed. Instead I took a punt and went south to a place called Æthelingæg, which lay at the southern edge of the swamp and was the biggest of all the marsh settlements. I was beginning to learn the swamp. Leofric stayed with Alfred, to protect the king and his family, but I explored, discovering scores of trackways through the watery void. The paths were called beamwegs and were made of logs that squelched underfoot, but by using them I could walk for miles. There were also rivers that twisted through the low land, and the biggest of those, the Pedredan, flowed close to Æthelingæg, which was an island, much of it covered with alders in which deer and wild goats lived, but there was also a large village on the island's highest spot and the headman had built himself a great hall there. It was not a royal hall, not even as big as the one I had made at Oxton, but a man could stand upright beneath its beams and the island was large enough to accommodate a small army.

A dozen beamwegs led away from Æthelingæg, but none led directly to the mainland. It would be a hard place for Guthrum to attack, because he would have to thread the swamp, but Svein, who we now knew commanded the Danes at Cynuit, at the Pedredan's mouth, would find it an easy place to approach for he could bring his ships up the river and, just north of Æthelingæg, he could turn south onto the river Thon, which flowed past the island. I took the punt into the center of the Thon and discovered, as I had feared, that it was more than deep enough to float the Dane's beast-headed ships.

I walked back to the place where the Thon flowed into the Pedredan. Across the wider river was a sudden hill, steep and high, which stood in the surrounding marshland like a giant's burial mound. It was a perfect place to make a fort, and if a bridge could be built across the Pedredan, then no Danish ship could pass upriver.

I walked back to the village where I discovered that the headman was a grizzled and stubborn old man called Haswold who was disinclined to help. I said I would pay good silver to have a bridge made across the Pedredan, but Haswold declared the war between Wessex and the Danes did not affect him. "There is madness over there," he said, waving vaguely at the eastern hills. "There's always madness over there, but here in the swamp we mind our own business. No one minds us and we don't mind them." He stank of fish and smoke. He wore otter skins that were greasy with fish oil and his graying beard was flecked by fish scales. He had small cunning eyes in an old cunning face, and he also had a half dozen wives, the youngest of whom was a child who could have been his own granddaughter, and he fondled her in front of me as if her existence proved his manhood. "I'm happy," he said, leering at me, "so why should I care for your happiness?"

"The Danes could end your happiness."

"The Danes?" He laughed at that, and the laugh turned into a cough. He spat. "If the Danes come," he went on, "then we go deep into the swamp and the Danes go." He grinned at me and I wanted to kill him, but that would have done no good. There were fifty or more men in the village and I would have lasted all of a dozen heartbeats, though the man I really feared was a tall, broad-shouldered, stooping man with a puzzled look on his face. What frightened me about him was that he carried a long hunting bow, not one of the short fowling bows that many of the marshmen possessed, but a stag killer, as tall as a man, and capable of shooting an arrow clean through a mail coat. Haswold must have sensed my fear of the bow for he summoned the man to stand beside him. The man looked confused by the summons, but obeyed. Haswold pushed a gnarled hand under the young girl's clothes, then stared at me as he fumbled, laughing at what he perceived as my impotence. "The Danes come," he said again, "and we go deep into the swamp and the Danes go away." He thrust his hand deeper into the girl's goatskin dress and mauled her breasts. "Danes can't follow us, and if they do follow us, then Eofer kills them." Eofer was the archer and, hearing his name, he looked startled, then worried. "Eofer's my man," Haswold boasted. "He puts arrows where I tell him to put them." Eofer nodded.

"Your king wants a bridge made," I said, "a bridge and a fort."

"King?" Haswold stared about the village. "I know no king. If any man is king here, 'tis me." He cackled with laughter at that and I looked at the villagers and saw nothing but dull faces. None shared Haswold's amusement. They were not, I thought, happy under his rule and perhaps he sensed what I was thinking, for he suddenly became angry, thrusting his girl bride away. "Leave us!" he shouted at me. "Just go away!"

I went away, returning to the smaller island where Alfred sheltered and where Edward lay dying. It was nightfall and the bishop's prayers to Saint Agnes had failed. Eanflæd told me how Alewold had persuaded Alfred to give up one of his most precious relics, a feather from the dove that Noah had released from the ark. Alewold cut the feather into two parts, returning one part to the king, while the other was scorched on a clean pan. When it was reduced to ash, the scraps were stirred into a cup of holy water, which Ælswith forced her son to drink. He had been wrapped in lambskin, for the lamb was the symbol of Saint Agnes who had been a child martyr in Rome.

But neither feather nor lambskin had worked. If anything, Eanflæd said, the boy was worse. Alewold was praying over him now. "He's given him the last rites," Eanflæd said. She looked at me with tears in her eyes. "Can Iseult help?"

"The bishop won't allow it," I said.

"He won't allow it?" she asked indignantly. "He's not the one who's dying!"

So Iseult was summoned, and Alfred came from the hut and Alewold, scenting heresy, came with him. Edward was coughing again, the sound terrible in the evening silence. Alfred flinched at the noise, then demanded to know if Iseult could cure his son's illness.

Iseult did not reply at once. Instead she turned and gazed across the swamp to where the moon rose above the mists. "The moon gets bigger," she said.

"Do you know a cure?" Alfred pleaded.

"A growing moon is good," Iseult said dully, then turned on him. "But there will be a price."

"Whatever you want!" he said.

"Not a price for me," she said, irritated that he had misunderstood her. "But there's always a price. One lives? Another must die."

"Heresy!" Alewold intervened.

I doubt Alfred understood Iseult's last three words, or did not care what she meant. He only snatched the tenuous hope that perhaps she could help. "Can you cure my son?" he demanded.

She paused, then nodded. "There is a way," she said.

"What way?"

"My way."

"Heresy!" Alewold warned again.

"Bishop!" Eanflæd said warningly, and the bishop looked abashed and fell silent.

"Now?" Alfred demanded of Iseult.

"Tomorrow night," Iseult said. "It takes time. There are things to do. If he lives till nightfall tomorrow I can help. You must bring him to me at moonrise."

"Not tonight?" Alfred pleaded.

"Tomorrow," Iseult said firmly.

"Tomorrow is the Feast of Saint Vincent," Alfred said, as though that might help, and somehow the child survived that night. Next day, Saint Vincent's Day, Iseult went with me to the eastern shore where we gathered lichen, burdock, celandine, and mistletoe. She would not let me use metal to scrape the lichen or cut the herbs, and before any was collected we had to walk three times around the plants, which, because it was winter, were poor and shriveled things. She also made me cut thorn boughs, and I was allowed to use a knife for that because the thorns were evidently not as important as the lichen or herbs. I watched the skyline as I worked, looking for any Danes, but if they patrolled the edge of the swamp none appeared that day. It was cold, a gusting wind clutching at our clothes. It took a long time to find the plants Iseult needed, but at last her pouch was full and I dragged the thornbushes back to the island and took them into the hut where she instructed me to dig two holes in the floor. "They must be as deep as the child is tall," she said, "and as far apart from each other as the length of your forearm."

She would not tell me what the pits were for. She was subdued, very close to tears. She hung the celandine and burdock from a roof beam, then pounded the lichen and the mistletoe into a paste that she moistened with spittle and urine, and she chanted long charms in her own language over the shallow wooden bowl. It all took a long time and sometimes she just sat exhausted in the darkness beyond the hearth and rocked to and fro. "I don't know that I can do it," she said once.

"You can try," I said helplessly.

"And if I fail," she said, "they will hate me more than ever."

"They don't hate you," I said.

"They think I am a sinner and a pagan," she said, "and they hate me."

"So cure the child," I said, "and they will love you."

I could not dig the pits as deep as she wanted, for the soil became ever wetter and, just a couple of feet down, the two holes were filling with brackish water. "Make them wider," Iseult ordered me, "wide enough so the child can crouch in them." I did as she said, and then she made me join the two holes by knocking a passage in the damp earth wall that divided them. That had to be done carefully to ensure that an arch of soil remained to leave a tunnel between the holes. "It is wrong," Iseult said, not talking of my excavation, but of the charm she planned to work. "Someone will die, Uhtred. Somewhere a child will die so this one will live."

"How do you know that?" I asked.

"Because my twin died when I was born," she said, "and I have his power. But if I use it he reaches from the dark world and takes the power back."

Darkness fell and the boy went on coughing, though to my ears it sounded feebler now, as though there was not enough life left in his small body. Alewold was praying still. Iseult crouched in the door of our hut, staring into the rain, and when Alfred came close she waved him away.

"He's dying," the king said helplessly.

"Not yet," Iseult said, "not yet."

Edward's breath rasped. We could all hear it, and we all thought every harsh breath would be his last, and still Iseult did not move, and then at last a rift showed in the rain clouds and a feeble wash of moonlight touched the marsh and she told me to fetch the boy.

Ælswith did not want Edward to go. She wanted him cured, but when I said Iseult insisted on working her charms alone, Ælswith wailed that she did not want her son to die apart from his mother. Her crying upset Edward, who began to cough again. Eanflæd stroked his forehead. "Can she do it?" she demanded of me.

"Yes," I said and did not know if I spoke the truth.

Eanflæd took hold of Ælswith's shoulders. "Let the boy go, my lady," she said. "Let him go."

"He'll die!"

"Let him go," Eanflæd said, and Ælswith collapsed into the whore's arms and I picked up Alfred's son, who felt as light as the feather that had not cured him. He was hot, yet shivering, and I wrapped him in a wool robe and carried him to Iseult.

"You can't stay here," she told me. "Leave him with me."

I waited with Leofric in the dark. Iseult insisted we could not watch through the hut's entrance, but I dropped my helmet outside the door and, by crouching under the eaves, I could just see a reflection of what happened inside. The small rain died and the moon grew brighter.

The boy coughed. Iseult stripped him naked and rubbed her herb paste on his chest, and then she began to chant in her own tongue, an endless chant it seemed, rhythmic, sad, and so monotonous that it almost put me to sleep. Edward cried once, and the crying turned to coughing and his mother screamed from her hut that she wanted him back, and Alfred calmed her and then came to join us and I waved him down so that he would not shadow the moonlight before Iseult's door.

I peered at the helmet and saw, in the small reflected firelight, that Iseult, naked herself now, was pushing the boy into one of the pits and then, still chanting, she drew him through the earth passage. Her chanting stopped and, instead, she began to pant, then scream, then pant again. She moaned, and Alfred made the sign of the cross, and then there was silence and I could not see properly, but suddenly Iseult cried aloud, a cry of relief, as if a great pain was ended, and I dimly saw her pull the naked boy out of the second pit. She laid him on her bed and he was silent as she crammed the thornbushes into the tunnel of earth. Then she lay beside the boy and covered herself with my large cloak.

There was silence. I waited and waited, and still there was silence. And the silence stretched until I understood that Iseult was sleeping, and the boy was sleeping, too, or else he was dead, and I picked up the helmet and went to Leofric's hut. "Shall I fetch him?" Alfred asked nervously.

"No."

"His mother—" he began.

"Must wait till morning, lord."

"What can I tell her?"

"That her son is not coughing, lord."

Ælswith screamed that Edward was dead, but Eanflæd and Alfred calmed her, and we all waited, and still there was silence, and in the end I fell asleep.

I woke in the dawn. It was raining as if the world was about to end, a torrential gray rain that swept in vast curtains from the Sæfern Sea, a rain that drummed on the ground and poured off the reed thatch and made streams on the small island where the little huts crouched. I went to the door of Leofric's shelter and saw Ælswith watching from her doorway. She looked desperate, like a mother about to hear that her child had died, and there was nothing but silence from Iseult's hut, and Ælswith began to weep, the terrible tears of a bereaved mother, and then there was a strange sound. At first I could not hear properly, for the seething rain was loud, but then I realized the sound was laughter. A child's laughter, and a heartbeat later Edward, still naked as an egg, and all muddy from his rebirth through the earth passage, ran from Iseult's hut and went to his mother.

"Dear God," Leofric said.

Iseult, when I found her, was weeping and would not be consoled. "I need you," I told her harshly.

She looked up at me. "Need me?"

"To build a bridge."

She frowned. "You think a bridge can be made with spells?"

"My magic this time," I said. "I want you healthy. I need a queen."

She nodded. And Edward, from that day forward, thrived.

The first men came, summoned by the priests I had sent onto the mainland. They came in ones and twos, struggling through the winter weather and the swamp, bringing tales of Danish raids, and when we had two days of sunshine they came in groups of six or seven so that the island became crowded. I sent them out on patrol, but ordered none to go too far west for I did not want to provoke Svein, whose men were camped beside the sea. He had not attacked us yet, which was foolish of him, for he could have brought his ships up the rivers and then struggled through the marsh, but I knew he would attack us when he was ready, and so I needed to make our defenses. And for that I needed Æthelingæg.

Alfred was recovering. He was still sick, but he saw God's favor in his son's recovery and it never occurred to him that it had been pagan magic that caused the recovery. Even Ælswith was generous and, when I asked her for the loan of her silver fox-fur cloak and what few jewels she possessed, she yielded them without fuss. The fur cloak was dirty, but Eanflæd brushed and combed it.

There were over twenty men on our island now, probably enough to capture Æthelingæg from its sullen headman, but Alfred did not want the marshmen killed. They were his subjects, he said, and if the Danes attacked they might yet fight for us, which meant the large island and its village must be taken by trickery and so, a week after Edward's rebirth, I took Leofric and Iseult south to Haswold's settlement. Iseult was dressed in the silver fur and had a silver chain in her hair and a great garnet brooch at her breast. I had brushed her hair till it shone and in that winter's gloom she looked like a princess come from the bright sky.

Leofric and I, dressed in mail and helmets, did nothing except walk around Æthelingæg, but after a while a man came from Haswold and said the chieftain wished to talk with us. I think Haswold expected us to go to his stinking hut, but I demanded he come to us instead. He could have taken from us whatever he wanted, of course, for there were only the three of us and he had his men, including Eofer the archer, but Haswold had at last understood that dire things were happening in the world beyond the swamp and that those events could pierce even his watery fastness, and so he chose to talk. He came to us at the settlement's northern gate, which was nothing more than a sheep hurdle propped against decaying fish traps and there, as I expected, he gazed at Iseult as though he had never seen a woman before. His small cunning eyes flickered at me and back to her. "Who is she?" he asked.

"A companion," I said carelessly. I turned to look at the sudden steep hill across the river where I wanted the fort made.

"Is she your wife?" Haswold asked.

"A companion," I said again. "I have a dozen like her," I added.

"I will pay you for her," Haswold said. A score of men were behind him, but only Eofer was armed with anything more dangerous than an eel spear.

I turned Iseult to face him; then I stood behind her and put my hands over her shoulders and undid the big garnet brooch. She shivered slightly and I whispered that she was safe and, when the brooch pin slid out of the heavy hide, I pulled her fur cloak apart. I showed her nakedness to Haswold and he dribbled into his fish-scaled beard and his dirty fingers twitched in his foul otter-skin furs, and then I closed the cloak and let Iseult fasten the brooch. "How much will you pay me?" I asked him.

"I can just take her," Haswold said, jerking his head at his men.

I smiled at that. "You could," I said, "but many of you will die before we die, and our ghosts will come back to kill your women and make your children scream. Have you not heard that we have a witch with us? You think your weapons can fight magic?"

None of them moved.

"I have silver," Haswold said.

"I don't need silver," I said. "What I want is a bridge and a fort." I turned and pointed to the hill across the river. "What is that hill called?"

He shrugged. "The hill," he said, "just the hill."

"It must become a fort," I said, "and it must have walls of logs and a gate of logs and a tower so that men can see a long way downriver. And then I want a bridge leading to the fort, a bridge strong enough to stop ships."

"You want to stop ships?" Haswold asked. He scratched his groin and shook his head. "Can't build a bridge."

"Why not?"

"Too deep." That was probably true. It was low tide now and the Pedredan flowed sullenly between steep and deep mud banks. "But I can block the river," Haswold went on, his eyes still on Iseult.

"Block the river," I said, "and build a fort."

"Give her to me," Haswold promised, "and you will have both."

"Do what I want," I said, "and you can have her, her sisters, and her cousins. All twelve of them."

Haswold would have drained the whole swamp and built a new Jerusalem for the chance to hump Iseult, but he had not thought beyond the end of his prick. But that was far enough for me, and I have never seen work done so quickly. It was done in days. He blocked the river first and did it cleverly by making a floating barrier of logs and felled trees, complete with their tangling branches, all of them lashed together with goat-hide ropes. A ship's crew could eventually dismantle such a barrier, but not if they were being assailed by spears and arrows from the fort on the hill that had a wooden palisade, a flooded ditch, and a flimsy tower made of alder logs bound together with leather ropes. It was all crude work, but the wall was solid enough, and I began to fear that the small fort would be finished before enough West Saxons arrived to garrison it, but the three priests were doing their job and the soldiers still came, and I put a score of them in Æthelingæg and told them to help finish the fort.

When the work was done, or nearly done, I took Iseult back to Æthelingæg and I dressed her as she had been dressed before, only this time she wore a deerskin tunic beneath the precious fur, and I stood her in the center of the village and said Haswold could take her. He looked at me warily, then looked at her. "She's mine?" he asked.

"All yours," I said, and stepped away from her.

"And her sisters?" he asked greedily. "Her cousins?"

"I shall bring them tomorrow."

He beckoned Iseult toward his hut. "Come," he said.

"In her country," I said, "it is the custom for the man to lead the woman to his bed."

He stared at Iseult's lovely, dark-eyed face above the swathing silver cloak. I stepped farther back, abandoning her, and he darted forward, reaching for her, and she brought her hands out from under the thick fur and she was holding Wasp-Sting and its blade sliced up into Haswold's belly. She gave a cry of horror and surprise as she brought the blade up, and I saw her hesitate, shocked by the effort required to pierce a man's belly and by the reality of what she had done. Then she gritted her teeth and ripped the blade hard, opening him up like a gutted carp, and he gave a strange mewing cry as he staggered back from her vengeful eyes. His intestines spilled into the mud, and I was beside her then with Serpent-Breath drawn. She was gasping, trembling. She had wanted to do it, but I doubted she would want to do it again.

"You were asked," I snarled at the villagers, "to fight for your king." Haswold was on the ground, twitching, his blood soaking his otter-skin clothes. He made a mewing noise and one of his filthy hands scrabbled among his own spilled guts. "For your king!" I repeated. "When you are asked to fight for your king it is not a request, but a duty! Every man here is a soldier and your enemy is the Danes and if you will not fight them, then you will fight against me!"

Iseult still stood beside Haswold, who jerked like a dying fish. I edged her away and stabbed Serpent-Breath down to slit his throat. "Take his head," she told me.

"His head?"

"Strong magic."

We mounted Haswold's head on the fort wall so that it stared toward the Danes, and in time eight more heads appeared there. They were the heads of Haswold's chief supporters, murdered by the villagers who were glad to be rid of them. Eofer, the archer, was not one of them. He was a simpleton, incapable of speaking sense, though he grunted and, from time to time, made howling noises. He could be led by a child, but when asked to use his bow he proved to have a terrible strength and uncanny accuracy. He was Æthelingæg's hunter, capable of dropping a full-grown boar at a hundred paces, and that was what his name meant: boar.

I left Leofric to command the garrison at Æthelingæg and took Iseult back to Alfred's refuge. She was silent and I thought her sunk in misery, but then she suddenly laughed. "Look!" She pointed at the dead man's blood matted and sticky in Ælswith's fur.

She still had Wasp-Sting. That was my short sword, a saxe, and it was a wicked blade in a close fight where men are so crammed together that there is no room to swing a long sword or an ax. She trailed the blade in the water, then used the hem of Ælswith's fur to scrub the diluted blood from the steel. "It is harder than I thought," she said, "to kill a man."

"It takes strength."

"But I have his soul now."

"Is that why you did it?'

"To give life," she said, "you must take it from somewhere else." She gave me back Wasp-Sting.

Alfred was shaving when we returned. He had been growing a beard, not for a disguise, but because he had been too low in spirits to bother about his appearance, but when Iseult and I reached his refuge he was standing naked to the waist beside a big wooden tub of heated water. His chest was pathetically thin, his belly hollow, but he had washed himself, combed his hair, and was now scratching at his stubble with an ancient razor he had borrowed from a marshman. His daughter, Æthelflaed, was holding a scrap of silver that served as a mirror. "I am feeling better," he told me solemnly.

"Good, lord," I said. "So am I."

"Does that mean you've killed someone?"

"She did." I jerked my head at Iseult.

He gave her a speculative look. "My wife," he said, dipping the razor in the water, "was asking whether Iseult is truly a queen."

"She was," I said, "but that means little in Cornwalum. She was queen of a dung heap."

"And she's a pagan?"

"It was a Christian kingdom," I said. "Didn't Brother Asser tell you that?"

"He said they were not good Christians."

"I thought that was for God to judge."

"Good, Uhtred, good!" He waved the razor at me, then stooped to the silver mirror and scraped at his upper lip. "Can she foretell the future?"

"She can."

He scraped in silence for a few heartbeats. Æthelflaed watched Iseult solemnly. "So tell me," Alfred said, "does she say I will be king in Wessex again?"

"You will," Iseult said tonelessly, surprising me.

Alfred stared at her. "My wife," he said, "says that we can look for a ship now that Edward is better. Look for a ship, go to Frankia, and perhaps travel on to Rome. There is a Saxon community in Rome." He scraped the blade against his jawbone. "They will welcome us."

"The Danes will be defeated," Iseult said, still tonelessly, but without a quiver of doubt in her voice.

Alfred rubbed his face. "The example of Boethius tells me she's right," he said.

"Boethius?" I asked. "Is he one of your warriors?"

"He was a Roman, Uhtred," Alfred said in a tone that chided me for not knowing, "and a Christian and a philosopher and a man rich in book learning. Rich indeed!" He paused, contemplating the story of Boethius. "When the pagan Alaric overran Rome," he went on, "and all civilization and true religion seemed doomed, Boethius alone stood against the sinners. He suffered, but he won through, and we can take heart from him. Indeed we can." He pointed the razor at me. "We must never forget the example of Boethius, Uhtred, never."

"I won't, lord," I said, "but do you think book learning will get you out of here?"

"I think," he said, "that when the Danes are gone, I shall grow a proper beard. Thank you, my sweet." This last was to Æthelflæd. "Give the mirror back to Eanflæd, will you?"

Æthelflæd ran off and Alfred looked at me with some amusement. "Does it surprise you that my wife and Eanflæd have become friends?"

"I'm glad of it, lord."

"So am I."

"But does your wife know Eanflæd's trade?" I asked.

"Not exactly," he said. "She believes Eanflæd was a cook in a tavern. Which is truth enough. So we have a fort at Æthelingæg?"

"We do. Leofric commands there and has forty-three men."

"And we have twenty-eight here. The very hosts of Midian!" He was evidently amused. "So we shall move there."

"Maybe in a week or two."

"Why wait?" he asked.

I shrugged. "This place is deeper in the swamp. When we have more men, when we know we can hold Æthelingæg, that is the time for you to go there."

He pulled on a grubby shirt. "Your new fort can't stop the Danes?"

"It will slow them, lord. But they could still struggle through the marsh." They would find it difficult, though, for Leofric was digging ditches to defend Æthelingæg's western edge.

"You're telling me Æthelingæg is more vulnerable than this place?"

"Yes, lord."

"Which is why I must go there," he said. "Men can't say their king skulked in an unreachable place, can they?" He smiled at me. "They must say he defied the Danes. That he waited where they could reach him, that he put himself into danger."

"And his family?" I asked.

"And his family," he said firmly. He thought for a moment. "If they come in force they could take all the swamp, isn't that true?"

"Yes, lord."

"So no place is safer than another. But how large a force does Svein have?"

"I don't know, lord."

"Don't know?" It was a reproof, gentle enough, but still a reproof.

"I haven't gone close to them, lord," I explained, "because till now we've been too weak to resist them, and so long as they leave us undisturbed then so long do we leave them undisturbed. There's no point in kicking a wild bees' nest, not unless you're determined to get the honey."

He nodded acceptance of that argument. "But we need to know how many bees there are, don't we?" he said. "So tomorrow we shall take a look at our enemy. You and me, Uhtred."

"No, lord," I said firmly. "I shall go. You shouldn't risk yourself."

"That is exactly what I need to do," he said, "and men must know I do it for I am the king, and why would men want a king who does not share their danger?" He waited for an answer, but I had none. "So let's say our prayers," he finished. "Then we shall eat."

It was fish stew. It was always fish stew.

And next day we went to find the enemy.

There were six of us: the man who poled the punt, Iseult and I, two of the newly arrived household troops, and Alfred. I tried once again to make him stay behind, but he insisted. "If anyone should stay," he said, "it is Iseult."

"She comes," I said.

"Evidently." He did not argue, and we all climbed into a large punt and went westward, and Alfred stared at the birds, thousands of birds. There were coots, moorhens, dabchicks, ducks, grebes, and herons, while off to the west, white against the sullen sky, was a cloud of gulls.

The marshman slid us silent and fast through secret channels. There were times when he seemed to be taking us directly into a bank of reeds or grass, yet the shallow craft would slide through into another stretch of open water. The incoming tide rippled through the gaps, bringing fish to the hidden nets and basket traps. Beneath the gulls, far off to the west, I could see the masts of Svein's fleet that had been dragged ashore on the coast.

Alfred saw them, too. "Why don't they join Guthrum?"

"Because Svein doesn't want to take Guthrum's orders," I said.

"You know that?"

"He told me so."

Alfred paused, perhaps thinking of my trial in front of the witan. He gave me a rueful look. "What sort of man is he?"

"Formidable."

"So why hasn't he attacked us here?"

I had been wondering the same thing. Svein had missed a golden chance to invade the swamp and hunt Alfred down. So why had he not even tried? "Because there's easier plunder elsewhere," I suggested, "and because he won't do Guthrum's bidding. They're rivals. If Svein takes Guthrum's orders, then he acknowledges Guthrum as his king."

Alfred stared at the distant masts that showed as small scratches against the sky. Then I mutely pointed toward a hill that reared steeply from the western water flats and the marshman obediently went that way, and when the punt grounded we clambered through thick alders and past some sunken hovels where sullen folk in dirty otter fur watched us pass. The marshman knew no name for the place, except to call it Brant, which meant steep, and it was steep. Steep and high, offering a view southward to where the Pedredan coiled like a great snake through the swamp's heart. And at the river's mouth, where sand and mud stretched into the Sæfern Sea, I could see the Danish ships.

They were grounded on the far bank of the Pedredan in the same place that Ubba had grounded his ships before meeting his death in battle. From there Svein could easily row to Æthelingæg, for the river was wide and deep, and he would meet no challenge until he reached the river barrier beside the fort where Leofric waited. I wanted Leofric and his garrison to have some warning if the Danes attacked, and this high hill offered a view of Svein's camp, but was far enough away so that it would not invite an attack from the enemy. "We should make a beacon here," I said to Alfred. A fire lit here would give Æthelingæg two or three hours' warning of a Danish attack.

He nodded but said nothing. He stared at the distant ships, but they were too far off to count. He looked pale, and I knew he had found the climb to the summit painful, so now I urged him downhill to where the hovels leaked smoke. "You should rest here, lord," I told him. "I'm going to count ships. But you should rest."

He did not argue and I suspected his stomach pains were troubling him again. I found a hovel that was occupied by a widow and her four children, and I gave her a silver coin and said her king needed warmth and shelter for the day, and I do not think she understood who he was, but she knew the value of a shilling and so Alfred went into her house and sat by the fire. "Give him broth," I told the widow, whose name was Elwide, "and let him sleep."

She scorned that. "Folk can't sleep while there's work!" she said. "There are eels to skin, fish to smoke, nets to mend, traps to weave."

"They can work," I said, pointing to the two household troops, and I left them all to Elwide's tender mercies while Iseult and I took the punt southward and, because the Pedredan's mouth was only three or four miles away and because Brant was such a clear land-mark, I left the marshman to help skin and smoke eels.

We crossed a smaller river and then poled through a long mere broken by marram grass and by now I could see the hill on the Pedredan's far bank where we had been trapped by Ubba, and I told

Iseult the story of the fight as I poled the punt across the shallows. The hull grounded twice and I had to push it into deeper water until I realized the tide was falling fast and so I tied the boat to a rotting stake. Then we walked across a drying waste of mud and sea lavender toward the Pedredan. I had grounded farther from the river than I had wanted, and it was a long walk into a cold wind, but we could see all we needed once we reached the steep bank at the river's edge. The Danes could also see us. I was not in mail, but I did have my swords, and the sight of me brought men to the farther shore, where they hurled insults across the swirling water. I ignored them. I was counting ships and saw twenty-four beast-headed boats hauled up on the strip of ground where we had defeated Ubba the year before. Ubba's burned ships were also there, their black ribs half buried in the sand where the men capered and shouted their insults.

"How many men can you see?" I asked Iseult.

There were a few Danes in the half-wrecked remnants of the monastery where Svein had killed the monks, but most were by the boats. "Just men?" she asked.

"Forget the women and children," I said. There were scores of women, mostly in the small village that was a small way upstream.

She did not know the English words for the bigger numbers, so she gave me her estimate by opening and closing her fingers six times. "Sixty?" I said, and nodded. "At most seventy. And there are twenty-four ships." She frowned, not understanding the point I made. "Twenty-four ships," I said, "means an army of what? Eight hundred? Nine hundred men? So those sixty or seventy men are the ship guards. And the others? Where are the others?" I asked the question of myself, watching as five of the Danes dragged a small boat to the river's edge. They planned to row across and capture us, but I did not intend to stay that long. "The others," I answered my own question, "have gone south. They've left their women behind and gone raiding. They're burning, killing, getting rich. They're raping Defnascir."

"They're coming," Iseult said, watching the five men clamber into the small boat.

"You want me to kill them?"

"You can?" She looked hopeful.

"No," I said, "so let's go."

We started back across the long expanse of mud and sand. It looked smooth, but there were runnels cutting through and the tide had turned and the sea was sliding back into the land with surprising speed. The sun was sinking, tangling with black clouds, and the wind pushed the flood up the Sæfern and the water gurgled and shivered as it filled the small creeks. I turned to see that the five Danes had abandoned their chase and gone back to the western bank where their fires looked delicate against the evening's fading light. "I can't see the boat," Iseult said.

"Over there," I said, but I was not certain I was right because the light was dimming and our punt was tied against a background of reeds, and now we were jumping from one dry spot to another, and the tide went on rising and the dry spots shrank and then we were splashing through the water and still the wind drove the tide inland.

The tides are big in the Sæfern. A man could make a house at low tide, and by high it would have vanished beneath the waves. Islands appear at low tide, islands with summits thirty feet above the water, and at the high tide they are gone, and this tide was pushed by the wind and it was coming fast and cold and Iseult began to falter so I picked her up and carried her like a child. I was struggling and the sun was behind the low western clouds and it seemed now that I was wading through an endless chill sea, but then, perhaps because the darkness was falling, or perhaps because Hoder, the blind god of the night, favored me, I saw the punt straining against its tether.

I dropped Iseult into the boat and hauled myself over the low side. I cut the rope, then collapsed, cold and wet and frightened, and let the punt drift on the tide.

"You must get back to the fire," Iseult chided me. I wished I had brought the marshman now for I had to find a route across the swamp and it was a long, cold journey in the day's last light. Iseult crouched beside me and stared far across the waters to where a hill reared up green and steep against the eastern land. "Eanflæd told me that hill is Avalon," she said reverentially.

"Avalon?"

"Where Arthur is buried."

"I thought you believed he was sleeping."

"He does sleep," she said fervently. "He sleeps in his grave with his warriors." She gazed at the distant hill that seemed to glow because it had been caught by the day's last errant shaft of sunlight spearing from the west beneath the furnace-glowing clouds. "Arthur," she said in a whisper. "He was the greatest king who ever lived. He had a magic sword." She told me tales of Arthur, how he had pulled his sword from a stone, and how he had led the greatest warriors to battle, and I thought that his enemies had been us, the English Saxons, yet Avalon was now in England, and I wondered if, in a few years, the Saxons would recall their lost kings and claim they were great and all the while the Danes would rule us. When the sun vanished Iseult was singing softly in her own tongue, but she told me the song was about Arthur and how he had placed a ladder against the moon and netted a swath of stars to make a cloak for his queen, Guinevere. Her voice carried us across the twilit water, sliding between reeds, and behind us the fires of the Danish ship guards faded in the encroaching dark, and far off a dog howled and the wind sighed cold and a spattering of rain shivered the black mere.

Iseult stopped singing as Brant loomed. "There's going to be a great fight," she said softly and her words took me by surprise and I thought she was still thinking of Arthur and imagining that the sleeping king would erupt from his earthy bed in gouts of soil and steel. "A fight by a hill," she went on, "a steep hill, and there will be a white horse and the slope will run with blood and the Danes will run from the Sais."

The Sais were us, the Saxons. "You dreamed this?" I asked.

"I dreamed it," she said.

"So it is true?"

"It is fate," she said, and I believed her, and just then the bow of the punt scraped on the island's shore.

It was pitch dark, but there were fish-smoking fires on the beach, and by their dying light we found our way to Elwide's house. It was made of alder logs thatched with reeds and I found Alfred sitting by the central hearth where he stared absently into the flames. Elwide, the two soldiers, and the marshman were all skinning eels at the hut's farther end where three of the widow's children were plaiting willow withies into traps and the fourth was gutting a big pike.

I crouched by the fire, wanting its warmth to bring life to my frozen legs.

Alfred blinked as though he was surprised to see me. "The Danes?" he asked.

"Gone inland," I said. "Left sixty or seventy men as ship guards." I crouched by the fire, shivering, wondering if I would ever be warm again.

"There's food here," Alfred said vaguely.

"Good," I said, "because we're starving."

"No, I mean there's food in the marshes," he said. "Enough food to feed an army. We can raid them, Uhtred, gather men and raid them. But that isn't enough. I have been thinking. All day, I've been thinking." He looked better now, less pained, and I suspected he had wanted time to think and had found it in this stinking hovel. "I'm not going to run away," he said firmly. "I'm not going to Frankia."

"Good," I said, though I was so cold I was not really listening to him.

"We're going to stay here," he said, "raise an army, and take Wessex back."

"Good," I said again. I could smell burning. The hearth was surrounded by flat stones and Elwide had put a dozen oat bannocks on the stones to cook and the edges nearest the flames were blackening. I moved one of them, but Alfred frowned and gestured for me to stop for fear of distracting him. "The problem," he said, "is that I cannot afford to fight a small war."

I did not see what other war he could fight, but kept silent.

"The longer the Danes stay here," he said, "the firmer their grip. Men will start giving Guthrum their allegiance. I can't have that."

"No, lord."

"So they have to be defeated." He spoke grimly. "Not beaten, Uhtred, but defeated!"

I thought of Iseult's dream but said nothing. Then I thought how often Alfred had made peace with the Danes instead of fighting them, and still I said nothing.

"In spring," he went on, "they'll have new men and they'll spread through Wessex until, by summer's end, there'll be no Wessex. So we have to do two things." He was not so much telling me as just thinking aloud. "First," he held out one long finger, "we have to stop them from dispersing their armies. They have to fight us here. They have to be kept together so they can't send small bands across the country and take estates." That made sense. Right now, from what we heard from the land beyond the swamp, the Danes were raiding all across Wessex. They were going fast, snatching what plunder they could before other men could take it, but in a few weeks they would start looking for places to live. By keeping their attention on the swamp, Alfred hoped to stop that process. "And while they look at us," he said, "the fyrd must be gathered."

I stared at him. I had supposed he would stay in the swamp until either the Danes overwhelmed us or we gained enough strength to take back a shire, and then another shire, a process of years, but his vision was much grander. He would assemble the army of Wessex under the Danish noses and take everything back at once. It was like a game of dice and he had decided to take everything he had, little as it was, and risk it all on one throw. "We shall make them fight a great battle," he said grimly, "and with God's help we shall destroy them."

There was a sudden scream. Alfred, as if startled from a reverie, looked up, but too late, because Elwide was standing over him, screaming that he had burned the oatcakes. "I told you to watch them!" she shouted and, in her fury, she slapped the king with a skinned eel. The blow made a wet sound as it struck and had enough force to knock Alfred sideways. The two soldiers jumped up, hands going to their swords, but I waved them back as Elwide snatched the burned cakes from the stones. "I told you to watch them!" she shrieked, and Alfred lay where he had fallen and I thought he was crying, but then I saw he was laughing. He was helpless with laughter, weeping with laughter, as happy as ever I saw him.

Because he had a plan to take back his kingdom.

Æthelingæg's garrison now had seventy-three men. Alfred moved there with his family and sent six of Leofric's men to Brant armed with axes and orders to make a beacon. He was at his best in those days, calm and confident, the panic and despair of the first weeks of January swept away by his irrational belief that he would regain his kingdom before summer touched the land. He was immensely cheered, too, by the arrival of Father Beocca who came limping from the landing stage, face beaming, to fall prostrate at the king's feet. "You live, lord!" Beocca said, clutching the king's ankles. "God be praised, you live!"

Alfred raised him and embraced him and both men wept, and next day, a Sunday, Beocca preached a sermon that I could not help hearing because the service was held in the open air, under a clear cold sky, and Æthelingæg's island was too small to escape the priest's voice. Beocca said how David, King of Israel, had been forced to flee his enemies, how he had taken refuge in the cave of Adullam, and how God had led him back into Israel and to the defeat of his enemies. "This is our Adullam!" Beocca said, waving his good hand at Æthelingæg's thatched roofs. "And this is our David!" He pointed to the king. "And God will lead us to victory!"

"It's a pity, father," I said to Beocca afterward, "that you weren't this belligerent two months ago."

"I rejoice," he said loftily, "to find you in the king's good graces."

"He's discovered the value," I said, "of murderous bastards like me, so perhaps he'll learn to distrust the advice of sniveling bastards like you who told him the Danes could be defeated by prayer."

He sniffed at that insult, then looked disapprovingly at Iseult. "You have news of your wife?"

"None."

Beocca had some news, though none of Mildrith. He had fled south in front of the invading Danes, getting as far as Dornwaraceaster in Thornsæta where he had found refuge with some monks. The Danes had come, but the monks had received warning of their approach and hid in an ancient fort that lay near the town. The Danes had sacked Dornwaraceaster, taking silver, coins, and women. Then they had moved eastward and shortly after that Huppa, Ealdorman of Thornsæta, had come to the town with fifty warriors. Huppa had set the monks and townspeople to mending the old Roman walls. "The folk there are safe for the moment," Beocca told me, "but there is not sufficient food if the Danes return and lay siege." Then Beocca had heard that Alfred was in the great swamps and Beocca had traveled alone, though on his last day of walking he had met six soldiers going to Alfred and so he had finished his journey with them. He brought no news of Wulfhere, but he had been told that Odda the Younger was somewhere on the upper reaches of Uisc in an ancient fort built by the old people. Beocca had seen no Danes on his journey. "They raid everywhere," he said gloomily, "but God be praised, we saw none of them."

"Is Dornwaraceaster a large place?" I asked.

"Large enough. It had three fine churches, three!"

"A market?"

"Indeed, it was prosperous before the Danes came."

"Yet the Danes didn't stay there?"

"Nor were they at Gifle," he said, "and that's a goodly place."

Guthrum had surprised Alfred, defeated the forces at Cippanhamm, and driven the king into hiding, but to hold Wessex he needed to take all her walled towns, and if Beocca could walk three days across country and see no Danes, then it suggested Guthrum did not have the men to hold all he had taken. He could bring more men from Mercia or East Anglia, but then those places might rise against their weakened Danish overlords, so Guthrum had to be hoping that more ships would come from Denmark. In the meantime, we learned, he had garrisons in Baðum, Readingum, Mærlebeorg, and Andefera, and doubtless he held other places, and Alfred suspected, rightly as it turned out, that most of eastern Wessex was in Danish hands, but great stretches of the country were still free of the enemy. Guthrum's men were making raids into those stretches, but they did not have sufficient force to garrison towns like Wintanceaster, Gifle, or Dornwaraceaster. In the early summer, Alfred knew, more ships would bring more Danes, so he had to strike before then, to which end, on the day after Beocca arrived, he summoned a council.

There were now enough men on Æthelingæg for a royal formality to prevail. I no longer found Alfred sitting outside a hut in the evening, but instead had to seek an audience with him. On the Monday of the council he gave orders that a large house was to be made into a church, and the family that lived there was evicted and some of the newly arrived soldiers were ordered to make a great cross for the gable and to carve new windows in the walls. The council itself met in what had been Haswold's hall, and Alfred had waited till we were assembled before making his entrance, and we had all stood as he came in and waited as he took one of the two chairs on the newly made dais. Ælswith sat beside him, her pregnant belly swathed in the silver fur cloak that was still stained with Haswold's blood.

We were not allowed to sit until the Bishop of Exanceaster said a prayer, and that took time, but at last the king waved us down. There were six priests in the half circle and six warriors. I sat beside Leofric, while the other four soldiers were newly arrived men who had served in Alfred's household troops. One of those was a gray-bearded man called Egwine who told me he had led a hundred men at Æsc's Hill and plainly thought he should now lead all the troops gathered in the swamp. I knew he had urged his case with the king and with Beocca who sat just below the dais at a rickety table on which he was trying to record what was said at the council. Beocca was having difficulties for his ink was ancient and faded, his quill kept splitting, and his parchments were wide margins torn from a missal, so he was unhappy, but Alfred liked to reduce arguments to writing.

The king formally thanked the bishop for his prayer, then announced, sensibly enough, that we could not hope to deal with Guthrum until Svein was defeated. Svein was the immediate threat for, though most of his men had gone south to raid Defnascir, he still had the ships with which to enter the swamp. "Twenty-four ships," Alfred said, raising an eyebrow at me.

"Twenty-four, lord," I confirmed.

"So, when his men are assembled, he can muster near a thousand men." Alfred let that figure linger awhile. Beocca frowned as his split quill spattered ink on his tiny patch of parchment.

"But two days ago," Alfred went on, "there were only seventy ship guards at the mouth of the Pedredan."

"Around seventy," I said. "There could be more we didn't see."

"Fewer than a hundred, though?"

"I suspect so, lord."

"So we must deal with them," Alfred said, "before the rest return to their ships." There was another silence. All of us knew how weak we were. A few men arrived every day, like the half dozen who had come with Beocca, but they came slowly, either because the news of Alfred's existence was spreading slowly, or else because the weather was cold and men do not like to travel on wet, cold days. Nor were there any thegns among the newcomers, not one. Thegns were noblemen, men of property, men who could bring scores of well-armed followers to a fight, and every shire had its thegns who ranked just below the reeve and ealdorman, who were themselves thegns. Thegns were the power of Wessex, but none had come to Æthelingæg. Some, we heard, had fled abroad, while others tried to protect their property. Alfred, I was certain, would have felt more comfortable if he had a dozen thegns about him, but instead he had me and Leofric and Egwine. "What are our forces now?" Alfred asked us.

"We have over a hundred men," Egwine said brightly.

"Of whom only sixty or seventy are fit to fight," I said. There had been an outbreak of sickness, men vomiting and shivering and hardly able to control their bowels. Whenever troops gather, such sickness seems to strike.

"Is that enough?" Alfred asked.

"Enough for what, lord?" Egwine was not quick-witted.

"Enough to get rid of Svein, of course," Alfred said, and again there was silence because the question was absurd.

Then Egwine straightened his shoulders. "More than enough, lord!"

Ælswith bestowed a smile on him.

"And how would you propose doing it?" Alfred asked.

"Take every man we have, lord," Egwine said, "every fit man, and attack them. Attack them!"

Beocca was not writing. He knew when he was hearing nonsense and he was not going to waste scarce ink on bad ideas.

Alfred looked at me. "Can it be done?"

"They'll see us coming," I said. "They'll be ready."

"March inland," Egwine said. "Come from the hills."

Again Alfred looked at me. "That will leave Æthelingæg undefended," I said, "and it will take at least three days, at the end of which our men will be cold, hungry, and tired, and the Danes will see us coming when we emerge from the hills, and that'll give them time to put on armor and gather weapons. And at best it will be equal numbers. At worst?" I just shrugged. After three or four days the rest of Svein's forces might have returned and our seventy or eighty men would be facing a horde.

"So how do you do it?" Alfred asked.

"We destroy their boats," I said.

"Go on."

"Without boats," I said, "they can't come up the rivers. Without boats, they're stranded."

Alfred nodded. Beocca was scratching away again. "So how do you destroy the boats?" the king asked.

I did not know. We could take seventy men to fight their seventy, but at the end of the fight, even if we won, we would be lucky to have twenty men still standing. Those twenty could burn the boats, of course, but I doubted we would survive that long. There were scores of Danish women at Cynuit and, if it came to a fight, they would join in and the odds were that we would be defeated. "Fire," Egwine said enthusiastically. "Carry fire in punts and throw the fire from the river."

"They are ship guards," I said tiredly, "and they'll be throwing spears and axes, sending arrows, and you might burn one boat, but that's all."

"Go at night," Egwine said.

"It's almost a full moon," I said, "and they'll see us coming. And if the moon is clouded we won't see their fleet."

"So how do you do it?" Alfred demanded again.

"God will send fire from heaven," Bishop Alewold said, and no one responded.

Alfred stood. We all got to our feet. Then he pointed at me. "You will destroy Svein's fleet," he said, "and I would know how you plan to do it by this evening. If you cannot do it then you"—he pointed to Egwine—"will travel to Defnascir, find Ealdorman Odda, and tell him to bring his forces to the river mouth and do the job for us."

"Yes, lord," Egwine said.

"By tonight," Alfred said to me coldly, and then he walked out.

He left me angry. He had meant to leave me angry. I stalked up to the newly made fort with Leofric and stared across the marshes to where the clouds heaped above the Sæfern. "How are we to burn twenty-four ships?" I demanded.

"God will send fire from heaven," Leofric said, "of course."

"I'd rather he sent a thousand troops."

"Alfred won't summon Odda," Leofric said. "He just said that to annoy you."

"But he's right, isn't he?" I said grudgingly. "We have to get rid of Svein."

"How?"

I stared at the tangled barricade that Haswold had made from felled trees. The water, instead of flowing downstream, was coming upstream because the tide was on the flood and so the ripples ran eastward from the tangled branches. "I remember a story," I said, "from when I was a small child." I paused, trying to recall the tale, which, I assume, had been told to me by Beocca. "The Christian god divided a sea, isn't that right?"

"Moses did," Leofric said.

"And when the enemy followed," I said, "they were drowned."

"Clever," Leofric said.

"So that's how we'll do it," I said.

"How?"

But instead of telling him I summoned the marshmen and talked with them, and by that night I had my plan and, because it was taken from the scriptures, Alfred approved it readily. It took another day to get everything ready. We had to gather sufficient punts to carry forty men and I also needed Eofer, the simpleminded archer. He was unhappy, not understanding what I wanted, and he gibbered at us and looked terrified, but then a small girl, perhaps ten or eleven years old, took his hand and explained that he had to go hunting with us. "He trusts you?" I asked the child.

"He's my uncle," she said. Eofer was holding her hand and he was calm again.

"Does Eofer do what you tell him?"

She nodded, her small face serious, and I told her she must come with us to keep her uncle happy.

We left before the dawn. We were twenty marshmen, skilled with boats, twenty warriors, a simpleminded archer, a child, and Iseult. Alfred, of course, did not want me to take Iseult, but I ignored him and he did not argue. Instead he watched us leave, then went to Æthelingæg's church that now boasted a newly made cross of alder wood nailed to its gable.

And low in the sky above the cross was the full moon. She was

low and ghostly pale, and as the sun rose she faded even more, but as
the ten punts drifted down the river I stared at her and said a silent
prayer to Hoder because the moon is his woman and it was she who
must give us victory. Because, for the first time since Guthrum had
struck in a winter's dawn, the Saxons were fighting back.

EIGHT

Before the Pedredan reaches the sea it makes a great curve through the swamp, a curve that is almost three-quarters of a circle and on the inside of the bank where the curve begins there was another tiny settlement: just a half dozen hovels built on stilts sunk into a slight rise in the ground. The settlement was called Palfleot, which means "the place with the stakes," for the folk who had once lived there had staked eel and fish traps in the nearby streams, but the Danes had driven those folk away and burned their houses, so that Palfleot was now a place of charred pilings and blackened mud. We landed there, shivering in the dawn. The tide was falling, exposing the great banks of sand and mud across which Iseult and I had struggled, while the wind was coming from the west, cold and fresh, hinting of rain, though for now there was a slanting sunlight throwing long shadows of marram grass and reeds across the marshes. Two swans flew south and I knew they were a message from the gods, but what their message was I could not tell.

The punts pushed away, abandoning us. They were now going north and east, following intricate waterways known only to the marshmen. We stayed for a while in Palfleot, doing nothing in particular, but doing it energetically so that the Danes, a long way off across the great bend in the river, would be sure to see us. We pulled down the blackened timbers and Iseult, who had acute eyesight, watched the place where the Danish ships' masts showed as scratches against the western clouds. "There's a man up a mast," she said after a while, and I stared, saw the man clinging to the mast top, and knew we had been spotted. The tide was falling, exposing more mud and sand, and now that I was sure we had been seen we walked across the drying expanse that was cradled by the river's extravagant bend.

As we drew closer I could see more Danes in their ships' rigging. They were watching us, but would not yet be worried for they out-numbered my few forces and the river lay between us and them, but whoever commanded in the Danish camp would also be ordering his men to arm themselves. He would want to be ready for whatever hap-pened, but I also hoped he would be clever. I was laying a trap for him, and for the trap to work he had to do what I wanted him to do, but at first, if he was clever, he would do nothing. He knew we were impotent, separated from him by the Pedredan, and so he was content to watch as we closed on the river's bank opposite his grounded ships and then slipped and slid down the steep muddy bluff that the ebbing tide had exposed. The river swirled in front of us, gray and cold.

There were close to a hundred Danes watching now. They were on their grounded boats, shouting insults. Some were laughing for it seemed clear to them that we had walked a long way to achieve noth-ing, but that was because they did not know Eofer's skills. I called the big bowman's niece to my side. "What I want your uncle Eofer to do," I explained to the small girl, "is kill some of those men."

"Kill them?" She stared up at me with wide eyes.

"They're bad men," I said, "and they want to kill you."

She nodded solemnly, then took the big man by the hand and led him to the water's edge where he sank up to his calves in the mud. It was a long way across the river and I wondered, pessimistically, if it was too far for even his massive bow, but Eofer strung the great stave and then waded into the Pedredan until he found a shallow spot, which meant he could go even farther into the river, and there he took an arrow from his sheaf, put it on the string, and hauled it back. He

made a grunting noise as he released and I watched the arrow twitch
off the cord. Then the fledging caught the air and the arrow soared
across the stream and plunged into a group of Danes standing on the
steering platform of a ship. There was a cry of anger as the arrow cut
down. It did not hit any of the group, but Eofer's next arrow struck a
man in his shoulder, and the Danes hurried back from their vantage
point by the ship's sternpost. Eofer, who was compulsively nodding
his shaggy head and making small animal noises, turned his aim to
another ship. He had extraordinary strength. The distance was too
great for any accuracy, but the danger of the long white-fledged arrows
drove the Danes back and it was our turn to jeer them. One of the
Danes fetched a bow and tried to shoot back, but his arrow sliced into
the river twenty yards short and we taunted them, laughed at them,
and capered up and down as Eofer's arrows slammed into ships' timbers. Only the one man had been wounded, but we had driven them
backward and that was humiliating to them. I let Eofer loose twenty
arrows. Then I waded into the river and took hold of his bow. I stood
in front of him so the Danes could not see what I was doing.

"Tell him not to worry," I told the girl, and she soothed Eofer who
was frowning at me and trying to remove his bow from my grasp.

I drew a knife and that alarmed him even more. He growled at
me, then plucked the bow from my hand. "Tell him it's all right," I
told the girl, and she soothed her uncle who then let me half sever
the woven hemp bowstring. I stepped away from him and pointed at
a group of Danes. "Kill them," I said.

Eofer did not want to draw the bow. Instead he fumbled under his
greasy woolen cap and produced a second bowstring, but I shook my
head and the small girl persuaded him he must use the half-severed
cord and so he pulled it nervously back and, just before it reached the
full draw, the string snapped and the arrow spun crazily into the sky
to float away on the river.

The tide had turned and the water was rising. "We go!" I shouted
to my men.

It was now the Danes' turn to jeer us. They thought we were

retreating because our one bowstring had broken, and so they shouted insults as we clambered back up the muddy bluff, and then I saw two men running along the far beach and I hoped they were carrying the orders I wanted.

They were. The Danes, released from the threat of Eofer's terrible bow, were going to launch two of their smaller ships. We had stung them, laughed at them, and now they would kill us.

All warriors have pride. Pride and rage and ambition are the goads to a reputation, and the Danes did not want us to think we had stung them without being punished for our temerity. They wanted to teach us a lesson. But they also wanted more. Before we left Æthelingæg I had insisted that my men be given every available coat of mail. Egwine, who had stayed behind with the king, had been reluctant to give up his precious armor, but Alfred had ordered it and so sixteen of my men were dressed in chain mail. They looked superb, like an elite group of warriors, and the Danes would win renown if they defeated such a group and captured the precious armor. Leather offers some protection, but chain mail over leather is far better and far more expensive, and by taking sixteen coats of mail to the river's edge, I had given the Danes an irresistible lure.

And they snapped at it.

We were going slowly, deliberately seeming to struggle in the soft ground as we headed back toward Palfleot. The Danes were also struggling, shoving their two ships down the riverbank's thick mud, but at last the boats were launched and then, on the hurrying flood tide, the Danes did what I had hoped they would do.

They did not cross the river. If they had crossed, then they would merely have found themselves on the Pedredan's eastern bank and we would have been half a mile ahead and out of reach, so instead the commander did what he thought was the clever thing to do. He tried to cut us off. They had seen us land at Palfleot and they reckoned our boats must still be there, and so they rowed their ships upriver to find those boats and destroy them.

Except our punts were not at Palfleot. They had been taken north and east, so that they were waiting for us in a reed-fringed dike, but now was not the time to use them. Instead, as the Danes went ashore at Palfleot, we made a huddle on the sand, watching them, and they thought we were trapped, and now they were on the same side of the river as us and the two ships' crews outnumbered us by over two to one, and they had all the confidence in the world as they advanced from the burned pilings of Palfleot to kill us in the swamp.

They were doing exactly what I wanted them to do.

And we now retreated. We went back raggedly, sometimes running to open a distance between us and the confident Danes. I counted seventy-six of them and we were only thirty strong because some of my men were with the hidden punts, and the Danes knew we were dead men and they hurried across the sand and creeks, and we had to go faster, ever faster, to keep them away from us. It began to rain, the drops carried on the freshening west wind, and I kept looking into the rain until at last I saw a silver bar of light glint and spill across the swamp's edge and knew the incoming tide was beginning its long fast race across the barren flats.

And still we went back, and still the Danes pursued us, but they were tiring now. A few shouted at us, daring us to stand and fight, but others had no breath to shout, just a savage intent to catch and kill us, but we were slanting eastward now toward a line of buckthorn and reeds, and there, in a flooding creek, were our punts.

We dropped into the boats, exhausted, and the marshmen poled us back down the creek that was a tributary of the river Bru, which barred the northern part of the swamp, and the flat-bottomed craft took us fast south, against the current, hurrying us past the Danes who could only watch from a quarter mile away and do nothing to stop us, and the farther we went from them, the more isolated they looked in that wide, barren place where the rain fell and the tide seethed as it flowed into the creek beds. The wind-driven water was running deep into the swamp now, a tide made bigger by the full moon, and suddenly the Danes saw their danger and turned back toward Palfleot.

But Palfleot was a long way off, and we had already left the stream and were carrying the punts to a smaller creek, one that ran down to the Pedredan, and that stream took us to where the blackened pilings leaned against the weeping sky, and where the Danes had tied their two ships. The two craft were guarded by only four men, and we came from the punts with a savage shout and drawn swords and the four men ran. The other Danes were still out in the swamp, only now it was not a swamp but a tidal flat, and they were wading through water.

And I had two ships. We hauled the punts aboard, and then the marshmen, divided between the ships, took the oars, and I steered one and Leofric took the other, and we rowed against that big tide toward Cynuit where the Danish ships were now unguarded except for a few men and a crowd of women and children who watched the two ships come and did not know they were crewed by their enemy. They must have wondered why so few oars bit the water, but how could they imagine that forty Saxons would defeat nearly eighty Danes? And so none opposed us as we ran the ships into the bank, and there I led my warriors ashore. "You can fight us," I shouted at the few ship guards left, "or you can live."

I was in chain mail, with my new helmet. I was a warlord. I banged Serpent-Breath against the big shield and stalked toward them. "Fight if you want!" I shouted. "Come and fight us!"

They did not. They were too few and so they retreated south and could only watch as we burned their ships. It took most of the day to ensure that the ships burned down to their keels, but burn they did, and their fires were a signal to the western part of Wessex that Svein had been defeated. He was not at Cynuit that day, but somewhere to the south, and as the ships burned I watched the wooded hills in fear that he would come with hundreds of men, but he was still far off and the Danes at Cynuit could do nothing to stop us. We burned twenty-three ships, including the *White Horse*, and the twenty-fourth, which was one of the two we had captured, carried us away as evening fell. We took good plunder from the Danish camp: food, rigging ropes, hides, weapons, and shields.

There were a score of Danes stranded on the low island of Palfleot. The rest had died in the rising water. The survivors watched us pass but did nothing to provoke us, and I did nothing to hurt them. We rowed on toward Æthelingæg and behind us, under a darkening sky, the water sheeted the swamp where white gulls cried above the drowned men and where, in the dusk, two swans flighted northward, their wings like drumbeats in the sky.

The smoke of the burned boats drifted to the clouds for three days, and on the second day Egwine took the captured ship downstream with forty men and they landed on Palfleot and killed all the surviving Danes, except for six who were taken prisoner, and five of those six were stripped of their armor and lashed to stakes in the river at low tide so that they drowned slowly on the flood. Egwine lost three men in that fight, but brought back mail, shields, helmets, weapons, arm rings, and one prisoner who knew nothing except that Svein had ridden toward Exanceaster. That prisoner died on the third day, the day that Alfred had prayers said in thanks to God for our victory. For now we were safe. Svein could not attack us for he had lost his ships, Guthrum had no way of penetrating the swamp, and Alfred was pleased with me.

"The king is pleased with you," Beocca told me. Two weeks before, I thought, the king would have told me that himself. He would have sat with me by the water's edge and talked, but now a court had formed and the king was hedged with priests.

"He should be pleased," I said. I had been practicing weapon-craft when Beocca sought me out. We practiced every day, using stakes instead of swords, and some men grumbled that they did not need to play at fighting, and those I opposed myself and, when they had been beaten down to the mud, I told them they needed to play more and complain less.

"He's pleased with you," Beocca said, leading me down the path beside the river, "but he thinks you are squeamish."

"Me! Squeamish?"

"For not going to Palfleot and finishing the job."

"The job was finished," I said. "Svein can't attack us without ships."

"But not all the Danes drowned," Beocca said.

"Enough died," I said. "Do you know what they endured? The terror of trying to outrun the tide?" I thought of my own anguish in the swamp, the inexorable tide, the cold water spreading, and the fear gripping the heart. "They had no ships! Why kill stranded men?"

"Because they are pagans," Beocca said, "because they are loathed by God and by men, and because they are Danes."

"And only a few weeks ago," I said, "you believed they would become Christians and all our swords would be beaten into ard points to plow fields."

Beocca shrugged that off. "So what will Svein do now?" he wanted to know.

"March around the swamp," I said, "and join Guthrum."

"And Guthrum is in Cippanhamm." We were fairly certain of that. New men were coming to the swamp and they all brought news. Much of it was rumor, but many had heard that Guthrum had strengthened Cippanhamm's walls and was wintering there. Large raiding parties still ravaged parts of Wessex, but they avoided the bigger towns in the south of the country where West Saxon garrisons had formed. There was one such garrison at Dornwaraceaster and another at Wintanceaster, and Beocca believed Alfred should go to one of those towns, but Alfred refused, reckoning that Guthrum would immediately besiege him. He would be trapped in a town, but the swamp was too big to be besieged and Guthrum could not hope to penetrate the marshes. "You have an uncle in Mercia, don't you?" Beocca asked, changing the subject abruptly.

"Æthelred. He's my mother's brother, and an ealdorman."

He heard the flat tone of my voice. "You're not fond of him?"

"I hardly know him." I had spent some weeks in his house, just long enough to quarrel with his son who was also called Æthelred.

"Is he a friend of the Danes?"

I shook my head. "They suffer him to live and he suffers them."

"The king has sent messengers to Mercia," Beocca said.

I grimaced. "If he wants them to rise against the Danes, they won't. They'll get killed."

"He'd rather they brought men south in the springtime," Beocca said and I wondered how a few Mercian warriors were supposed to get past the Danes to join us, but said nothing. "We look to the spring-time for our salvation," Beocca went on, "but in the meantime the king would like someone to go to Cippanhamm."

"A priest," I asked sourly, "to talk to Guthrum?"

"A soldier," Beocca said, "to gauge their numbers."

"So send me," I offered.

Beocca nodded, then limped along the riverbank where the willow fish traps had been exposed by the falling tide. "It's so different from Northumbria," he said wistfully.

I smiled at that. "You miss Bebbanburg?"

"I would like to end my days at Lindisfarena," he said. "I would like to say my dying prayer on that island." He turned and gazed at the eastern hills. "The king would go to Cippanhamm himself," he said, almost as an afterthought.

I thought I had misheard, then realized I had not. "That's madness," I protested.

"It's kingship," he said.

"Kingship?"

"The witan chooses the king," Beocca said sternly, "and the king must have the trust of the people. If Alfred goes to Cippanhamm and walks among his enemies, then folk will know he deserves to be king."

"And if he's captured," I said, "then folk will know he's a dead king."

"So you must protect him," he said. I said nothing. It was indeed madness, but Alfred was determined to show he deserved to be king. He had, after all, usurped the throne from his nephew, and in those early years of his reign he was ever mindful of that. "A small group will travel," Beocca said, "you, some other warriors, a priest, and the king."

"Why the priest?"

"To pray, of course."

I sneered at that. "You?"

Beocca patted his lamed leg. "Not me. A young priest."

"Better to send Iseult," I said.

"No."

"Why not? She's keeping the king healthy." Alfred was in sudden good health, better than he had been in years, and it was all because of the medicines that Iseult made. The celandine and burdock she had gathered on the mainland had taken away the agony in his arse, while other herbs calmed the pains in his belly. He walked confidently, had bright eyes, and looked strong.

"Iseult stays here," Beocca said.

"If you want the king to live," I said, "send her with us."

"She stays here," Beocca said, "because we want the king to live." It took me a few heartbeats to understand what he had said, and when I did realize his meaning I turned on him with such fury that he stumbled backward. I said nothing, for I did not trust myself to speak, or perhaps I feared that speech would turn to violence. Beocca tried to look severe, but only looked fearful. "These are difficult times," he said plaintively, "and the king can only put his trust in men who serve God. In men who are bound to him by their love of Christ."

I kicked at an eel trap, sending it spinning over the bank into the river. "For a time," I said, "I almost liked Alfred. Now he's got his priests back and you're dripping poison into him."

"He—" Beocca began.

I turned on him, silencing him. "Who rescued the bastard? Who burned Svein's ships? Who, in the name of your luckless god, killed Ubba? And you still don't trust me?"

Beocca was trying to calm me now, making flapping gestures. "I fear you are a pagan," he said, "and your woman is assuredly a pagan."

"My woman healed Edward," I snarled. "Does that mean nothing?"

"It could mean," he said, "that she did the devil's work."

I was astonished into silence by that.

"The devil does his work in the land," Beocca said earnestly, "and it would serve the devil well if Wessex was to vanish. The devil wants the king dead. He wants his own pagan spawn all across England! There is a greater war, Uhtred. Not the fight between Saxon and Dane, but between God and the devil, between good and evil! We are part of it!"

"I've killed more Danes than you can dream of," I told him.

"But suppose," he said, pleading with me now, "that your woman has been sent by the devil? That the evil one allowed her to heal Edward so that the king would trust her? And then, when the king, in all innocence, goes to spy on the enemy, she betrays him!"

"You think she would betray him?" I asked sourly. "Or do you mean I might betray him?"

"Your love of the Danes is well known," Beocca said stiffly, "and you spared the men on Palfleot."

"So you think I can't be trusted?"

"I trust you," he said, without conviction. "But other men?" He waved his palsied hand in an impotent gesture. "But if Iseult is here—" He shrugged, not ending the thought.

"So she's to be a hostage," I said.

"A surety, rather."

"I gave the king my oath," I pointed out.

"And you have sworn oaths before, and you are known as a liar, and you have a wife and child, yet live with a pagan whore, and you love the Danes as you love yourself, and do you really think we can trust you?" This all came out in a bitter rush. "I have known you, Uhtred," he said, "since you crawled on Bebbanburg's rush floors. I baptized you, taught you, chastised you, watched you grow, and I know you better than any man alive and I do not trust you." Beocca stared at me belligerently. "If the king does not return, Uhtred, then your whore will be given to the dogs." He had delivered his message now, and he seemed to regret the force of it for he shook his head.

"The king should not go. You're right. It's a madness. It is stupidity! It is"—he paused, searching for a word, and came upon one of the worst condemnations in his vocabulary—"it is irresponsible! But he insists, and if he goes then you must also go, for you're the only man here who can pass as a Dane. But bring him back, Uhtred, bring him back, for he is dear to God and to all Saxons."

Not to me, I thought, he was not dear to me. That night, brooding on Beocca's words, I was tempted to flee the swamp, to go away with Iseult, find a lord, give Serpent-Breath a new master, but Ragnar had been a hostage and so I had no friend among my enemies, and if I fled I would break my oath to Alfred and men would say Uhtred of Bebban- burg could never be trusted again and so I stayed. I tried to persuade Alfred not to go to Cippanhamm. It was, as Beocca had said, irrespon- sible, but Alfred insisted. "If I stay here," he said, "men will say I hid from the Danes. Others face them, but I hide? No. Men must see me, must know that I live, and know that I fight." For once Ælswith and I were in agreement, and we both tried to keep him in Æthelingæg, but Alfred would not be dissuaded. He was in a strange mood, suffused with happiness, utterly confident that God was on his side, and, because his sickness had abated, he was full of energy and confidence.

He took six companions. The priest was a young man called Adel- bert who carried a small harp wrapped in leather. It seemed ridiculous to take a harp to the enemy, but Adelbert was famed for his music and Alfred blithely said that we should sing God's praises while we were among the Danes. The other four were all experienced warriors who had been part of his royal guard. They were called Osferth, Wul- frith, Beorth, and Egwine, the last of whom swore to Ælswith that he would bring the king home, which made Ælswith throw a bitter glance at me. Whatever favor I had gained by Iseult's cure of Edward had evaporated under the influence of the priests.

We dressed for war in mail and helmets, while Alfred insisted on wearing a fine blue cloak, trimmed with fur, which made him con- spicuous, but he wanted folk to see a king. The best horses were selected, one for each of us and three spare mounts, and we swam them across the river, then followed log roads until we came at last to firm ground close to the island where Iseult said Arthur was buried. I had left Iseult with Eanflæd who shared quarters with Leofric.

It was February now. There had been a spell of fine weather after the burning of Svein's fleet and I had thought we should travel then, but Alfred insisted on waiting until the eighth day of February, because that was the Feast of Saint Cuthman, a Saxon saint from East Anglia, and Alfred reckoned that must be a propitious day. Perhaps he was right, for the day turned out wet and bitterly cold, and we were to discover that the Danes were reluctant to leave their quarters in the worst weather. We went at dawn and by midmorning we were in the hills overlooking the swamp, which was half hidden by a mist thickened by the smoke from the cooking fires of the small villages. "Are you familiar with Saint Cuthman?" Alfred asked me cheerfully.

"No, lord."

"He was a hermit," Alfred said. We were riding north, keeping on the high ground with the swamp to our left. "His mother was crippled and so he made her a wheelbarrow."

"A wheelbarrow? What could a cripple do with a wheelbarrow?"

"No, no, no! He pushed her about in it! So she could be with him as he preached. He pushed her everywhere."

"She must have liked that."

"There's no written life of him that I know of," Alfred said, "but we must surely compose one. He could be a saint for mothers."

"Or for wheelbarrows, lord."

We saw our first evidence of the Danes just after midday. We were still on the high ground, but in a valley that sloped to the marshes we saw a substantial house with limewashed walls and thick thatch. Smoke came from the roof, while in a fenced apple orchard were a score of horses. No Dane would ever leave such a place unplundered, which suggested the horses belonged to them and that the farm was garrisoned. "They're there to watch the swamp," Alfred suggested.

"Probably." I was cold. I had a thick woolen cloak, but I was still cold.

"We shall send men here," Alfred said, "and teach them not to steal apples."

We stayed that night in a small village. The Danes had been there and the folk were frightened. At first, when we rode up the rutted track between the houses, they hid, thinking we were Danes, but when they heard our voices they crept out and stared at us as if we had just ridden down from the moon. Their priest was dead, killed by the pagans, so Alfred insisted that Adelbert hold a service in the burned-out remnants of the church. Alfred himself acted as precentor, accompanying his chanting with the priest's small harp. "I learned to play as a child," he told me. "My stepmother insisted, but I'm not very good."

"You're not," I agreed, which he did not like.

"There is never enough time to practice," he complained.

We lodged in a peasant's house. Alfred, reckoning that the Danes would have taken the harvest from wherever we visited, had laden the spare horses with smoked fish, smoked eels, and oatcakes, so we provided most of the food and, after we had eaten, the peasant couple knelt to me and the woman tentatively touched the skirt of my mail coat. "My children," she whispered, "there are two of them. My daughter is about seven years old and my boy is a little older. They are good children."

"What of them?" Alfred intervened.

"The pagans took them, lord," the woman said. She was crying. "You can find them, lord?" she said, tugging my mail. "You can find them and bring them back? My little ones? Please?"

I promised to try, but it was an empty promise for the children would have long gone to the slave market and, by now, would either be working on some Danish estate or, if they were pretty, sent overseas where heathen men pay good silver for Christian children.

We learned that the Danes had come to the village shortly after Twelfth Night. They had killed, captured, stolen, and ridden on southward. A few days later they had returned, going back northward, driving a band of captives and a herd of captured horses laden with plunder. Since then the villagers had seen no Danes except for the few on the swamp's edge. Those Danes, they said, caused no trouble, perhaps because they were so few and dared not stir up the enmity of the country about them. We heard the same tale in other villages. The Danes had come, they had pillaged, then had gone back north.

But on our third day we at last saw a force of the enemy riding on the Roman road that cuts straight westward across the hills from Baðum. There were close to sixty of them, and they rode hard in front of dark clouds and the gathering night. "Going back to Cippanhamm," Alfred said. It was a foraging party, and their packhorses carried nets stuffed with hay to feed their warhorses, and I remembered my childhood winter in Readingum, when the Danes first invaded Wessex, and how hard it had been to keep horses and men alive in the cold. We had cut feeble winter grass and pulled down thatch to feed our horses that still became skeletal and weak. I have often listened to men declare that all that is needed to win a war is to assemble men and march against the enemy, but it is never that easy. Men and horses must be fed, and hunger can defeat an army much faster than spears. We watched the Danes go north, then turned aside to a half-ruined barn that offered us shelter for the night.

It began to snow that night, a relentless soft snow, silent and thick, so that by dawn the world was white under a pale blue sky. I suggested we waited till the snow had thawed before we rode further, but Egwine, who came from this part of the country, said we were only two or three hours south of Cippanhamm and Alfred was impatient. "We go," he insisted. "We go there, look at the town, and ride away."

So we rode north, our hooves crunching the newly fallen snow, riding through a world made new and clean. Snow clung to every twig and branch while ice skimmed the ditches and ponds. I saw a fox's trail crossing a field and thought that the spring would bring a plague of the beasts, for there would have been no one to hunt them, and the lambs would die bloodily and the ewes would bleat pitifully.

We came in sight of Cippanhamm before midday, though the

great pall of smoke, made by hundreds of cooking fires, had shown in the sky all morning. We stopped south of the town, just where the road emerged from a stand of oaks, and the Danes must have noticed us, but none came from the gates to see who we were. It was too cold for men to stir themselves. I could see guards on the walls, though none stayed there long, retreating to whatever warmth they could find between their short forays along the wooden ramparts. Those ramparts were bright with round shields painted blue and white and bloodred and, because Guthrum's men were there, black. "We should count the shields," Alfred said.

"It won't help," I said. "They carry two or three shields each and hang them on the walls to make it look as if they have more men."

Alfred was shivering and I insisted we find some shelter. We turned back into the trees, following a path that led to the river, and a mile or so upstream we came across a mill. The millstone had been taken away, but the building itself was whole and it was well made, with stone walls and a turf roof held up by stout rafters. There was a hearth in a room where the miller's family had lived, but I would not let Egwine light a fire in case the trickle of smoke brought curious Danes from the town. "Wait till dark," I said.

"We'll freeze by then," he grumbled.

"Then you shouldn't have come," I snapped.

"We have to get closer to the town," Alfred said.

"You don't," I said. "I do." I had seen horses paddocked to the west of the walls and I reckoned I could take our best horse and ride about the town's western edge and count every horse I saw. That would give a rough estimate of the Danish numbers, for almost every man would have a horse. Alfred wanted to come, but I shook my head. It was pointless for more than one man to go, and sensible that the one man who did go should speak Danish, so I told him I would see him back in the mill before nightfall and then I rode north. Cippanhamm was built on a hill that was almost encircled by the river, so I could not ride clear around the town, but I went as close to the walls as I dared and stared across the river and saw no horses on its farther bank, which suggested that the Danes were keeping their beasts on the western side of the town. I went there, keeping in the snowy woods, and though the Danes must have seen me, they could not be bothered to ride into the snow to chase one man, and so I was able to find the paddocks where their horses shivered. I spent the day counting. Most of the horses were in fields beside the royal compound and there were hundreds of them. By late afternoon I had estimated that there were twelve hundred, and those were only the ones I could see, and the best horses would be in the town, but my reckoning was good enough. It would give Alfred an idea of how large Guthrum's force was. Say two thousand men? And elsewhere in Wessex, in the towns the Danes had occupied, there must be another thousand. That was a strong force, but not quite strong enough to capture all the kingdom. That would have to wait until spring when reinforcements would come from Denmark or from the three conquered kingdoms of England. I rode back to the watermill as dusk fell. There was a frost and the air was still. Three rooks flew across the river as I dismounted. I reckoned one of Alfred's men could rub my horse down; all I wanted was to find some warmth and it was plain Alfred had risked lighting a fire, for smoke was pouring out of the hole in the turf roof.

They were all crouched about the small fire and I joined them, stretching my hands to the flames. "Two thousand men," I said, "more or less."

No one answered.

"Didn't you hear me?" I asked, and looked around the faces.

There were five faces. Only five.

"Where's the king?" I asked.

"He went," Adelbert said helplessly.

"He did what?"

"He went to the town," the priest said. He was wearing Alfred's rich blue cloak and I assumed Alfred had taken Adelbert's plain garment.

I stared at him. "You let him go?"

"He insisted," Egwine said.

"How could we stop him?" Adelbert pleaded. "He's the king!"

"You hit the bastard, of course," I snarled. "You hold him down till the madness passes. When did he go?"

"Just after you left," the priest said miserably, "and he took my harp," he added.

"And when did he say he'd be back?"

"By nightfall."

"It is nightfall," I said. I stood and stamped out the fire. "You want the Danes to come and investigate the smoke?" I doubted the Danes would come, but I wanted the damned fools to suffer. "You." I pointed to one of the four soldiers. "Rub my horse down. Feed it."

I went back to the door. The first stars were bright and the snow glinted under a sickle moon.

"Where are you going?" Adelbert had followed me.

"To find the king, of course."

If he lived. And if he did not, then Iseult was dead.

I had to beat on Cippanhamm's western gate, provoking a disgruntled voice from the far side demanding to know who I was.

"Why aren't you up on the ramparts?" I asked in return.

The bar was lifted and the gate opened a few inches. A face peered out, then vanished as I pushed the gate hard inward, banging it against the suspicious guard. "My horse went lame," I said, "and I've walked here."

He recovered his balance and pushed the gate shut. "Who are you?" he asked again.

"Messenger from Svein."

"Svein!" He lifted the bar and dropped it into place. "Has he caught Alfred yet?"

"I'll tell Guthrum that news before I tell you."

"Just asking," he said.

"Where is Guthrum?" I asked. I had no intention of going anywhere near the Danish chieftain for, after my insults to his dead mother, the best I could hope for was a swift death, and the likelihood was a very slow one.

"He's in Alfred's hall," the man said, and pointed south. "That side of town, so you've still farther to walk." It never occurred to him that any messenger from Svein would never ride alone through Wessex, that such a man would come with an escort of fifty or sixty men, but he was too cold to think, and besides, with my long hair and my thick arm rings I looked like a Dane. He retreated into the house beside the gate where his comrades were clustered around a hearth and I walked on into a town made strange. Houses were missing, burned in the first fury of the Danish assault, and the large church by the marketplace on the hilltop was nothing but blackened beams touched white by the snow. The streets were frozen mud, and only I moved there, for the cold was keeping the Danes in the remaining houses. I could hear singing and laughter. Light leaked past shutters or glowed through smoke holes in low roofs. I was cold and I was angry. There were men here who could recognize me, and men who might recognize Alfred, and his stupidity had put us both in danger. Would he have been mad enough to go back to his own hall? He must have guessed that was where Guthrum would be living and he would surely not risk being recognized by the Danish leader, which suggested he would be in the town rather than the royal compound.

I was walking toward Eanflæd's old tavern when I heard the roars. They were coming from the east side of town and I followed the sound, which led me to the nunnery by the river wall. I had never been inside the convent, but the gate was open and the courtyard inside was lit by two vast fires that offered some warmth to the men nearest the flames. And there were at least a hundred men in the courtyard, bellowing encouragement and insults at two other men who were fighting in the mud and melted snow between the fires. They were fighting with swords and shields, and every clash of blade against blade or of blade against wood brought raucous shouts. I glanced briefly at the fighters, then searched for faces in the crowd. I was looking for Haesten, or anyone else who might recognize me, but I saw no one, though it was hard to distinguish faces in the flickering shadows. There was no sign of any nuns and I assumed they had fled, were dead, or had been taken away for the conquerors' amusement.

I slunk along the courtyard wall. I was wearing my helmet and its faceplate was an adequate disguise, but some men threw me curious glances, for it was unusual to see a helmeted man off a battlefield. In the end, seeing no one I recognized, I took the helmet off and hung it from my belt. The nunnery church had been turned into a feasting hall, but there was only a handful of drunks inside, oblivious to the noise outside. I stole a half loaf of bread from one of the drunks and took it back outside and watched the fighting.

Steapa Snotor was one of the two men. He no longer wore his mail armor, but was in a leather coat, and he fought with a small shield and a long sword. Around his waist was a chain that led to the courtyard's northern side where two men held it and, whenever Steapa's opponent seemed to be in danger, they yanked on the chain to pull the huge Saxon off balance. He was being made to fight as Haesten had been fighting when I first discovered him, and doubtless Steapa's captors were making good money from fools who wanted to try their prowess against a captured warrior. Steapa's current opponent was a thin, grinning Dane who tried to dance around the huge man and slide his sword beneath the small shield, doing what I had done when I had fought Steapa, but Steapa was doggedly defending himself, parrying each blow and, when the chain allowed him, counterattacking fast. Whenever the Danes jerked him backward the crowd jeered and once, when the men yanked the chain too hard and Steapa turned on them, only to be faced by three long spears, the crowd gave him a great cheer. He whipped back to parry the next attack, then stepped backward, almost to the spear points, and the thin man followed fast, thinking he had Steapa at a disadvantage, but Steapa suddenly checked, slammed the shield down onto his opponent's blade, and brought his left hand around, sword hilt foremost, to hit the man on the head. The Dane went down, Steapa reversed the sword to stab, and the chain dragged him off his feet and the spears threatened him with death if he finished the job. The crowd liked it. He had won.

Money changed hands. Steapa sat by the fire, his grim face showing nothing, and one of the men holding the chain shouted for another opponent. "Ten pieces of silver if you wound him! Fifty if you kill him!"

Steapa, who probably did not understand a word, just stared at the crowd, daring another man to take him on, and sure enough a half-drunken brute came grinning from the crowd. Bets were made as Steapa was prodded to his feet. It was like a bull baiting, except Steapa was being given only one opponent at a time. They would doubtless have set three or four men on him, except that the Danes who had taken him prisoner did not want him dead so long as there were still fools willing to pay to fight him.

I was sidling around the courtyard's edge, still looking at faces. "Six pennies?" a voice said behind me and I turned to see a man grinning beside a door. It was one of a dozen similar doors, evenly spaced along the lime-washed wall.

"Six pennies?" I asked, puzzled.

"Cheap," he said, and he pushed back a small shutter on the door and invited me to look inside.

I did. A tallow candle lit the tiny room, which must have been where a nun had slept, and inside was a low bed and on the bed was a naked woman who was half covered by a man who had dropped his breeches. "He won't be long," the man said.

I shook my head and moved away from the shutter.

"She was a nun here," the man said. "Nice and young. Pretty, too. Screams like a pig usually."

"No," I said.

"Four pennies? She won't put up a fight. Not now she won't."

I walked on, convinced I was wasting my time. Had Alfred been and gone? More likely, I thought grimly, the fool had gone back to his hall and I wondered if I dared go there, but the thought of Guthrum's revenge deterred me. The new fight had started. The Dane was crouching low, trying to cut Steapa's feet from under him, but Steapa was swatting his blows easily enough and I sidled past the men holding his chains and saw another room off to my left, a large room, perhaps where the nuns had eaten, and a glint of gold in the light of its dying fire drew me inside.

The gold was not metal. It was the gilding on the frame of a small harp that had been stamped on so hard that it broke. I looked around the shadows and saw a man lying in a heap at the far end and went to him. It was Alfred. He was barely conscious, but he was alive and, so far as I could see, unwounded, but he was plainly stunned and I dragged him to the wall and sat him up. He had no cloak and his boots were gone. I left him there, went back to the church, and found a drunk to befriend. I helped him to his feet, put my arm around his shoulders, and persuaded him I was taking him to his bed, then took him through the back door to the latrine yard of the nunnery where I punched him three times in the belly and twice in the face, then carried his hooded cloak and tall boots back to Alfred.

The king was conscious now. His face was bruised. He looked up at me without showing any surprise, then rubbed his chin. "They didn't like the way I played," he said.

"That's because the Danes like good music," I said. "Put these on." I threw the boots beside him, draped him with the cloak, and made him pull the hood over his face. "You want to die?" I asked him angrily.

"I want to know about my enemies," he said.

"And I found out for you," I said. "There are roughly two thousand of them."

"That's what I thought," he said, then grimaced. "What's on this cloak?"

"Danish vomit," I said.

He shuddered. "Three of them attacked me," he sounded surprised. "They kicked and punched me."

"I told you, the Danes like good music," I said, helping him to his feet. "You're lucky they didn't kill you."

"They thought I was Danish," he said, then spat blood that trickled from his swollen lower lip.

"Were they drunk?" I asked. "You don't even look like a Dane."

"I pretended I was a musician who couldn't speak." He mouthed silently at me, then grinned bloodily, proud of his deception. I did not grin back and he sighed. "They were very drunk, but I need to know their mood, Uhtred. Are they confident? Are they readying to attack?" He paused to wipe more blood from his lips. "I could only find that out by coming to see them for myself. Did you see Steapa?"

"Yes."

"I want to take him back with us."

"Lord," I said savagely, "you are a fool. He's in chains. He's got half a dozen guards."

"Daniel was in a lion's den, yet he escaped. Saint Paul was imprisoned, yet God freed him."

"Then let God look after Steapa," I said. "You're coming back with me. Now."

He bent to relieve a pain his belly. "They punched me in the stomach," he said as he straightened. In the morning, I thought, he would have a rare black eye to display. He flinched as a huge cheer sounded from the courtyard and I guessed Steapa had either died or downed his last opponent. "I want to see my hall," Alfred said stubbornly.

"Why?"

"I'm a man who would look at his own home. You can come or stay."

"Guthrum's there! You want to be recognized? You want to die?"

"Guthrum will be inside, and I just want to look at the outside."

He would not be dissuaded and so I led him through the courtyard to the street, wondering if I should simply pick him up and carry him away, but in his obdurate mood he would probably struggle and shout until men came to find out the cause of the noise. "I wonder what happened to the nuns," he said as we left the nunnery.

"One of them is being whored in there for pennies," I said.

"Oh, dear God." He made the sign of the cross and turned back and I knew he was thinking of rescuing the woman, so I dragged him onward. "This is madness!" I protested.

"It is a necessary madness," he said calmly, then stopped to lecture me. "What does Wessex believe? It thinks I am defeated, it thinks the Danes have won, it readies itself for the spring and the coming of more Danes. So they must learn something different. They must learn that the king lives, that he walked among his enemies, and that he made fools of them."

"That he got given a bloody nose and a black eye," I said.

"You won't tell them that," he said, "any more than you'll tell folk about that wretched woman who hit me with an eel. We must give men hope, Uhtred, and in the spring that hope will blossom into victory. Remember Boethius, Uhtred, remember Boethius! Never give up hope."

He believed it. He believed that God was protecting him, that he could walk among his enemies without fear or harm, and to an extent he was right, for the Danes were well supplied with ale, birch wine, and mead, and most were much too drunk to care about a bruised man carrying a broken harp.

No one stopped us going into the royal compound, but there were six black-cloaked guards at the hall door and I refused to let Alfred get close to them. "They'll take one look at your bloodied face," I said, "and finish what the others began."

"Then let me at least go to the church."

"You want to pray?" I asked sarcastically.

"Yes," he said simply.

I tried to stop him. "If you die here," I said, "then Iseult dies."

"That wasn't my doing," he said.

"You're the king, aren't you?"

"The bishop thought you would join the Danes," he said, "and others agreed."

"I have no friends left among the Danes," I said. "They were your hostages and they died."

"Then I shall pray for their pagan souls," he said, and pulled away from me and went to the church door where he instinctively pushed the hood off his head to show respect. I snatched it back over his hair, shadowing his bruises. He did not resist, but just pushed the door open and made the sign of the cross.

The church was being used to shelter more of Guthrum's men. There were straw mattresses, heaps of chain mail, stacks of weapons, and a score of men and women gathered around a newly made hearth in the nave. They were playing dice and none took any particular interest in our arrival until someone shouted that we should shut the door.

"We're leaving," I said to Alfred. "You can't pray here."

He did not answer. He was gazing reverently to where the altar had been, and where a half dozen horses were now tethered.

"We're leaving!" I insisted again.

And just then a voice hailed me. It was a voice full of astonishment and I saw one of the dice players stand and stare at me. A dog ran from the shadows and began to jump up and down, trying to lick me, and I saw the dog was Nihtgenga and that the man who had recognized me was Ragnar. Earl Ragnar, my friend.

Who I had thought was dead.

NINE

R agnar embraced me. There were tears in both our eyes and for a
moment neither of us could speak, though I retained enough
sense to look behind me to make sure Alfred was safe. He was squatting
beside the door, deep in the shadow of a bale of wool, with his cloak's
hood drawn over his face. "I thought you were dead!" I said to Ragnar.

"I hoped you would come," he said at the same moment, and for
a time we both talked and neither listened, and then Brida walked from
the back of the church and I watched her, seeing a woman instead of a
girl, and she laughed to see me and gave me a decorous kiss.

"Uhtred." She said my name as a caress. We had been lovers once,
though we had been little more than children then. She was Saxon,
but she had chosen the Danish side to be with Ragnar. The other
women in the hall were hung with silver, garnet, jet, amber, and gold,
but Brida wore no jewelry other than an ivory comb that held her
thick black hair in a pile. "Uhtred," she said again.

"Why aren't you dead?" I asked Ragnar. He had been a hostage,
and the hostages' lives had been forfeit the moment Guthrum
crossed the frontier.

"Wulfhere liked us," Ragnar said. He put an arm around my
shoulder and drew me to the central hearth where the fire blazed.
"This is Uhtred," he announced to the dice players, "a Saxon, which

makes him scum, of course, but he is also my friend and my brother. Ale." He pointed to jars. "Wine. Wulfhere let us live."

"And you let him live?"

"Of course we did! He's here. Feasting with Guthrum."

"Wulfhere? Is he a prisoner?"

"He's an ally!" Ragnar said, thrusting a pot into my hand and pulling me down beside the fire. "He's with us now." He grinned at me, and I laughed for the sheer joy of finding him alive. He was a big man, golden-haired, open-faced, and as full of mischief, life, and kindness as his father had been. "Wulfhere used to talk to Brida," Ragnar went on, "and through her to me. We liked each other. Hard to kill a man you like."

"You persuaded him to change sides?"

"Didn't need a great deal of persuasion," Ragnar said. "He could see we were going to win, and by changing sides he keeps his land, doesn't he? Are you going to drink that ale or just stare at it?"

I pretended to drink, letting some of the ale drip down my beard, and I remembered Wulfhere telling me that when the Danes came we must all make what shifts we could to survive. But Wulfhere? Alfred's cousin and the Ealdorman of Wiltunscir? He had changed sides? So how many other thegns had followed his example and now served the Danes?

"Who's that?" Brida asked. She was staring at Alfred. He was in shadow, but there was something oddly mysterious about the way he squatted alone and silent.

"A servant," I said.

"He can come by the fire."

"He cannot," I said harshly. "I'm punishing him."

"What did you do?" Brida called to him in English. His face came up and he stared at her, but the hood still shadowed him.

"Speak, you bastard," I said, "and I'll whip you till your bones show." I could just see his eyes in the hood's shadow. "He insulted me," I spoke in Danish again, "and I've sworn him to silence, and for every word he utters he receives ten blows of the whip."

That satisfied them. Ragnar forgot the strange hooded servant and told me how he had persuaded Wulfhere to send a messenger to Guthrum, promising to spare the hostages, and how Guthrum had warned Wulfhere when the attack would come to make sure that the ealdorman had time to remove the hostages from Alfred's revenge. That, I thought, was why Wulfhere had left so early on the morning of the attack. He had known the Danes were coming. "You call him an ally," I said. "Does that make him just a friend? Or a man who will fight for Guthrum?"

"He's an ally," Ragnar said, "and he's sworn to fight for us. At least he's sworn to fight for the Saxon king."

"The Saxon king?" I asked, confused. "Alfred?"

"Not Alfred, no. The true king. The boy who was the other one's son."

Ragnar meant Æthelwold, who had been heir to Alfred's brother, King Æthelred, and of course the Danes would want Æthelwold. Whenever they captured a Saxon kingdom they appointed a Saxon as king, and that gave their conquest a cloak of legality, though the Saxon never lasted long. Guthrum, who already called himself King of East Anglia, wanted to be King of Wessex, too, but by putting Æthelwold on the throne he might attract other West Saxons who could convince themselves they were fighting for the true heir. And once the fight was over and Danish rule established, Æthelwold would quietly be killed.

"But Wulfhere will fight for you?" I persisted.

"Of course he will! If he wants to keep his land," Ragnar said, then grimaced. "But what fighting? We just sit here like sheep and do nothing!"

"It's winter."

"Best time to fight. Nothing else to do." He wanted to know where I had been since Yule and I said I had been deep inside Defnascir. He assumed I had been making sure my family was safe, and he also assumed I had now come to Cippanhamm to join him. "You're not sworn to Alfred, are you?" he asked.

"Who knows where Alfred is?" I evaded the question.

"You were sworn to him," he said reproachfully.

"I was sworn to him," I said, truthfully enough, "but only for a year, and that year has long ended." That was no lie; I just did not tell Ragnar I had sworn myself to Alfred once again.

"So you can join me?" he asked eagerly. "You'll give me your oath?"

I took the question lightly, though in truth it worried me. "You want my oath," I asked, "just so I can sit here like a sheep doing nothing?"

"We make some raids," Ragnar said defensively, "and men are guarding the swamp. That's where Alfred is. In the swamps. But Svein will dig him out." So Guthrum and his men had yet to hear that Svein's fleet was ashes beside the sea.

"So why are you just sitting here?" I asked.

"Because Guthrum won't divide his army," Ragnar said. I half smiled at that because I remembered Ragnar's grandfather advising Guthrum never to divide an army again. Guthrum had done that at Æsc's Hill and that had been the first victory of the West Saxons over the Danes. He had done it again when he abandoned Werham to attack Exanceaster, and the part of his army that went by sea was virtually destroyed by the storm. "I've told him," Ragnar said, "that we should split the army into a dozen parts. Take a dozen more towns and garrison them. All those places in southern Wessex, we should capture them, but he won't listen."

"Guthrum holds the north and east," I said, as if I was defending him.

"And we should have the rest! But instead we're waiting till spring in hope more men will join us. Which they will. There's land here, good land. Better than the land up north." He seemed to have forgotten the matter of my oath. I knew he would want me to join him, but instead he talked of what happened in Northumbria, how our enemies, Kjartan and Sven, thrived in Dunholm, and how that father and son dared not leave the fortress for fear of Ragnar's revenge. They had taken his sister captive and, so far as Ragnar knew, they held her still,

and Ragnar, like me, was sworn to kill them. He had no news of Bebbanburg other than that my treacherous uncle still lived and held the
fortress. "When we've finished with Wessex," Ragnar promised me,
"we shall go north. You and I together. We'll carry swords to Dunholm."

"Swords to Dunholm," I said and raised my pot of ale.

I did not drink much, or if I did it seemed to have little effect. I was
thinking, sitting there, that with one sentence I could finish Alfred forever. I could betray him; I could have him dragged in front of Guthrum
and then watch as he died. Guthrum would even forgive me the
insults to his mother if I gave him Alfred, and thus I could finish Wessex, for without Alfred there was no man about whom the fyrd would
muster. I could stay with my friend Ragnar, I could earn more arm
rings, I could make a name that would be celebrated wherever Northmen sailed their long ships, and all it would take was one sentence.

And I was so tempted that night in Cippanhamm's royal church.
There is such joy in chaos. Stow all the world's evils behind a door
and tell men that they must never, ever, open the door, and it will be
opened because there is pure joy in destruction. At one moment,
when Ragnar was bellowing with laughter and slapping my shoulder
so hard that it hurt, I felt the words form on my tongue. That is
Alfred, I would have said, pointing at him, and all my world would
have changed and there would have been no more England. Yet, at
the last moment, when the first word was on my tongue, I choked it
back. Brida was watching me, her shrewd eyes calm, and I caught her
gaze and I thought of Iseult. In a year or two, I thought, Iseult would
look like Brida. They had the same tense beauty, the same dark color-
ing, and the same smoldering fire in the soul. If I spoke, I thought,
Iseult would be dead, and I could not bear that. And I thought of
Æthelflaed, Alfred's daughter, and knew she would be enslaved, and
also knew that wherever the remnants of the Saxons gathered about
their fires of exile my name would be cursed. I would be Uhtredærwe
forever, the man who destroyed a people.

"What were you about to say?" Brida asked.

"That we have never known such a hard winter in Wessex."

She gazed at me, not believing my answer. Then she smiled. "Tell me, Uhtred," she spoke in English, "if you thought Ragnar was dead, then why did you come here?"

"Because I don't know where else to be," I said.

"So you came here? To Guthrum? Whom you insulted?"

So they knew about that. I had not expected them to know and I felt a surge of fear. I said nothing.

"Guthrum wants you dead," Brida said, speaking in Danish now.

"He doesn't mean it," Ragnar said.

"He does mean it," Brida insisted.

"Well, I won't let him kill Uhtred," Ragnar said. "You're here now!" He slapped me on the back again and glared at his men as if daring any of them to betray my presence to Guthrum. None of them moved, but they were nearly all of them drunk and some already asleep.

"You're here now," Brida said, "yet not so long ago you were fighting for Alfred and insulting Guthrum."

"I was on my way to Defnascir," I said, as if that explained anything.

"Poor Uhtred," Brida said. Her right hand fondled the black-and-white fur at the back of Nihtgenga's neck. "And I thought you'd be a hero to the Saxons."

"A hero? Why?"

"The man who killed Ubba?"

"Alfred doesn't want heroes," I said, loudly enough for him to hear, "only saints."

"So tell us about Ubba!" Ragnar demanded, and so I had to describe Ubba's death and the Danes, who love a good story of a fight, wanted every detail. I told the tale well, making Ubba into a great hero who had almost destroyed the West Saxon army, and I said he had been fighting like a god, and told how he had broken our shield wall with his great ax. I described the burning ships, their smoke drifting over the battle slaughter like a cloud from the netherworld, and I said I had found myself facing Ubba in his victory charge. That was not true, of course, and the Danes knew it was not true. I had not just found myself opposing Ubba, but had sought him out. But when a story is told it must be seasoned with modesty and the listeners, understanding that custom, murmured approval. "I have never known such fear," I said, and I told how we had fought, Serpent-Breath against Ubba's ax, and how he had chopped my shield into firewood, and then I described, truthfully, how he had lost his footing in the spilled guts of a dead man. The Danes about the fire sighed with disappointment. "I cut the tendons of his arm," I said, chopping my left hand into the crook of my right elbow to show where I had cut him, "and then beat him down."

"He died well?" a man asked anxiously.

"As a hero," I said, and I told how I had put the ax back into his dying hand so that he would go to Valhalla. "He died very well," I finished.

"He was a warrior," Ragnar said. He was drunk now. Not badly drunk, but tired drunk. The fire was dying, thickening the shadows at the western end of the church where Alfred sat. More stories were told, the fire died, and the few candles guttered. Men were sleeping, and still I sat until Ragnar lay back and began to snore. I waited longer, letting the room go to sleep, and only then did I go back to Alfred. "We go now," I said. He did not argue.

No one appeared to notice as we went into the night, closing the door quietly behind us. "Who were you talking with?" Alfred asked me.

"Earl Ragnar."

He stopped, puzzled. "Wasn't he one of the hostages?"

"Wulfhere let them live," I said.

"He let them live?" he asked, astonished.

"And Wulfhere is now on Guthrum's side." I gave him the bad news. "He's here, in the hall. He's agreed to fight for Guthrum."

"Here?" Alfred could scarce believe what I said. Wulfhere was his cousin, he had married Alfred's niece, he was family. "He's here?"

"He's on Guthrum's side," I said harshly.

He just stared at me. "No." He mouthed the word, rather than said it. "And Æthelwold?" he asked.

"He's a prisoner," I said.

"A prisoner!" He asked the question sharply, and no wonder, for Æthelwold had no value to the Danes as a prisoner unless he had agreed to become their token king on the West Saxon throne.

"A prisoner," I said. It was not true, of course, but I liked Æthelwold and I owed him a favor. "He's a prisoner," I went on, "and there's nothing we can do about it, so let's get away from here." I pulled him toward the town, but too late, for the church door opened and Brida came out with Nihtgenga.

She told the dog to stay at her heels as she walked toward me. Like me she was not drunk, though she must have been very cold for she wore no cloak over her plain blue woolen dress. The night was brittle with frost, but she did not shiver. "You're going?" She spoke in English. "You're not staying with us?"

"I have a wife and child," I said.

She smiled at that. "Whose names you have not mentioned all evening, Uhtred. So what happened?" I gave no answer and she just stared at me, and there was something very unsettling in her gaze. "So what woman is with you now?" she asked.

"Someone who looks like you," I admitted.

She laughed at that. "And she would have you fight for Alfred?"

"She sees the future," I said, evading the question. "She dreams it."

Brida stared at me. Nihtgenga whined softly and she put down a hand to calm him. "And she sees Alfred surviving?"

"More than surviving," I said. "She sees him winning." Beside me Alfred stirred and I hoped he had the sense to keep his head lowered.

"Winning?"

"She sees a green hill of dead men," I said, "a white horse, and Wessex living again."

"Your woman has strange dreams," Brida said, "but you never answered my first question, Uhtred. If you thought Ragnar was dead, why did you come here?"

I had no ready answer so made none.

"Who did you expect to find here?" she asked.

"You?" I suggested glibly.

She shook her head, knowing I lied. "Why did you come?" I still had no answer and Brida smiled sadly. "If I was Alfred," she said, "I would send a man who spoke Danish to Cippanhamm, and that man would go back to the swamp and tell all he had seen."

"If you think that," I said, "then why don't you tell them?" I nodded toward Guthrum's black-cloaked men guarding the hall door.

"Because Guthrum is a nervous fool," she said savagely. "Why help Guthrum? And when Guthrum fails, Ragnar will take command."

"Why doesn't he command now?"

"Because he is like his father. He's decent. He gave his word to Guthrum and he won't break his word. And tonight he wanted you to give him an oath, but you didn't."

"I do not want Bebbanburg to be a gift of the Danes," I answered.

She thought about that, and understood it. "But do you think," she asked scornfully, "that the West Saxons will give you Bebbanburg? It's at the other end of Britain, Uhtred, and the last Saxon king is rotting in a swamp."

"This will give it to me," I said, pulling back my cloak to show Serpent-Breath's hilt.

"You and Ragnar can rule the north," she said.

"Maybe we will," I said. "So tell Ragnar that when this is all finished, when all is decided, I shall go north with him. I shall fight Kjartan. But in my own time."

"I hope you live to keep that promise," she said, then leaned forward and kissed my cheek. Then, without another word, she turned and walked back to the church.

Alfred let out a breath. "Who is Kjartan?"

"An enemy," I said shortly. I tried to lead him away, but he stopped me.

He was staring at Brida who was nearing the church. "That is the girl who was with you at Wintanceaster?"

"Yes." He was talking of the time when I had first come to Wessex and Brida had been with me.

"And does Iseult truly see the future?'

"She has not been wrong yet."

He made the sign of the cross, then let me lead him back through the town that was quieter now, but he would not go with me to the western gate, insisting we return to the nunnery where, for a moment, we both crouched near one of the dying fires in the courtyard to get what warmth we could from the embers. Men slept in the nunnery church, but the courtyard was now deserted and quiet, and Alfred took a piece of half-burning wood and, using it as a torch, went to the row of small doors that led to the nuns' sleeping cells. One door had been fastened with two hasps and a short length of thick chain and Alfred paused there.

"Draw your sword," he ordered me.

When Serpent-Breath was naked he unwound the chain from the hasps and pushed the door inward. He entered cautiously, pushing the hood back from his face. He held the torch high, and in its light I saw the big man huddled on the floor.

"Steapa!" Alfred hissed.

Steapa was only pretending to be asleep and he uncoiled from the floor with wolflike speed, lashing out at Alfred, and I rammed the sword toward his breast, but then he saw Alfred's bruised face and he froze, oblivious of the blade. "Lord?"

"You're coming with us," Alfred said.

"Lord!" Steapa fell to his knees in front of his king.

"It's cold out there," Alfred said. It was freezing inside the cell as well. "You can sheath your sword, Uhtred." Steapa looked at me and seemed vaguely surprised to find I was the man he had been fighting when the Danes came. "The two of you will be friends," Alfred said sternly, and the big man nodded. "And we have one other person to fetch," Alfred said, "so come."

"One other person?" I asked.

"You spoke of a nun," Alfred said.

So I had to find the nun's cell, and she was still there, lying crushed against the wall by a Dane who was snoring flabbily. The flamelight showed a small, frightened face half hidden by the Dane's beard. His beard was black and her hair was gold, pale gold, and she was awake and, seeing us, gasped, and that woke the Dane who blinked in the flamelight and then snarled at us as he tried to throw off the thick cloaks serving as blankets. Steapa hit him and it was like the sound of a bullock being clubbed, wet and hard at the same time. The man's head snapped back and Alfred pulled the cloaks away and the nun tried to hide her nakedness. Alfred hurriedly put the cloaks back. He had been embarrassed and I had been impressed for she was young and very beautiful and I wondered why such a woman would waste her sweetness on religion. "You know who I am?" Alfred asked her. She shook her head. "I am your king," he said softly, "and you will come with us, sister."

Her clothes were long gone, so we swathed her in the heavy cloaks. The Dane was dead by now, his throat cut by Wasp-Sting, and I had found a pouch of coins strung around his neck on a leather thong. "That money goes to the church," Alfred said.

"I found it," I said, "and I killed him."

"It is the money of sin," he said patiently, "and must be redeemed." He smiled at the nun. "Are there any other sisters here?" he asked.

"Only me," she said in a small voice.

"And now you are safe, sister." He straightened. "We can go."

Steapa carried the nun who was called Hild. She clung to him, whimpering, either from the cold or, more likely, from the memory of her ordeal.

We could have captured Cippanhamm that night with a hundred men. It was so bitterly cold that no guards stood on the ramparts. The gate sentries were in a house by the wall, crouched by the fire, and all the notice they took of the bar being lifted was to shout a bad-tempered question wanting to know who we were. "Guthrum's men," I called back, and they did not bother us further. A half hour later we were in the water mill, reunited with Father Adelbert, Egwine, and the three soldiers.

"We should give thanks to God for our deliverance," Alfred said to Father Adelbert, who had been aghast to see the blood and bruises on the king's face. "Say a prayer, father," Alfred ordered.

Adelbert prayed, but I did not listen. I just crouched by the fire, thought I would never be warm again, and then slept.

It snowed all next day. Thick snow. We made a fire, careless that the Danes might see the smoke, for no Dane was going to struggle through the bitter cold and deepening snow to investigate one small, far-off trickle of gray against a gray sky.

Alfred brooded. He spoke little that day, though once he frowned and asked me if it could really be true about Wulfhere. "We didn't see him with Guthrum," he added plaintively, desperately hoping that the ealdorman had not betrayed him.

"The hostages lived," I said.

"Dear God," he said, convinced by that argument, and leaned his head against the wall. He watched the snow through one of the small windows. "He's family!" he said after a while, then fell silent again.

I fed the horses the last of the hay we had brought with us, then sharpened my swords for lack of anything else to do. Hild wept. Alfred tried to comfort her, but he was awkward and had no words, and oddly it was Steapa who calmed her. He talked to her softly, his voice a deep grumble, and when Serpent-Breath and Wasp-Sting were as sharp as I could make them, and as the snow sifted endlessly onto a silent world, I brooded like Alfred.

I thought of Ragnar wanting my oath. I thought of him wanting my allegiance.

The world began in chaos and it will end in chaos. The gods brought the world into existence, and they will end it when they fight among themselves, but in between the chaos of the world's birth and the chaos of the world's death is order, and order is made by oaths, and oaths bind us like the buckles of a harness.

I was bound to Alfred by an oath, and before I gave that oath I had wanted to bind myself to Ragnar, but now I felt affronted that he had even asked me. That was pride growing in me and changing me. I was Uhtred of Bebbanburg, slayer of Ubba, and while I would give an oath to a king, I was reluctant to make an oath to an equal. The oath-giver is subservient to the man who accepts the oath. Ragnar would have said I was a friend, he would be generous, he would treat me like a brother, but his assumption that I would give him an oath demonstrated that he still believed I was his follower. I was a lord of Northumbria, but he was a Dane, and to a Dane all Saxons are lesser men, and so he had demanded an oath. If I gave him an oath, then he would be generous but I would be expected to show gratitude, and I could only ever hold Bebbanburg because he allowed me to hold it. I had never thought it all through before, but suddenly, on that cold day, I understood that among the Danes I was as important as my friends, and without friends I was just another landless, masterless warrior. But among the Saxons I was another Saxon, and among the Saxons I did not need another man's generosity.

"You look thoughtful, Uhtred." Alfred interrupted my reverie.

"I was thinking, lord," I said, "that we need warm food." I fed the fire, then went outside to the stream where I knocked away the skim of ice and scooped water into a pot. Steapa had followed me outside, not to talk, but to piss, and I stood behind him. "At the witanegemot," I said, "you lied about Cynuit."

He tied the scrap of rope that served as a belt and turned to look at me. "If the Danes had not come," he said in his growling voice, "I would have killed you."

I did not argue with that, for he was probably right. "At Cynuit," I said instead, "when Ubba died, where were you?"

"There."

"I didn't see you," I said. "I was in the thick of the battle, but I didn't see you."

"You think I wasn't there?" He was angry.

"You were with Odda the Younger?" I asked, and he nodded. "You were with him," I guessed, "because his father told you to protect him?" He nodded again. "And Odda the Younger," I said, "stayed a long way from any danger. Isn't that right?"

He did not answer, but his silence told me I was right. He decided he had nothing more to say to me so he started back toward the mill, but I pulled on his arm to stop him. He was surprised by that. Steapa was so big and so strong and so feared that he was unused to men using force on him, and I could see the slow anger burning in him. I fed it. "You were Odda's nursemaid," I sneered. "The great Steapa Snotor was a nursemaid. Other men faced and fought the Danes and you just held Odda's hand."

He just stared at me. His face, so tight-skinned and expressionless, was like an animal's gaze, nothing there but hunger and anger and violence. He wanted to kill me, especially after I used his nickname, but I understood something more about Steapa Snotor. He was truly stupid. He would kill me if he was ordered to kill me, but without someone to instruct him he did not know what to do, so I thrust the pot of water at him. "Carry that inside," I told him. He hesitated. "Don't stand there like a dumb ox!" I snapped. "Take it! And don't spill it." He took the pot. "It has to go on the fire," I told him, "and next time we fight the Danes you'll be with me."

"You?"

"Because we are warriors," I said, "and our job is to kill our enemies, not be nursemaids to weaklings."

I collected firewood, then went inside to find Alfred staring at nothing and Steapa sitting beside Hild who now seemed to be consoling him rather than being consoled. I crumbled oatcakes and dried fish into the water and stirred the mess with a stick. It was a gruel of sorts and tasted horrible, but it was hot.

That night it stopped snowing and next morning we went home.

Alfred need not have gone to Cippanhamm. Anything he learned there he could have discovered by sending spies, but he had insisted on going himself and he came back more worried than before. He had learned some good things, that Guthrum did not have the men to subjugate all Wessex and so was waiting for reinforcements, but he had also learned that Guthrum was trying to turn the nobility of Wessex to his side. Wulfhere was sworn to the Danes. Who else?

"Will the fyrd of Wiltunscir fight for Wulfhere?" he asked us.

Of course they would fight for Wulfhere. Most of the men in Wiltunscir were loyal to their lord, and if their lord ordered them to follow his banner to war, then they would march. Those men who were in the parts of the shire not occupied by the Danes might go to Alfred, but the rest would do what they always did, follow their lord. And other ealdormen, seeing that Wulfhere had not lost his estates, would reckon that their own future, and their family's safety, lay with the Danes. The Danes had ever worked that way. Their armies were too small and too disorganized to defeat a great kingdom so they recruited lords of the kingdom, flattered them, even made them into kings, and only when they were secure did they turn on those Saxons and kill them.

So back in Æthelingæg Alfred did what he did best. He wrote letters. He wrote letters to all his nobility, and messengers were sent into every corner of Wessex to find ealdormen, thegns, and bishops, and deliver the letters. I am alive, the scraps of parchment said, and after Easter I shall take Wessex from the pagans, and you will help me. We waited for the replies.

"You must teach me to read," Iseult said when I told her about the letters.

"Why?"

"It is a magic," she said.

"What magic? So you can read psalms?"

"Words are like breath," she said. "You say them and they're gone. But writing traps them. You could write down stories, poems."

"Hild will teach you," I said, and the nun did, scratching letters in the mud. I watched them sometimes and thought they could have been taken for sisters except one had hair black as a raven's wing and the other had hair of pale gold.

So Iseult learned her letters and I practiced the men with their weapons and shields until they were too tired to curse me, and we also made a new fortress. We restored one of the beamwegs that led south to the hills at the edge of the swamp, and where that log road met dry land we made a strong fort of earth and tree trunks. None of

Guthrum's men tried to stop the work, though we saw Danes watching us from the higher hills, and by the time Guthrum understood what we were doing the fort was finished. In late February a hundred Danes came to challenge it, but they saw the thorn palisade protecting the ditch, saw the strength of the log wall behind the ditch, saw our spears thick against the sky, and rode away.

Next day I took sixty men to the farm where we had seen the Danish horses. They were gone, and the farm was burned out. We rode inland, seeing no enemy. We found newborn lambs slaughtered by foxes, but no Danes, and from that day on we rode ever deeper into Wessex, carrying the message that the king lived and fought, and some days we met Danish bands, but we only fought if we outnumbered them for we could not afford to lose men.

Ælswith gave birth to a daughter whom she and Alfred called Æthelgifu. Ælswith wanted to leave the swamp. She knew that Huppa of Thornsæta was holding Dornwaraceaster, for the ealdorman had replied to Alfred's letter saying that the town was secure and, as soon as Alfred demanded it, the fyrd of Thornsæta would march to his aid. Dornwaraceaster was not so large as Cippanhamm, but it had Roman walls and Ælswith was tired of living in the marshes, tired of the endless damp, of the chill mists, and she said her newborn baby would die of the cold, and that Edward's sickness would come back, and Bishop Alewold supported her. He had a vision of a large house in Dornwaraceaster, of warm fires and priestly comfort, but Alfred refused. If he moved to Dornwaraceaster then the Danes would immediately abandon Cippanhamm and besiege Alfred and starvation would soon threaten the garrison, but in the swamp there was food. In Dornwaraceaster Alfred would be a prisoner of the Danes, but in the swamp he was free, and he wrote more letters, telling Wessex he lived, that he grew stronger, and that after Easter but before Pentecost, he would strike the pagans.

It rained that late winter. Rain and more rain. I remember standing on the muddy parapet of the new fort and watching the rain just falling and falling. Mail coats rusted, fabrics rotted, and food went moldy. Our boots fell apart and we had no men skilled in making new ones. We slid and splashed through greasy mud, our clothes were never dry, and still gray swathes of rain marched from the west. Thatch dripped, huts flooded, the world was sullen. We ate well enough, though as more men came to Æthelingæg, the food became scarcer, but no one starved and no one complained except Bishop Alewold who grimaced whenever he saw another fish stew. There were no deer left in the swamp—all had been netted and eaten—but at least we had fish, eels, and wildfowl, while outside the swamp, in those areas the Danes had plundered, folk starved. We practiced with our weapons, fought mock battles with staves, watched the hills, and welcomed the messengers who brought news. Burgweard, the fleet commander, wrote from Hamtun saying that the town was garrisoned by Saxons but that Danish ships were off the coast. "I don't suppose he's fighting them," Leofric remarked glumly when he heard that news.

"He doesn't say so," I said.

"Doesn't want to get his nice ships dirty," Leofric guessed.

"At least he still has the ships."

A letter came from a priest in distant Kent saying that Vikings from Lundene had occupied Contwaraburg and others had settled on the Isle of Sceapig, and that the ealdorman had made his peace with the invaders. News came from Suth Seaxa of more Danish raids, but also a reassurance from Arnulf, Ealdorman of Suth Seaxa, that his fyrd would gather in the spring. He sent a gospel book to Alfred as a token of his loyalty, and for days Alfred carried the book until the rain soaked into the pages and made the ink run. Wiglaf, Ealdorman of Sumorsæte, appeared in early March and brought seventy men. He claimed to have been hiding in the hills south of Baðum and Alfred ignored the rumors that said Wiglaf had been negotiating with Guthrum. All that mattered was that the ealdorman had come to Æthelingæg and Alfred gave him command of the troops that continually rode inland to shadow the Danes and to ambush their forage parties. Not all the news was so encouraging. Wilfrith of Hamptonscir

had fled across the water to Frankia, as had a score of other ealdormen and thegns.

But Odda the Younger, Ealdorman of Defnascir, was still in Wessex. He sent a priest who brought a letter reporting that the ealdorman was holding Exanceaster. "God be praised," the letter read, "but
there are no pagans in the town."

"So where are they?" Alfred asked the priest. We knew that Svein,
despite losing his ships, had not marched to join Guthrum, which
suggested he was still skulking in Defnascir.

The priest, a young man who seemed terrified of the king, shrugged,
hesitated, then stammered that Svein was close to Exanceaster.

"Close?" the king asked.

"Nearby," the priest managed to say.

"They besiege the town?" Alfred asked.

"No, lord."

Alfred read the letter a second time. He always had great faith in
the written word and he was trying to find some hint of the truth
that had escaped him in the first reading. "They are not in
Exanceaster," he concluded, "but the letter does not say where they
are. Nor how many they are. Nor what they're doing."

"They are nearby, lord," the priest said hopelessly. "To the west, I
think."

"The west?"

"I think they're to the west."

"What's to the west?" Alfred asked me.

"The high moor," I said.

Alfred threw the letter down in disgust. "Maybe you should go to
Defnascir," he told me, "and find out what the pagans are doing."

"Yes, lord," I said.

"It will be a chance to discover your wife and child," Alfred said.

There was a sting there. As the winter rains fell the priests hissed
their poison into Alfred's ears and he was willing enough to hear their
message, which was that the Saxons would only defeat the Danes if
God willed it. And God, the priests said, wanted us to be virtuous.

And Iseult was a pagan, as was I, and she and I were not married, while I had a wife, and so the accusation was whispered about the swamp that it was Iseult who stood between Alfred and victory. No one said it openly, not then, yet Iseult sensed it. Hild was her protector in those days, because Hild was a nun, a Christian, and a victim of the Danes, but many thought Iseult was corrupting Hild. I pretended to be deaf to the whispers until Alfred's daughter told me of them.

Æthelflaed was almost seven and her father's favorite child. Ælswith was fonder of Edward, and in those wet winter days she worried about her son's health and the health of her newborn child, which gave Æthelflaed a deal of freedom. She would stay at her father's side much of the time, but she also wandered about Æthelingæg where she was spoiled by soldiers and villagers. She was a bright ripple of sunlight in those rain-sodden days. She had golden hair, a sweet face, blue eyes, and no fear. One day I found her at the southern fort, watching a dozen Danes who had come to watch us. I told her to go back to Æthelingæg and she pretended to obey me, but an hour later, when the Danes had gone, I found her hiding in one of the turf-roofed shelters behind the wall. "I hoped the Danes would come," she told me.

"So they could take you away?"

"So I could watch you kill them."

It was one of the rare days when it was not raining. There was sunshine on the green hills and I sat on the wall, took Serpent-Breath from her fleece-lined scabbard, and began sharpening her two edges with a whetstone. Æthelflaed insisted on trying the whetstone and she laid the long blade on her lap and frowned in concentration as she drew the stone down the sword. "How many Danes have you killed?" she asked.

"Enough."

"Mama says you don't love Jesus."

"We all love Jesus," I said evasively.

"If you loved Jesus," she said seriously, "then you could kill more Danes. What's this?" She had found the deep nick in one of Serpent-Breath's edges.

"It's where she hit another sword," I said. It had happened at Cippanhamm during my fight with Steapa and his huge sword had bitten deep into Serpent-Breath.

"I'll make her better," she said, and worked obsessively with the whetstone, trying to smooth the nick's edges. "Mama says Iseult is an aglæcwif." She stumbled over the word, then grinned in triumph because she had managed to say it. I said nothing. An aglæcwif was a fiend, a monster. "The bishop says it, too," Æthelflaed said earnestly. "I don't like the bishop."

"You don't?"

"He dribbles." She tried to demonstrate and managed to spit onto Serpent-Breath. She rubbed the blade. "Is Iseult an aglæcwif?"

"Of course not. She made Edward better."

"Jesus did that, and Jesus sent me a baby sister." She scowled because all her efforts had made no impression on the nick in Serpent-Breath.

"Iseult is a good woman," I said.

"She's learning to read. I can read."

"You can?"

"Almost. If she reads then she can be a Christian. I'd like to be an aglæcwif."

"You would?" I asked, surprised.

For answer she growled at me and crooked a small hand so that her fingers looked like claws. Then she laughed. "Are those Danes?" She had seen some horsemen coming from the south.

"That's Wiglaf," I said.

"He's nice."

I sent her back to Æthelingæg on Wiglaf's horse and I thought of what she had said and wondered, for the thousandth time, why I was among Christians who believed I was an offense to their god. They called my gods dwolgods, which meant false gods, so that made me Uhtredærwe, living with an aglæcwif and worshipping dwolgods. I flaunted it, though, always wearing my hammer amulet openly, and that night Alfred, as ever, flinched when he saw it. He had summoned me to his hall where I found him bent over a tafl board. He was play-ing against Beocca, who had the larger set of pieces. It seems a simple game, tafl, where one player has a king and a dozen other pieces, and the other has double the pieces, but no king, and then you move the pieces about the checkered board until one or other player has all his wooden pieces surrounded. I had no patience for it, but Alfred was fond of the game, though when I arrived he seemed to be losing and so was relieved to see me. "I want you to go to Defnascir," he said.

"Of course, lord."

"I fear your king is threatened, lord," Beocca said happily.

"Never mind," Alfred said irritably. "You're to go to Defnascir," he said, turning back to me, "but Iseult must stay here."

I bridled at that. "She's to be a hostage again?" I asked.

"I need her medicines," Alfred said.

"Even though they're made by an aglæcwif?"

He gave me a sharp look. "She is a healer," he said, "and that means she is God's instrument, and with God's help she will come to the truth. Besides, you must travel fast and don't need a woman for company. You will go to Defnascir and find Svein, and once you've found him you will instruct Odda the Younger to raise the fyrd. Tell him Svein must be driven from the shire, and once Odda has achieved that, he is to come here with his household troops. He com-mands my bodyguard. He should be here."

"You want me to give Odda orders?" I asked, partly in surprise, partly with scorn.

"I do," Alfred said, "and I order you to make your peace with him."

"Yes, lord," I replied.

He heard the sarcasm in my voice. "We are all Saxons, Uhtred, and now, more than ever, is the time to heal our wounds."

Beocca, realizing that defeating Alfred at tafl would not help the king's mood, was taking the pieces from the board. "A house divided against itself," he interjected, "will be destroyed. Saint Matthew said that."

"Praise God for that truth," Alfred said, "and we must be rid of

Svein." That was a greater truth. Alfred wanted to march against Guthrum after Easter, but he could scarcely do that if Svein's forces were behind him. "You find Svein," the king told me, "and Steapa will accompany you."

"Steapa!"

"He knows the country," Alfred said, "and I have told him he is to obey you."

"It's best that two of you go," Beocca said earnestly. "Remember that Joshua sent two spies against Jericho."

"You're delivering me to my enemies," I said bitterly, though when I thought about it I decided that using me as a spy made sense. The Danes in Defnascir would be looking out for Alfred's scouts, but I could speak the enemy's language and could pass for one of them and so I was safer than anyone else in Alfred's force. As for Steapa, he was from Defnascir, he knew the country, and he was Odda's sworn man, so he was best suited for carrying a message to the ealdorman.

And so the two of us rode south from Æthelingæg on a day of driving rain.

Steapa did not like me and I did not like him and so we had nothing to say to each other except when I suggested what path we take, and he never disagreed. We kept close to the large road, the road the Romans had made, though I went cautiously for such roads were much used by Danish bands seeking forage or plunder. This was also the route Svein must take if he marched to join Guthrum, but we saw no Danes. We saw no Saxons either. Every village and farm on the road had been pillaged and burned so that we journeyed through a land of the dead.

On the second day Steapa headed westward. He did not explain the sudden change of direction, but doggedly pushed up into the hills and I followed him because he knew the countryside and I supposed he was taking the small paths that would lead to the high bleakness of Dærentmora. He rode urgently, his hard face grim, and I called to him once that we should take more care in case there were Danish forage parties in the small valleys, but he ignored me. Instead, almost at a gallop, he rode down into one of those small valleys until he came in sight of a farmstead.

Or what had been a farmstead. Now it was wet ashes in a green place, a deep green place where narrow pastures were shadowed by tall trees on which the very first haze of spring was just showing. Flowers were thick along the pasture edges, but there were none where the few small buildings had stood. There were only embers and the black smear of ash in mud, and Steapa, abandoning his horse, walked among the ashes. He had lost his great sword when the Danes captured him at Cippanhamm, so now he carried a huge war ax and he prodded the wide blade into the dark piles.

I rescued his horse, tied both beasts to the scorched trunk of an ash that had once grown by the farmyard, and watched him. I said nothing, for I sensed that one word would release all his fury. He crouched by the skeleton of a dog and just stared at the fire-darkened bones for a few minutes, then reached out and stroked the bared skull. There were tears on Steapa's face, or perhaps it was the rain that fell softly from low clouds.

A score of people had once lived there. A larger house had stood at the southern end of the settlement and I explored its charred remains, seeing where the Danes had dug down by the old posts to find hidden coins. Steapa watched me. He was by one of the smaller patches of charred timbers and I guessed he had grown up there, in a slave hovel. He did not want me near him, and I pointedly stayed away, wondering if I dared suggest to him that we ride on. But he began digging instead, hacking the damp red soil with his huge war ax and scooping the earth out with bare hands until he had made a shallow grave for the dog. It was a skeleton now. There were still patches of fur on the old bones, but the flesh had been eaten away so that the ribs were scattered, so this had all happened weeks before. Steapa gathered the bones and laid them tenderly in the grave.

That was when the people came. You can ride through a land-scape of the dead and see no one, but they will see you. Folk hide when enemies come. They go up into the woods and they wait there, and now three men came from the trees.

"Steapa," I said. He turned on me, furious that I had interrupted him, then saw I was pointing westward.

He gave a roar of recognition and the three men, who were holding spears, ran toward him. They dropped their weapons and they hugged the huge man, and for a time they all spoke together, but then they calmed down and I took one aside and questioned him. The Danes had come soon after Yule, he told me. They had come suddenly, before anyone was even aware that there were pagans in Defnascir. These men had escaped because they had been felling a beech tree in a nearby wood, and they had heard the slaughter. Since then they had been living in the forests, scared of the Danes who still rode about Defnascir in search of food. They had seen no Saxons.

They had buried the folk of the farm in a pasture to the south, and Steapa went there and knelt in the wet grass. "His mother died," the man told me. He spoke English with such a strange accent that I continually had to ask him to repeat himself, but I understood those three words. "Steapa was good to his mother," the man said. "He brought her money. She was no slave any more."

"His father?"

"He died long time back. Long time."

I thought Steapa was going to dig up his mother, so I crossed and stood in front of him. "We have a job to do," I said.

He looked up at me, his harsh face expressionless.

"There are Danes to kill," I said. "The Danes who killed folk here must be killed themselves."

He nodded abruptly, then stood, towering over me again. He cleaned the blade of his ax and climbed into his saddle. "There are Danes to kill," he said and, leaving his mother in her cold grave, we went to find them.

TEN

We rode south. We went cautiously, for folk said the Danes were still seen in this part of the shire, though we saw none. Steapa was silent until, in a river meadow, we rode past a ring of stone pillars, one of the mysteries left behind by the old people. Such rings stand all across England and some are huge, though this one was a mere score of lichen-covered stones, none taller than a man, standing in a circle some fifteen paces wide. Steapa glanced at them, then astonished me by speaking. "That's a wedding," he said.

"A wedding?"

"They were dancing," he growled, "and the devil turned them to stone."

"Why did the devil do that?" I asked cautiously.

"Because they wed on a Sunday, of course. Folk never should wed on a Sunday, never! Everyone knows that." We rode on in silence, then, surprising me again, he began to talk about his mother and father and how they had been serfs of Odda the Elder. "But life was good for us," he said.

"It was?"

"Plowing, sowing, weeding, harvest, threshing."

"But Ealdorman Odda didn't live back there," I said, jerking my thumb toward Steapa's destroyed homestead.

"No! Not him!" Steapa was amused I should even ask such a question. "He wouldn't live there, not him! Had his own big hall. Still does. But he had a steward there. Man to give us orders. He was a big man! Very tall!"

I hesitated. "But your father was short?"

Steapa looked surprised. "How did you know that?"

"I just guessed."

"He was a good worker, my father."

"Did he teach you to fight?"

"He didn't, no. No one did. I just learned myself."

The land was less damaged the farther south we went. And that was strange, for the Danes had come this way. We knew that, for folk said the Danes were still in the southern part of the shire, but life suddenly seemed normal. We saw men spreading dung on fields, and other men ditching or hedging. There were lambs in the pastures. To the north the foxes had become fat on dead lambs, but here the shepherds and their dogs were winning that ceaseless battle.

And the Danes were in Cridianton.

A priest told us that in a village hard under a great oak-covered hill beside a stream. The priest was nervous because he had seen my long hair and arm rings and he presumed I was a Dane, and my northern accent did not persuade him otherwise, but he was reassured by Steapa. The two talked, and the priest gave his opinion that it would be a wet summer.

"It will," Steapa agreed. "The oak greened before the ash."

"Always a sign," the priest said.

"How far is Cridianton?" I broke into the conversation.

"A morning's walk, lord."

"You've seen the Danes there?" I asked.

"I seen them, lord, I have," he said.

"Who leads them?"

"Don't know, lord."

"They have a banner?" I asked.

He nodded. "It hangs on the bishop's hall, lord. It shows a white horse."

So it was Svein. I did not know who else it could have been, but the white horse confirmed that Svein had stayed in Defnascir rather than try to join Guthrum. I twisted in the saddle and looked at the priest's village that was unscarred by war. No thatch had been burned, no granaries emptied, and the church was still standing. "Have the Danes come here?" I asked.

"Oh yes, lord, they came. Came more than once."

"Did they rape? Steal?"

"No, lord. But they bought some grain. Paid silver for it."

Well-behaved Danes. That was another strange thing. "Are they besieging Exanceaster?" I asked. That would have made a sort of sense. Cridianton was close enough to Exanceaster to give most of the Danish troops shelter while the rest invested the larger town.

"No, lord," the priest said, "not that I know of."

"Then what are they doing?" I asked.

"They're just in Cridianton, lord."

"And Odda is in Exanceaster?"

"No, lord. He's in Ocmundtun. He's with Lord Harald."

I knew the shire reeve's hall was in Ocmundtun, which lay beneath the northern edge of the great moor. But Ocmundtun was also a long journey from Cridianton and no place to be if a man wanted to harry the Danes.

I believed the priest when he said Svein was at Cridianton, but we still rode there to see for ourselves. We used wooded, hilly tracks and came to the town at midafternoon and saw the smoke rising from cooking fires, then saw the Danish shields hanging from the palisade. Steapa and I were hidden in the high woods and could see men guarding the gate, and other men standing watch in a pasture where forty or fifty horses were grazing on the first of the spring grass. I could see Odda the Elder's hall where I had been reunited with Mildrith after the fight at Cynuit, and I could also see a triangular Danish banner flying above the larger hall that was the bishop's home. The western gate was open, though well guarded, and despite the sentries and the shields on the wall the town looked like a place at peace, not at war. There should be Saxons on this hill, I thought, Saxons watching the enemy, ready to attack. Instead the Danes were living undisturbed. "How far to Ocmundtun?" I asked Steapa.

"We can make it by nightfall."

I hesitated. If Odda the Younger was at Ocmundtun, then why go there? He was my enemy and sworn to my death. Alfred had given me a scrap of parchment on which he had written words commanding Odda to greet me peaceably, but what force did writing have against hatred?

"He won't kill you," Steapa said, surprising me again. He had evidently guessed my thoughts. "He won't kill you," he said again.

"Why not?"

"Because I won't bide him killing you," Steapa said and turned his horse west.

We reached Ocmundtun at dusk. It was a small town built along a river and guarded by a high spur of limestone on which a stout palisade offered a refuge if attackers came. No one was on the limestone spur now and the town, which had no walls, looked placid. There might be war in Wessex, but Ocmundtun, like Cridianton, was evidently at peace. Harald's hall was close to the fort on its hill and no one challenged us as we rode into the forecourt where servants recognized Steapa. They greeted him warily, but then a steward came from the hall door and, seeing the huge man, clapped his hands twice in a sign of delight. "We heard you were taken by the pagans," the steward said.

"I was."

"They let you go?"

"My king freed me," Steapa growled as though he resented the question. He slid from his horse and stretched. "Alfred freed me."

"Is Harald here?" I asked the steward.

"My lord is inside." The steward was offended that I had not called the reeve "lord."

"Then so are we," I said, and led Steapa into the hall. The steward flapped at us because custom and courtesy demanded that he seek his lord's permission for us to enter the hall, but I ignored him.

A fire burned in the central hearth and dozens of rushlights stood on the platforms at the hall's edges. Boar spears were stacked against the wall on which hung a dozen deerskins and a bundle of valuable pine marten pelts. A score of men were in the hall, evidently waiting for supper, and a harpist played at the far end. A pack of hounds rushed to investigate us and Steapa beat them off as we walked to the fire to warm ourselves. "Ale," Steapa said to the steward.

Harald must have heard the noise of the hounds, for he appeared at a door leading from the private chamber at the back of the hall. He blinked when he saw us. He had thought the two of us were enemies, then he had heard that Steapa was captured, yet here we were, side by side. The hall fell silent as he limped toward us. It was only a slight limp, the result of a spear wound in some battle that had also taken two fingers of his sword hand. "You once chided me," he said, "for carrying weapons into your hall. Yet you bring weapons into mine."

"There was no gatekeeper," I said.

"He was having a piss, lord," the steward explained.

"There are to be no weapons in the hall," Harald insisted.

That was customary. Men get drunk in the hall and can do enough damage to one another with the knives we use to cut meat, and drunken men with swords and axes can turn a supper table into a butcher's yard. We gave the steward our weapons. I hauled off my mail coat and told the steward to hang it on a frame to dry, then have a servant clean its links.

Harald formally welcomed us when our weapons were gone. He said the hall was ours and that we should eat with him as honored guests. "I would hear your news," he said, beckoning a servant who brought us pots of ale.

"Is Odda here?" I demanded.

"The father is, yes. Not the son."

I swore. We had come here with a message for Ealdorman Odda, Odda the Younger, only to discover that it was the wounded father, Odda the Elder, who was in Ocmundtun. "So where is the son?" I asked.

Harald was offended by my brusqueness, but he remained courteous. "The ealdorman is in Exanceaster."

"Is he besieged there?"

"No."

"And the Danes are in Cridianton?"

"They are."

"And are they besieged?" I knew the answer to that, but wanted to hear Harald admit it.

"No," he said.

I let the ale pot drop. "We come from the king," I said. I was supposedly speaking to Harald, but I strode down the hall so that the men on the platforms could hear me. "We come from Alfred," I said, "and Alfred wishes to know why there are Danes in Defnascir. We burned their ships, we slaughtered their ship guards, and we drove them from Cynuit, yet you allow them to live here? Why?"

No one answered. There were no women in the hall, for Harald was a widower who had not remarried, and so the supper guests were all his warriors or else thegns who led men of their own. Some looked at me with loathing, for my words imputed cowardice to them, while others looked down at the floor. Harald glanced at Steapa as if seeking the big man's support, but Steapa just stood by the fire, his savage face showing nothing. I turned back to stare at Harald. "Why are there Danes in Defnascir?" I demanded.

"Because they are welcome here," a voice said behind me.

I turned to see an old man standing in the door. White hair showed beneath the bandage that swathed his head, and he was so thin and so weak that he had to lean on the door frame for support. At first I did not recognize him, for when I had last spoken to him he had been a big man, well built and vigorous, but Odda the Elder had taken an ax blow to the skull at Cynuit and he should have died from such a wound, yet somehow he had lived, and here he was, though now he was skeletal, pale, haggard, and feeble. "They are here," Odda said, "because they are welcome. As are you, Lord Uhtred, and you, Steapa."

A woman was tending Odda the Elder. She had tried to pull him away from the door and take him back to his bed, but now she edged past him into the hall and stared at me. Then, seeing me, she did what she had done the very first time she saw me. She did what she had done when she came to marry me. She burst into tears.

It was Mildrith.

Mildrith was robed like a nun in a pale gray dress, belted with rope, over which she wore a large wooden cross. She had a close-fitting gray bonnet from which strands of her fair hair escaped. She stared at me, burst into tears, made the sign of the cross, and vanished. A moment later Odda the Elder followed her, too frail to stand any longer, and the door closed.

"You are indeed welcome here," Harald said, echoing Odda's words.

"But why are the Danes welcome here?" I asked.

Because Odda the Younger had made a truce. Harald explained it as we ate. No one in this part of Defnascir had heard how Svein's ships had been burned at Cynuit. They only knew that Svein's men, and their women and children, had marched south, burning and plundering, and Odda the Younger had taken his troops to Exanceaster, which he had prepared for a siege, but instead Svein had offered to talk. The Danes, quite suddenly, had stopped raiding. Instead they had settled in Cridianton and sent an embassy to Exanceaster, and Svein and Odda had made their private peace.

"We sell them horses," Harald said, "and they pay well for them. Twenty shillings a stallion, fifteen a mare."

"You sell them horses," I said flatly.

"So they will go away," Harald explained.

Servants threw a big birch log onto the fire. Sparks exploded outward, scattering the hounds who lay just beyond the ring of hearth stones.

"How many men does Svein lead?" I asked.

"Many," Harald said.

"Eight hundred?" I asked. "Nine?" Harald shrugged. "They came in twenty-four ships," I went on, "only twenty-four. So how many men can he have? No more than a thousand, and we killed a few, and others must have died in the winter."

"We think he has eight hundred," Harald said reluctantly.

"And how many men in the fyrd? Two thousand?"

"Of which only four hundred are seasoned warriors," Harald said. That was probably true. Most men of the fyrd are farmers, while every Dane is a sword warrior, but Svein would never have pitted his eight hundred men against two thousand. Not because he feared losing, but because he feared that in gaining victory he would lose a hundred men. That was why he had stopped plundering and made his truce with Odda, because in southern Defnascir he could recover from his defeat at Cynuit. His men could rest, feed, make weapons, and get horses. Svein was husbanding his men and making them stronger. "It was not my choice," Harald said defensively. "The ealdorman ordered it."

"And the king," I retorted, "ordered Odda to drive Svein out of Defnascir."

"What do we know of the king's orders?" Harald asked bitterly, and it was my turn to give him news, to tell how Alfred had escaped Guthrum and was in the great swamp.

"And sometime after Easter," I said, "we shall gather the shire fyrds and we shall cut Guthrum into pieces." I stood. "There will be no more horses sold to Svein." I said it loudly so that every man in the big hall could hear me.

"But—" Harald began, then shook his head. He had doubtless been about to say that Odda the Younger, Ealdorman of Defnascir, had ordered the horses to be sold, but his voice trailed away.

"What are the king's orders?" I demanded of Steapa.

"No more horses," he thundered.

There was silence until Harald irritably gestured at the harpist who struck a chord and began playing a melancholy tune. Someone began singing, but no one joined in and his voice trailed away. "I must look to the sentinels," Harald said, and he threw me an inquisitive look that I took as an invitation to join him, and so I buckled on my swords and then walked with him down Ocmundtun's long street to where three spearmen stood guard beside a wooden hut. Harald talked to them for a moment, then led me farther east, away from the light of the sentinels' fire. A moon silvered the valley, lighting the empty road until the track vanished among trees. "I have thirty fighting men," Harald said suddenly.

He was telling me he was too weak to fight. "How many men does Odda have in Exanceaster?" I asked.

"A hundred? Hundred and twenty?"

"The fyrd should have been raised."

"I had no orders," Harald said.

"Did you seek any?"

"Of course I did." He was angry with me now. "I told Odda we should drive Svein away, but he wouldn't listen."

"Did he tell you the king ordered the fyrd raised?"

"No." Harald paused, staring down the moonlit road. "We heard nothing of Alfred, except that he'd been defeated and was hiding. And we heard the Danes were all across Wessex, and that more were gathering in Mercia."

"Odda didn't think to attack Svein when he landed?"

"He thought to protect himself," Harald said, "and sent me to the Tamur."

The Tamur was the river that divided Wessex from Cornwalum. "The Britons are quiet?" I asked.

"Their priests are telling them not to fight us."

"But priests or no priests," I said, "they'll cross the river if the Danes look like winning."

"Aren't they winning already?" Harald asked bitterly.

"We're still free men," I said.

He nodded at that. Behind us, in the town, a dog began howling and he turned as if the noise indicated trouble, but the howling stopped with a sharp yelp. He kicked a stone in the road. "Svein frightens me," he admitted suddenly.

"He's a frightening man," I agreed.

"He's clever," Harald said, "clever, strong, and savage."

"A Dane," I said drily.

"A ruthless man," Harald went on.

"He is," I agreed, "and do you think that after you have fed him, supplied him with horses and given him shelter, he will leave you alone?"

"No," he said, "but Odda believes that."

Then Odda was a fool. He was nursing a wolf cub that would tear him to shreds when it was strong enough. "Why didn't Svein march north to join Guthrum?" I asked.

"I wouldn't know."

But I knew. Guthrum had been in England for years now. He had tried to take Wessex before, and he had failed, but now, on the very brink of success, he had paused. Guthrum the Unlucky, he was called, and I suspected he had not changed. He was wealthy, led many men, but he was cautious. Svein, though, came from the Norsemen's settlements in Ireland and was a very different creature. He was younger than Guthrum, less wealthy than Guthrum, and led fewer men, but he was undoubtedly the better warrior. Now, bereft of his ships, he was weakened, but he had persuaded Odda the Younger to give him refuge and he gathered his strength so that when he did meet Guthrum he would not be a defeated leader in need of help, but a spear Dane of power. Svein, I thought, was a far more dangerous man than Guthrum, and Odda the Younger was only making him more dangerous.

"Tomorrow," I said, "we must start raising the fyrd. Those are the king's orders."

Harald nodded. I could not see his face in the darkness, but I sensed he was not happy, yet he was a sensible man and must have known that Svein had to be driven out of the shire. "I shall send the

messages," he said, "but Odda might stop the fyrd assembling. He's
made his truce with Svein and he won't want me breaking it. Folk
will obey him before they obey me."

"And what of his father?" I asked. "Will they obey him?"

"They will," he said, "but he's a sick man. You saw that. It's a miracle he lives at all."

"Maybe because my wife nurses him?"

"Yes," he said, and fell silent. There was something odd in the air
now, something unexpressed, a discomfort. "Your wife nurses him
well," he finished awkwardly.

"He's her godfather," I said.

"So he is."

"It is good to see her," I said, not because I meant it, but because
it was the proper thing to say and I could think of nothing else. "And
it will be good to see my son," I added with more warmth.

"Your son," Harald said flatly.

"He's here, isn't he?"

"Yes." Harald flinched. He turned away to look at the moon and I
thought he would say no more, but then he summoned his courage
and looked back to me. "Your son, Lord Uhtred," he said, "is in the
churchyard."

It took a few heartbeats for that to make sense, and then it did
not make any sense at all, but left me confused. I touched my hammer amulet. "In the churchyard?"

"It is not my place to tell you."

"But you will tell me," I said, and my voice sounded like Steapa's
growl.

Harald stared at the moon-touched river, silver white beneath the
black trees. "Your son died," he said. He waited for my response, but I
neither moved nor spoke. "He choked to death."

"Choked?"

"A pebble," Harald said. "He was just a baby. He must have picked
the pebble up and swallowed it."

"A pebble?" I asked.

"A woman was with him, but . . ." Harald's voice trailed away. "She tried to save him, but she could do nothing. He died."

"On Saint Vincent's Day," I said.

"You knew?"

"No," I said, "I didn't know." But Saint Vincent's Day had been the day when Iseult drew Alfred's son, the Ætheling Edward, through the earth. And somewhere, Iseult had told me, a child must die so that the king's heir, the Ætheling, could live.

And it had been my child. Uhtred the Younger. Whom I had hardly known. Edward had been given breath and Uhtred had twitched and fought and gasped and died.

"I'm sorry," Harald said. "It was not my place to tell you, but you needed to know before you saw Mildrith again."

"She hates me," I said bleakly.

"Yes," he said, "she does." He paused. "I thought she would go mad with grief, but God has preserved her. She would like—"

"Like what?"

"To join the sisters at Cridianton. When the Danes leave. They have a nunnery there, a small house."

I did not care what Mildrith did. "And my son is buried here?"

"Under the yew tree"—he turned and pointed—"beside the church."

So let him stay there, I thought. Let him rest in his short grave to wait the chaos of the world's ending.

"Tomorrow," I said, "we raise the fyrd."

Because there was a kingdom to save.

Priests were summoned to Harald's hall and the priests wrote the summons for the fyrd. Most thegns could not read, and many of their priests would probably struggle to decipher the few words, but the messengers would tell them what the parchments said. They were to arm their men and bring them to Ocmundtun, and the wax seal on the summons was the authority for those orders. The seal showed Odda the Elder's badge of a stag. "It will take a week," Harald warned me, "for most of the fyrd to reach here, and the ealdorman will try to stop it happening at all."

"What will he do?"

"Tell the thegns to ignore it, I suppose."

"And Svein? What will he do?"

"Try to kill us."

"And he has eight hundred men who can be here tomorrow," I said.

"And I have thirty men," Harald said bleakly.

"But we do have a fortress," I said, pointing to the limestone ridge with its palisade.

I did not doubt that the Danes would come. By summoning the fyrd we threatened their safety, and Svein was not a man who would take a threat lightly, and so, while the messages were carried north and south, the townsfolk were told to take their valuables up to the fort beside the river. Some men were set to strengthening the palisade. Others took livestock up onto the moor so the beasts could not be taken by the Danes. Steapa went to every nearby settlement and demanded that men of fighting age go to Ocmundtun with any weapon they possessed, so that by that afternoon the fort was manned by over eighty men. Few were warriors, most had no weapons other than an ax, but from the foot of the hill they looked formidable enough. Women carried food and water to the fort, and most of the town reckoned to sleep up there, despite the rain, for fear that the Danes would come in the night.

Odda the Elder refused to go to the fort. He was too sick, he said, and too feeble, and if he was supposed to die then he would die in Harald's hall. Harald and I tried to persuade him, but he would not listen. "Mildrith can go," he said.

"No," she said. She sat by Odda's bed, her hands clutched tight under the sleeves of her gray robe. She stared at me, challenge in her eyes, daring me to give her an order to abandon Odda and go to the fortress.

"I am sorry," I said to her.

"Sorry?"

"About our son."

"You were not a father to him," she accused me. Her eyes glistened. "You wanted him to be a Dane! You wanted him to be a pagan! You didn't even care for his soul!"

"I cared for him," I said, but she ignored that. I had not sounded convincing, even to myself.

"His soul is safe," Harald said gently. "He is in the Lord Jesus's arms. He is happy."

Mildrith looked at him and I saw how Harald's words had comforted her, though she still began crying. She caressed her wooden cross. Then Odda the Elder reached out and patted her arm. "If the Danes come, lord," I said to him, "I shall send men for you." I turned then and went from the sickroom. I could not cope with Mildrith crying or with the thought of a dead son. Such things are difficult, much more difficult than making war, and so I buckled on my swords, picked up my shield, and put on my splendid wolf-crested helmet so that, when Harald came from Odda's chamber, he checked to see me standing like a warlord by his hearth. "If we make a big fire at the eastern end of town," I said, "we'll see the Danes come. It will give us time to carry Lord Odda to the fort."

"Yes." He looked up at the great rafters of his hall, and perhaps he was thinking that he would never see it thus again, for the Danes would come and the hall would burn. He made the sign of the cross.

"Fate is inexorable," I told him. What else was there to say? The Danes might come, the hall might burn, but they were small things in the balance of a kingdom, and so I went to order the fire that would illuminate the eastern road, but the Danes did not come that night. It rained softly all through the darkness, so that in the morning the folk in the fort were wet, cold, and unhappy. Then, in the dawn, the first men of the fyrd arrived. It might take days for the farther parts of the shire to receive their summons and to arm men and despatch them to Ocmundtun, but the nearer places sent men straight away so that by late morning there were close to three hundred beneath the fort. No more than seventy of those could be called war-riors, men who had proper weapons, shields, and at least a leather coat. The rest were farm laborers with hoes or sickles or axes.

Harald sent foraging parties to find grain. It was one thing to gather a force, quite another to feed it, and none of us knew how long we would have to keep the men assembled. If the Danes did not come to us, then we would have to go to them and force them from Cridianton, and for that we would need the whole fyrd of Defnascir. Odda the Younger, I thought, would never allow that to happen.

Nor did he. For, as the rain ended and the noontime prayers were said, Odda himself came to Ocmundtun and he did not come alone, but rode with sixty of his warriors in chain mail and as many Danes in their war glory. The sun came out as they appeared from the east-ern trees and it shone on mail and on spear points, on bridle chains and stirrup irons, on polished helmets and bright shield bosses. They spread into the pastures on either side of the road and advanced on Ocmundtun in a wide line, and at its center were two standards. One, the black stag, was the banner of Defnascir, while the other was a Danish triangle and displayed the white horse.

"There'll be no fight," I told Harald.

"There won't?"

"Not enough of them. Svein can't afford to lose men, so he's come to talk."

"I don't want to meet them here." He gestured at the fort. "We should be in the hall."

He ordered that the best armed men should go down to the town, and there we filled the muddy street outside the hall as Odda and the Danes came from the east. The horsemen had to break their line to enter the town, making a column instead, and the column was led by three men. Odda was in the center and he was flanked by two Danes, one of them Svein of the White Horse.

Svein looked magnificent, a silver-white warrior. He rode a white horse, wore a white woolen cloak, and his mail and boar-snouted hel-met had been scrubbed with sand until they glowed silver in the

watery sunlight. His shield bore a silvered boss around which a white
horse had been painted. The leather of his bridle, saddle, and scabbard
had been bleached pale. He saw me, but showed no recognition, just
looked along the line of men barring the street and seemed to dismiss
them as useless. His banner of the white horse was carried by the second horseman who had the same darkened face as his master, a face
hammered by sun and snow, ice and wind.

"Harald." Odda the Younger had ridden ahead of the two Danes.
He was sleek as ever, gleaming in mail, and with a black cloak draping
his horse's rump. He smiled as though he welcomed the meeting. "You
have summoned the fyrd. Why?"

"Because the king commanded it," Harald said.

Odda still smiled. He glanced at me, appeared not to notice I was
present, then looked to the hall door where Steapa had just appeared.
The big man had been talking with Odda the Elder, and now he
stared at Odda the Younger with astonishment. "Steapa!" Odda the
Younger said. "Loyal Steapa! How good to see you!"

"You too, lord."

"My faithful Steapa," Odda said, plainly pleased to be reunited
with his erstwhile bodyguard. "Come here!" he commanded, and
Steapa pushed past us and knelt in the mud by Odda's horse and reverently kissed his master's boot. "Stand," Odda said, "stand. With you
beside me, Steapa, who can hurt us?"

"No one, lord."

"No one," Odda repeated, then smiled at Harald. "You said the
king ordered the fyrd summoned? There is a king in Wessex?"

"There is a king in Wessex," Harald said firmly.

"There is a king skulking in the marshes!" Odda said, loudly
enough for all Harald's men to hear. "He is the king of frogs, perhaps?
A monarch of eels? What kind of king is that?"

I answered for Harald, only I answered in Danish. "A king who
ordered me to burn Svein's boats. Which I did. All but one, which I
kept and still have."

Svein took off his boar-snouted helmet and looked at me and

again there was no recognition. His gaze was like that of the great serpent of death that lies at the foot of Yggdrasil. "I burned the *White
Horse*," I told him, "and warmed my hands on its flames." Svein spat
for answer. "And the man beside you"—I spoke to Odda now, using
English—"is the man who burned your church at Cynuit, the man
who killed the monks. The man who is cursed in heaven, in hell, and
in this world, yet now he is your ally?"

"Does that goat turd speak for you?" Odda demanded of Harald.

"These men speak for me," Harald said, indicating the warriors
behind him.

"But by what right do you raise the fyrd?" Odda asked. "I am an
ealdorman!"

"And who made you ealdorman?" Harald asked. He paused, but
Odda gave no answer. "The king of frogs?" Harald asked. "The
monarch of eels? If Alfred has no authority, then you have lost yours
with his."

Odda was plainly surprised by Harald's defiance, and he was probably irritated by it, but he gave no sign of annoyance. He just went on
smiling. "I do believe," he said to Harald, "that you have misunderstood what happens in Defnascir."

"Then explain to me," Harald said.

"I shall," Odda said, "but we shall talk with ale and food." He
looked up at the sky. The brief sun was gone behind cloud and a chill
wind was gusting the thatch of the street. "And we should talk under
a roof," Odda suggested, "before it rains again."

There were matters to be agreed first, though that was done soon
enough. The Danish horsemen would withdraw to the eastern end of
the town while Harald's men would retreat to the fort. Each side
could take ten men into the hall, and all of those men were to leave
their weapons heaped in the street where they were to be guarded by
six Danes and as many Saxons.

Harald's servants brought ale, bread, and cheese. There was no
meat offered, for it was the season of Lent. Benches were placed on
either side of the hearth. Svein crossed to our side of the fire as the

benches were brought and at long last deigned to recognize me. "It was really you who burned the ships?" he asked.

"Including yours."

"The *White Horse* took a year and a day to build," he said, "and she was made of trees from which we'd hung Odin's sacrifices. She was a good ship."

"She's all ash on the seashore now," I said.

"Then one day I shall repay you," he retorted, and though he spoke mildly, there was a world of threat in his voice. "And you were wrong," he added.

"Wrong?" I asked. "Wrong to burn your ships?"

"There was no altar of gold at Cynuit."

"Where you burned the monks," I said.

"I burned them alive," he agreed, "and warmed my hands on their flames." He smiled at that memory. "You could join me again?" he suggested. "I shall forgive you burning my ship and you and I can fight side by side once more? I need good men. I pay well."

"I am sworn to Alfred."

"Ah." He nodded. "So be it. Enemies." He went back to Odda's benches.

"You would see your father before we talk?" Harald asked Odda, gesturing toward the door at the hall's end.

"I shall see him," Odda said, "when our friendship is repaired. And you and I must be friends." He said the last words loudly and they prompted men to sit on the benches. "You summoned the fyrd," he spoke to Harald, "because Uhtred brought you orders from Alfred?"

"He did."

"Then you did the right thing," Odda said, "and that is to be praised." Svein, listening to the translation that was provided by one of his own men, stared flatly at us. "And now you will do the right thing again," Odda continued, "and send the fyrd home."

"The king has ordered otherwise," Harald said.

"What king?" Odda asked.

"Alfred, who else?"

"But there are other kings in Wessex," Odda said. "Guthrum is King of East Anglia, and he is in Wessex, and some say Æthelwold will be crowned king before the summer."

"Æthelwold?" Harald asked.

"You'd not heard?" Odda asked. "Wulfhere of Wiltunscir has sided with Guthrum, and both Guthrum and Wulfhere have said Æthelwold will be King of Wessex. And why not? Is not Æthelwold the son of our last king? Should he not be king?"

Harald, uncertain, looked at me. He had not heard of Wulfhere's defection, and it was hard news for him. I nodded. "Wulfhere is with Guthrum," I said.

"So Æthelwold, son of Æthelred, will be king in Wessex," Odda said, "and Æthelwold has thousands of swords at his command. Ælfrig of Kent is with the Danes. There are Danes in Lundene, on Sceapig, and on the walls of Contwaraburg. All northern Wessex is in Danish hands. There are Danes here, in Defnascir. What, tell me, is Alfred king of?"

"Of Wessex," I said.

Odda ignored me, looking at Harald. "Alfred has our oaths," Harald said stubbornly.

"And I have your oath," Odda reminded him. He sighed. "God knows, Harald, no one was more loyal to Alfred than I. Yet he failed us! The Danes came and the Danes are here, and where is Alfred? Hiding! In a few weeks their armies will march! They will come from Mercia, from Lundene, from Kent! Their fleets will be off our coast. Armies of Danes and fleets of vikings! What will you do then?"

Harald shifted uneasily. "What will you do?" he retorted.

Odda gestured at Svein who, the question translated, spoke for the first time. I interpreted for Harald. Wessex is doomed, Svein said in his grating voice. By summer it will be swarming with Danes, with men newly come from the north, and the only Saxons who will live will be those men who aid the Danes now. Those who fight against the Danes, Svein said, will be dead, and their women will be whores and their children will be slaves and their homes will be lost and their names shall be forgotten like the smoke of an extinguished fire.

"And Æthelwold will be king?" I asked scornfully. "You think we will all bow to a whoring drunkard?"

Odda shook his head. "The Danes are generous," he said, and he drew back his cloak and I saw that he wore six golden arm rings. "To those who help them," he said, "there will be the reward of land, wealth, and honor."

"And Æthelwold will be king?" I asked again.

Odda again gestured at Svein. The big Dane seemed bored, but he stirred himself. "It is right," he said, "that Saxons should be ruled by a Saxon. We shall make a king here."

I scorned that. They had made Saxon kings in Northumbria and in Mercia and those kings were feeble, leashed to the Danes, and then I understood what Svein meant and I laughed aloud. "He's promised you the throne!" I accused Odda.

"I've heard more sense from a pig's fart," Odda retorted, but I knew I was right. Æthelwold was Guthrum's candidate for the throne of Wessex, but Svein was no friend of Guthrum and would want his own Saxon as king. Odda.

"King Odda," I said jeeringly, then spat into the fire.

Odda would have killed me for that, but we met under the terms of a truce and so he forced himself to ignore the insult. He looked at Harald. "You have a choice, Harald," he said, "You can die or you can live."

Harald was silent. He had not known about Wulfhere, and the news had appalled him. Wulfhere was the most powerful ealdorman in Wessex, and if he thought Alfred was doomed, then what was Harald to think? I could see the shire reeve's uncertainty. His decency wanted him to declare loyalty to Alfred, but Odda had suggested that nothing but death would follow such a choice. "I . . ." Harald began, then fell silent, unable to say what he thought for he did not know his own mind.

"The fyrd is raised," I spoke for him, "at the king's orders, and the king's orders are to drive the Danes from Defnascir."

Odda spat into the fire for answer.

"Svein has been defeated," I said. "His ships are burned. He is like a whipped dog and you give him comfort." Svein, when that was translated, gave me a look like the stroke of a whip. "Svein," I went on as though he was not present, "must be driven back to Guthrum."

"You have no authority here," Odda said.

"I have Alfred's authority," I said, "and a written order telling you to drive Svein from your shire."

"Alfred's orders mean nothing," Odda said, "and you croak like a swamp frog." He turned to Steapa. "You have unfinished business with Uhtred."

Steapa looked uncertain for a heartbeat, then understood what his master meant. "Yes, lord," he said.

"Then finish it now."

"Finish what now?" Harald asked.

"Your king"—Odda said the last word sarcastically—"ordered Steapa and Uhtred to fight to the death. Yet both live! So your king's orders have not been obeyed."

"There is a truce!" Harald protested.

"Either Uhtred stops interfering in the affairs of Defnascir," Odda said forcefully, "or I shall have Steapa kill Uhtred. You want to know who is right? Alfred or me? You want to know who will be king in Wessex, Æthelwold or Alfred? Then put it to the test, Harald. Let Steapa and Uhtred finish their fight and see which man God favors. If Uhtred wins then I shall support you, and if he loses?" He smiled. He had no doubt who would win.

Harald stayed silent. I looked at Steapa and, as on the first time I met him, saw nothing on his face. He had promised to protect me, but that was before he had been reunited with his master. The Danes looked happy. Why should they mind two Saxons fighting? Harald, though, still hesitated, and then the weary, feeble voice sounded from the doorway at the back of the hall. "Let them fight, Harald, let them fight." Odda the Elder, swathed in a wolf-skin blanket, stood at the door. He held a crucifix. "Let them fight," he said again, "and God will guide the victor's arm."

Harald looked at me. I nodded. I did not want to fight, but a man cannot back down from combat. What was I to do? Say that to expect God to indicate a course of action through a duel was nonsense? To appeal to Harald? To claim that everything Odda had said was wrong and that Alfred would win? If I had refused to fight I was granting the argument to Odda, and in truth he had half convinced me that Alfred was doomed, and Harald, I am sure, was wholly convinced. Yet there was more than mere pride making me fight in the hall that day. There was a belief, deep in my soul, that somehow Alfred would survive. I did not like him, I did not like his god, but I believed fate was on his side. So I nodded again, this time to Steapa. "I do not want to fight you," I said to him, "but I have given an oath to Alfred, and my sword says he will win and that Danish blood will dung our fields."

Steapa said nothing. He just flexed his huge arms, then waited as one of Odda's men went outside and returned with two swords. No shields, just swords. He had taken a pair of blades at random from the pile and he offered them to Steapa first, who shook his head, indicating that I should have the choice. I closed my eyes, groped, and took the first hilt that I touched. It was a heavy sword, weighted toward its tip. A slashing weapon, not a piercing blade, and I knew I had chosen wrong.

Steapa took the other and scythed it through the air so that the blade sang. Svein, who had betrayed little emotion so far, looked impressed, while Odda the Younger smiled. "You can put the sword down," he told me, "and thus yield the argument to me."

Instead I walked to the clear space beside the hearth. I had no intention of attacking Steapa, but would let him come to me. I felt weary and resigned. Fate is inexorable.

"For my sake," Odda the Elder spoke behind me, "make it fast."

"Yes, lord," Steapa said, and he took a step toward me and then turned as fast as a striking snake and his blade whipped in a slash that took Odda the Younger's throat. The sword was not as sharp as it could have been, so that the blow drove Odda down, but it also ripped his gullet open so that blood spurted a blade's length into the air, then splashed into the fire where it hissed and bubbled. Odda was on the floor rushes now, his legs twitching, his hands clutching at his throat that still pumped blood. He made a gargling noise, turned on his back, and went into a spasm so that his heels drummed against the floor and then, just as Steapa stepped forward to finish him, he gave a last jerk and was dead.

Steapa drove the sword into the floor, leaving it quivering there. "Alfred rescued me," he announced to the hall. "Alfred took me from the Danes. Alfred is my king."

"And he has our oaths," Odda the Elder added, "and my son had no business making peace with the pagans."

The Danes stepped back. Svein glanced at me, for I was still holding a sword. Then he looked at the boar spears leaning against the wall, judging whether he could snatch one before I attacked him. I lowered the blade. "We have a truce," Harald said loudly.

"We have a truce," I told Svein in Danish.

Svein spat on the bloody rushes. Then he and his standard-bearer took another cautious backward pace.

"But tomorrow," Harald said, "there will be no truce, and we shall come to kill you."

The Danes rode from Ocmundtun. And next day they also went from Cridianton. They could have stayed if they wished. There were more than enough of them to defend Cridianton and make trouble in the shire, but Svein knew he would be besieged and, man by man, worn down until he had no force at all, and so he went north, going to join Guthrum, and I rode to Oxton. The land had never looked more beautiful, the trees were hazed with green, and bullfinches were feasting on the first tight fruit buds, while anemones, stitchwort, and white violets glowed in sheltered spots. Lambs ran from the buck hares in the pastures. The sun shimmered the wide sea-reach of the Uisc and the sky was full of lark song beneath which the foxes took lambs, magpies and jays feasted on other birds' eggs, and plowmen impaled crows at the edges of the fields to ensure a good harvest.

"There'll be butter soon," a woman told me. She really wanted to know if I was returning to the estate, but I was not. I was saying farewell. There were slaves living there, doing their jobs, and I assured them Mildrith would appoint a steward sooner or later. Then I went to the hall and I dug beside the post and found my hoard untouched. The Danes had not come to Oxton. Wirken, the sly priest of Exanmynster, heard I was at the hall and rode a donkey up to the estate. He assured me he had kept a watchful eye on the place, and doubtless he wanted a reward. "It belongs to Mildrith now," I told him.

"The Lady Mildrith? She lives?"

"She lives," I said curtly, "but her son is dead."

"God rest his poor soul," Wirken said, making the sign of the cross. I was eating a scrap of ham and he looked at it hungrily, knowing I broke the rules of Lent. He said nothing, but I knew he was cursing me for being a pagan.

"And the Lady Mildrith," I went on, "would live a chaste life now. She says she will join the sisters in Cridianton."

"There are no sisters in Cridianton," Wirken said. "They're all dead. The Danes saw to that before they left."

"Other nuns will settle there," I said. Not that I cared, for the fate of a small nunnery was none of my business. Oxton was no longer my business. The Danes were my business, and the Danes had gone north and I would follow them.

For that was my life. That spring I was twenty-one years old and for half my life I had been with armies. I was not a farmer. I watched the slaves tearing the couch grass from the home fields and knew the tasks of farming bored me. I was a warrior, and I had been driven from my home of Bebbanburg to the southern edge of England and I think I knew, as Wirken babbled on about how he had guarded the storehouses through the winter, that I was now going north again. Ever north. Back home.

"You lived off these storehouses all winter," I accused the priest.

"I watched them all winter, lord."

"And you got fat as you watched," I said. I climbed into my saddle. Behind me were two bags, ripe with money, and they stayed there as I rode to Exanceaster and found Steapa in The Swan. Next morning, with six other warriors from Ealdorman Odda's guard, we rode north. Our way was marked by pillars of smoke, for Svein was burning and plundering as he went, but we had done what Alfred had wanted us to do. We had driven Svein back to Guthrum, so that now the two largest Danish armies were united. If Alfred had been stronger he might have left them separate and marched against each in turn, but Alfred knew he had only one chance to take back his kingdom, and that was to win one battle. He had to overwhelm all the Danes and destroy them in one blow, and his weapon was an army that existed only in his head. He had sent demands that the fyrd of Wessex would be summoned after Easter and before Pentecost, but no one knew whether it would actually appear. Perhaps we would ride from the swamp and find no one at the meeting place. Or perhaps the fyrd would come, and there would be too few men. The truth was that Alfred was too weak to fight, but to wait longer would only make him weaker. So he had to fight or lose his kingdom.

So we would fight.

ELEVEN

You will have many sons," Iseult told me. It was dark, though a half-moon was hazed by a mist. Somewhere to the northeast a dozen fires burned in the hills, evidence that a strong Danish patrol was watching the swamp. "But I am sorry about Uhtred," she said.

I wept for him then. I do not know why the tears had taken so long to come, but suddenly I was overwhelmed by the thought of his helplessness, his sudden smile, and the pity of it all. Both my half brothers and my half sister had died when they were babies and I do not remember my father crying, though perhaps he did. I do remember my stepmother shrieking in grief, and how my father, disgusted by the sound, had gone hunting with his hawks and hounds.

"I saw three kingfishers yesterday," Iseult said.

Tears were running down my cheeks, blurring the misted moon. I said nothing.

"Hild says the blue of the kingfisher's feathers is for the virgin and the red is for Christ's blood."

"And what do you say?"

"That your son's death is my doing."

"*Wyrd biþ ful āraed,*" I said. Fate is fate. It cannot be changed or cheated. Alfred had insisted I marry Mildrith so I would be tied to Wessex and would put roots deep into its rich soil, but I already had roots

in Northumbria, roots twisted into the rock of Bebbanburg, and perhaps my son's death was a sign from the gods that I could not make a
new home. Fate wanted me to go to my northern stronghold and until
I reached Bebbanburg I would be a wanderer. Men fear wanderers for
they have no rules. The Danes came as strangers, rootless and violent,
and that, I thought, was why I was always happier in their company.
Alfred could spend hours worrying about the righteousness of a law,
whether it concerned the fate of orphans or the sanctity of boundary
markers, and he was right to worry because folk cannot live together
without law, or else every straying cow would lead to bloodshed, but
the Danes hacked through the law with swords. It was easier that way,
though once they had settled a land they started to make their own
laws. "It was not your fault," I said. "You don't command fate."

"Hild says there is no such thing as fate," Iseult said.

"Then Hild is wrong."

"There is only the will of God," Iseult said, "and if we obey that
we go to heaven."

"And if we choose not to," I said, "isn't that fate?"

"That's the devil," she said. "We are sheep, Uhtred, and we
choose our shepherd, a good one or a bad one."

I thought Hild must have soured Iseult with Christianity, but I
was wrong. It was a priest who had come to Æthelingæg while I had
been in Defnascir who had filled her head with his religion. He was a
British priest from Dyfed, a priest who spoke Iseult's native tongue
and also knew both English and Danish. I was ready to hate him as I
hated Brother Asser, but Father Pyrlig stumbled into our hut next
morning booming that he had found five goose eggs and was dying
of hunger. "Dying! That's what I am, dying of starvation!" He looked
pleased to see me. "You're the famous Uhtred, eh? And Iseult tells
me you hate Brother Asser? Then you're a friend of mine. Why Abraham doesn't take Asser to his bosom I do not know, except maybe
Abraham doesn't want the little bastard clinging to his bosom. I
wouldn't. It would be like suckling a serpent, it would. Did I say I
was hungry?"

He was twice my age and a big man, big bellied and big-hearted. His hair stuck out in ungovernable clumps, he had a broken nose, only four teeth, and a broad smile. "When I was a child," he told me, "ever such a little child, I used to eat mud. Can you believe that? Do Saxons eat mud? Of course they do, and I thought I don't want to eat mud. Mud is for toads, it is. So eventually I became a priest. And you know why? Because I never saw a hungry priest! Never! Did you ever see a hungry priest? Nor me!" All this tumbled out without any introduction. Then he spoke earnestly to Iseult in her own tongue and I was sure he was pouring Christianity into her, but then he translated for me. "I'm telling her that you can make a marvelous dish with goose eggs. Break them up, stir them well, and add just a little crumbled cheese. So Defnascir is safe?"

"Unless the Danes send a fleet," I said.

"Guthrum has that in mind," Pyrlig said. "He wants the Danes in Lundene to send their ships to the south coast."

"You know that?"

"I do indeed, I do indeed! He told me! I've just spent ten days in Cippanhamm. I speak Danish, see, because I'm clever, and so I was an ambassador for my king. How about that! Me, who used to eat mud, an ambassador! Crumble the cheese finer, my love. That's right. I had to discover, you see, how much money Guthrum would pay us to bring our spearmen over the hills and start skewering Saxons. Now that's a fine ambition for a Briton, skewering Saxons, but the Danes are pagans, and God knows we can't have pagans loose in the world."

"Why not?"

"It's just a fancy of mine," he said, "just a fancy." He stabbed his finger into a tiny pot of butter, then licked it. "It isn't really sour," he told Iseult, "not very, so stir it in." He grinned at me. "What happens when you put two bulls to a herd of cows?"

"One bull dies."

"There you are! Gods are the same, which is why we don't want pagans here. We're cows and the gods are bulls."

"So we get humped?"

He laughed. "Theology's difficult. Anyway, God is my bull so here I am, telling the Saxons about Guthrum."

"Did Guthrum offer you money?" I asked.

"He offered me the kingdoms of the world! He offered me gold, silver, amber, and jet! He even offered me women, or boys if I had that taste, which I don't. And I didn't believe a single promise he made. Not that it mattered. The Britons aren't going to fight anyway. God doesn't want us to. No! My embassy was all a pretense. Brother Asser sent me. He wanted me to spy on the Danes, see? Then tell Alfred what I saw, so that's what I'm doing."

"Asser sent you?"

"He wants Alfred to win. Not because he loves the Saxons—even Brother Asser isn't that curdled—but because he loves God."

"And will Alfred win?"

"If God has anything to do with it, yes," Pyrlig said cheerfully, then gave a shrug. "But the Danes are strong in men. A big army! But they're not happy, I can tell you that. And they're all hungry. Not starving, mind you, but pulling their belts tighter than they'd like, and now Svein's there so there'll be even less food. Their own fault, of course. Too many men in Cippanhamm! And too many slaves! They have scores of slaves. But he's sending the slaves to Lundene, to sell them there. They need some baby eels, eh? That'll fatten them up." The elvers were swarming into the Sæfern Sea and slithering up the shallow waterways of the swamp where they were being netted in abundance. There was no hunger in Æthelingæg, not if you gorged on elvers. "I caught three basketfuls yesterday," Pyrlig said happily, "and a frog. It had a face just like Brother Asser so I gave it a blessing and threw it back. Don't just stir the eggs, girl! Beat them! I hear your son died?"

"Yes," I answered stiffly.

"I am sorry," he said with genuine feeling, "I am truly sorry, for to lose a child is a desperate hard thing. I sometimes think God must like children. He takes so many to him. I believe there's a garden in heaven, a green garden where children play all the time. He's got two sons of mine up there, and I tell you, the youngest must be making the angels scream. He'll be pulling the girls' hair and beating up the other boys like they were goose eggs."

"You lost two sons?"

"But I kept three others and four daughters. Why do you think I'm never home?" He grinned at me. "Noisy little things they are, children, and such appetites! Sweet Jesus, they'd eat a horse a day if they could! There are some folk who say priests shouldn't marry and there are times I think they're right. Do you have any bread, dearest?"

Iseult pointed to a net hanging from the roof. "Cut the mold off," she told me.

"I like to see a man obeying a woman," Father Pyrlig said as I fetched the loaf.

"Why's that?" I asked.

"Because it means I'm not alone in this sorry world. Good God, but that Ælswith was weaned on gall juice, wasn't she? Got a tongue in her like a starving weasel! Poor Alfred."

"He's happy enough."

"Good God, man, that's the last thing he is! Some folk catch God like a disease, and he's one of them. He's like a cow after winter, he is."

"He is?"

"You know when the late spring grass comes in? All green and new and rich? And you put the poor cow out to eat and she blows up like a bladder? She's nothing but shit and wind and then she gets the staggers and drops down dead if you don't take her off the grass for a while. That's Alfred. He got too much of the good green grass of God, and now he's sick on it. But he's a good man, a good man. Too thin, he is, but good. A living saint, no less. Ah, good girl, let's eat." He scooped some of the eggs with his fingers, then passed the pot to me. "Thank God it's Easter next week," he said with his mouth full so that scraps of egg lodged in his huge beard, "and then we can eat meat again. I'm wasting away without meat. You know Iseult will be baptized at Easter?"

"She told me," I said shortly.

"And you don't approve? Just think of it as a good wash. Then maybe you won't mind so much."

I was not in Æthelingæg for Iseult's baptism, nor did I wish to be, for I knew Easter with Alfred would be nothing but prayers and psalms and priests and sermons. Instead I took Steapa and fifty men up into the hills, going toward Cippanhamm, for Alfred had ordered that the Danes were to be harried mercilessly in the next few weeks. He had decided to assemble the fyrd of Wessex close to Ascension Day, which was just six weeks away, and those were the weeks in which Guthrum would be hoping to revive his hungry horses on the spring grass, and so we rode to ambush Danish forage parties. Kill one forage party and the next must be protected by a hundred extra horsemen, and that wearies the horses even more and so requires still more forage. It worked for a while, but then Guthrum began sending his foragers north into Mercia where they were not opposed.

It was a time of waiting. There were two smiths in Æthelingæg now and, though neither had all the equipment they wanted, and though fuel for their furnaces was scarce, they were making good spear points. One of my jobs was to take men to cut ash poles for the spear shafts. Alfred was writing letters, trying to discover how many men the shires could bring to battle, and he sent priests to Frankia to persuade the thegns who had fled there to return. More spies came from Cippanhamm confirming that Svein had joined Guthrum and that Guthrum was strengthening his horses and raising men from the Danish parts of England. He was ordering his West Saxon allies like Wulfhere to arm their men, and warning his garrisons in Wintanceaster, Readingum, and Baðum that they must be ready to abandon their ramparts and march to his aid. Guthrum had his own spies and must have known Alfred was planning to assemble an army, and I dare say he welcomed that news for such an army would be Alfred's last hope and, should Guthrum destroy the fyrd, Wessex would fall, never to rise again.

Æthelingæg seethed with rumor. Guthrum, it was said, had five thousand men. Ships had come from Denmark and a new army of

Norsemen had sailed from Ireland. The Britons were marching. The fyrd of Mercia was on Guthrum's side, and it was said the Danes had set up a great camp at Cracgelad on the river Temes where thousands of Mercian troops, both Danish and Saxon, were assembling. The rumors of Guthrum's strength crossed the sea and Wilfrith of Hamptonscir wrote from Frankia begging Alfred to flee Wessex. "Take ship to this coast," he wrote, "and save your family."

Leofric rarely rode on patrols with us but stayed in Æthelingæg, for he had been named commander of the king's bodyguard. He was proud of that, as he should have been, for he had been peasant born and he could neither read nor write, and Alfred usually insisted that his commanders were literate. Eanflæd's influence was behind the appointment, for she had become a confidante of Ælswith. Alfred's wife went nowhere without Eanflæd—even in church the onetime whore sat just behind Ælswith—and when Alfred held court, Eanflæd was always there. "The queen doesn't like you," Eanflæd told me one rare day when I found her alone.

"She's not a queen," I said. "Wessex doesn't have queens."

"She should be a queen," she said indignantly. "It would be right and proper." She was carrying a heap of plants and I noticed her forearms were a pale green. "Dyeing," she explained brusquely, and I followed her to where a great cauldron was bubbling on a fire. She threw in the plants and began stirring the mess in the pot. "We're making green linen," she said.

"Green linen?"

"Alfred must have a banner," she said indignantly. "He can't fight without a banner." The women were making two banners. One was the great green dragon flag of Wessex, while the other bore the cross of Christianity. "Your Iseult's working on the cross," Eanflæd told me.

"I know."

"You should have been at her baptism."

"I was killing Danes."

"But I'm glad she's baptized. Come to her senses, she has."

In truth, I thought, Iseult had been battered into Christianity. For weeks she had endured the rancor of Alfred's churchmen, had been accused of witchcraft and of being the devil's instrument, and it had worn her down. Then came Hild with her gentler Christianity, and Pyrlig who spoke of God in Iseult's tongue, and Iseult had been persuaded. That meant I was the only pagan left in the swamp and Eanflæd glanced pointedly at my hammer amulet. She said nothing of it, instead asking me whether I truly believed we could defeat the Danes.

"Yes," I said confidently, though of course I did not know.

"How many men will Guthrum have?"

I knew the questions were not Eanflæd's but Ælswith's. Alfred's wife wanted to know if her husband had any chance of survival or whether they should take the ship we had captured from Svein and sail to Frankia. "Guthrum will lead four thousand men," I said, "at least."

"At least?"

"Depends how many come from Mercia," I said, then thought for a heartbeat, "but I expect four thousand."

"And Wessex?"

"The same," I said. I was lying. With enormous luck we could assemble three thousand, but I doubted it. Two thousand? Not likely, but possible. My real fear was that Alfred would raise his banner and no one would come, or that only a few hundred men would arrive. We could lead three hundred from Æthelingæg, but what could three hundred do against Guthrum's great army?

Alfred also worried about numbers, and he sent me to Hamptonscir to discover how much of the shire was occupied by the Danes. I found them well entrenched in the north, but the south of the shire was free of them and in Hamtun, where Alfred's fleet was based, the war ships were still drawn up on the beach. Burgweard, the fleet's commander, had over a hundred men in the town, all that was left of his crews, and he had them manning the palisade. He claimed he could not leave Hamtun for fear that the Danes would attack and capture the ships, but I had Alfred's scrap of parchment with his dragon seal on it, and I used it to order him to keep thirty men to protect the ships and bring the rest to Alfred.

"When?" he asked gloomily.

"When you're summoned," I said, "but it will be soon. And you're to raise the local fyrd, too. Bring them."

"And if the Danes come here?" he asked. "If they come by sea?"

"Then we lose the fleet," I said, "and we build another."

His fear was real enough. Danish ships were off the south coast again. For the moment, rather than attempt an invasion, they were being Vikings. They landed, raided, raped, burned, stole, and went to sea again, but they were numerous enough for Alfred to worry that a whole army might land somewhere on the coast and march against him. We were harassed by that fear and by the knowledge that we were few and the enemy numerous, and that the enemy's horses were fattening on the new grass.

"Ascension Day," Alfred announced on the day I returned from Hamtun.

That was the day we should be ready in Æthelingæg, and on the Sunday after, which was the Feast of Saint Monica, we would gather the fyrd, if there was a fyrd. Reports said the Danes were readying to march and it was plain they would launch their attack south toward Wintanceaster, the town that was the capital of Wessex, and to protect it, to bar Guthrum's road south, the fyrd would gather at Egbert's Stone. I had never heard of the place, but Leofric assured me it was an important spot, the place where King Egbert, Alfred's grandfather, had given judgments. "It isn't one stone," he said, "but three."

"Three?"

"Two big pillars and another boulder on top. The giants made it in the old days."

And so the summons were issued. Bring every man, the parchments instructed, bring every weapon, and say your prayers, for what is left of Wessex will meet at Egbert's Stone to carry battle to the Danes, and no sooner was the summons sent than disaster struck. It came just a week before the fyrd was to gather.

Huppa, Ealdorman of Thornsæta, wrote that forty Danish ships were off his coast and that he dared not lead the fyrd away from their threat. Worse, because the Danes were so numerous, he had begged Harald of Defnascir to lend him men.

That letter almost destroyed Alfred's spirits. He had clung to his dream of surprising Guthrum by raising an unexpectedly powerful army, but all his hopes were now shredding away. He had always been thin, but suddenly he looked haggard and he spent hours in the church, wrestling with God, unable to understand why the Almighty had so suddenly turned against him. And two days after the news of the Danish fleet, Svein of the White Horse led three hundred mounted men in a raid against the hills on the edge of the swamp and, because scores of men from the Sumorsæte fyrd had gathered in Æthelingæg, Svein discovered and stole their horses. We had neither the room nor the forage to keep many horses in Æthelingæg itself, and so they were pastured beyond the causeway, and I watched from the fort as Svein, riding a white horse and wearing his white-plumed helmet and white cloak, rounded up the beasts and drove them away. There was nothing I could do to stop him. I had twenty men in the fort and Svein was leading hundreds.

"Why were the horses not guarded?" Alfred wanted to know.

"They were," Wiglaf, Ealdorman of Sumorsæte, said, "and the guards died." He saw Alfred's anger, but not his despair. "We haven't seen a Dane here for weeks!" he pleaded. "How were we to know they'd come in force?"

"How many men died?"

"Only twelve."

"Only?" Alfred asked, wincing. "And how many horses lost?"

"Sixty-three."

On the night before Ascension Day Alfred walked beside the river. Beocca, faithful as a hound, followed him at a distance, wanting to offer the king God's reassurance, but instead Alfred called to me. There was a moon, and its light shadowed his cheeks and made his pale eyes look almost white. "How many men will we have?" he asked abruptly.

I did not need to think about the answer. "Two thousand."

He nodded. He knew that number as well as I did.

"Maybe a few more," I suggested.

He grunted at that. We would lead three hundred and fifty men from Æthelingæg and Wiglaf, Ealdorman of Sumorsæte, had promised a thousand, though in truth I doubted if that many would come. The fyrd of Wiltunscir had been weakened by Wulfhere's defection, but the southern part of the shire should yield five hundred men, and we could expect some from Hamptonscir, but beyond that we would depend on whatever few men made it past the Danish garrisons that now ringed the heartland of Wessex. If Defnascir and Thornsæta had sent their fyrds, then we would have numbered closer to four thousand, but they were not coming.

"And Guthrum?" Alfred asked. "How many will he have?"

"Four thousand."

"More like five," Alfred said. He stared at the river that was running low between the muddy banks. The water rippled about the wicker fish traps. "So should we fight?"

"What choice do we have?"

He smiled at that. "We have a choice, Uhtred," he assured me. "We can run away. We can go to Frankia. I could become a king in exile and pray that God brings me back."

"You think God will?"

"No," he admitted. If he ran away, then he knew he would die in exile.

"So we fight," I said.

"And on my conscience," he said, "I will forever bear the weight of all those men who died in a hopeless cause. Two thousand against five thousand? How can I justify leading so few against so many?"

"You know how."

"So I can be king?"

"So that we are not slaves in our own land," I said.

He pondered that for a while. An owl flew low overhead, a sudden surprise of white feathers and the rush of air across stubby wings. It was an omen, I knew, but of what kind? "Perhaps we are being punished," Alfred said.

"For what?"

"For taking the land from the Britons?"

That seemed nonsense to me. If Alfred's god wanted to punish him for his ancestors having taken the land from the Britons, then why send the Danes? Why not send the Britons? God could resurrect Arthur and let his people have their revenge, but why send a new people to take the land? "Do you want Wessex or not?" I asked harshly.

He said nothing for a while, then gave a sad smile. "In my conscience," he said, "I can find no hope for this fight, but as a Christian I must believe we can win it. God will not let us lose."

"Nor will this," I said, and I slapped Serpent-Breath's hilt.

"So simple?" he asked.

"Life is simple," I said. "Ale, women, sword, and reputation. Nothing else matters."

He shook his head and I knew he was thinking about God and prayer and duty, but he did not argue. "So if you were I, Uhtred," he said, "would you march?"

"You've already made up your mind, lord," I said, "so why ask me?"

He nodded. A dog barked in the village and he turned to stare at the cottages and the hall and the church he had made with its tall alder cross. "Tomorrow," he said, "you will take a hundred horsemen and patrol ahead of the army."

"Yes, lord."

"And when we meet the enemy," he went on, still staring at the cross, "you will choose fifty or sixty men from the bodyguard. The best you can find. And you will guard my banners."

He did not say more, nor did he need to. What he meant was that I was to take the best warriors, the most savage men, the dangerous warriors who loved battle, and I was to lead them in the place where the fight would be hardest, for an enemy loves to capture his foe's banners. It was an honor to be asked and, if the battle was lost, an almost certain death sentence. "I shall do it gladly, lord," I said, "but ask a favor of you in return."

"If I can," he said guardedly.

"If you can," I said, "don't bury me. Burn my body on a pyre, and put a sword in my hand."

He hesitated, then nodded, knowing he had agreed to a pagan funeral. "I never told you," he said, "that I am sorry about your son."

"So am I, lord."

"But he is with God, Uhtred, he is assuredly with God."

"So I'm told, lord, so I'm told."

And next day we marched. Fate is inexorable, and though numbers and reason told us we could not win, we dared not lose, and so we marched to Egbert's Stone.

We marched with ceremony. Twenty-three priests and eighteen monks formed our vanguard and chanted a psalm as they led Alfred's forces away from the fort guarding the southern trackway and east toward the heartland of Wessex.

They chanted in Latin so the words meant nothing to me, but Father Pyrlig had been given use of one of Alfred's horses and, dressed in a leather coat and with a great sword strapped to his side and with a stout-shafted boar spear on one shoulder, he rode alongside me and translated the words. " 'God,' " he said, " 'you have abandoned me, you have scattered us, you are angry with us, now turn to us again.' That sounds a reasonable request, doesn't it? You've kicked us in the face, so now give us a cuddle, eh?"

"It really means that?"

"Not the bit about kicks and cuddles. That was me." He grinned at me. "I do miss war. Isn't that a sin?"

"You've seen war?"

"Seen it? I was a warrior before I joined the church! Pyrlig the Fearless, they called me. I killed four Saxons in a day once. All by myself and I had nothing but a spear. And they had swords and shields, they did. Back home they made a song about me, but mind you, the Britons will sing about anything. I can sing you the song, if you like? It tells how I slaughtered three hundred and ninety-four Saxons in one day, but it's not entirely accurate."

"So how many did you kill?"

"I told you. Four." He laughed.

"So how did you learn English?"

"My mother was a Saxon, poor thing. She was taken in a raid on Mercia and became a slave."

"So why did you stop being a warrior?"

"Because I found God, Uhtred. Or God found me. And I was becoming too proud. Songs about yourself go to your head and I was wickedly proud of myself, and pride is a terrible thing."

"It's a warrior's weapon," I said.

"It is indeed," he agreed, "and that is why it is a terrible thing, and why I pray God purges me of it."

We were well ahead of the priests now, climbing toward the nearest hilltop to look north and east for the enemy, but the churchmen's voices followed us, their chant strong in the morning air. " 'Through God we shall do bravely,' " Father Pyrlig interpreted for me, " 'and God shall trample down our enemies.' Now there's a blessed thought for a fine morning, Lord Uhtred!"

"The Danes are saying their own prayers, father."

"But to what god, eh? No point in shouting at a deaf man, is there?" He curbed his horse at the hilltop and stared northward. "Not even a mouse stirring."

"The Danes are watching," I said. "We can't see them, but they can see us."

If they were watching, then what they could see was Alfred's three hundred and fifty men riding or walking away from the swamp, and in the distance another five or six hundred men who were the fyrd from the western part of Sumorsæte who had camped to the south of the swamp and now marched to join our smaller column. Most of the men from Æthelingæg were real soldiers, trained to stand in the shield wall, but we also had fifty of the marshmen. I had wanted Eofer, the strong bowman, to come with us, but he could not fight without his niece telling him what to do and I had no intention of taking a child to war and so we had left Eofer behind. A good number of women and chil-

dren were following the column, though Alfred had sent Ælswith and
his children south to Scireburnan under a guard of forty men. We
could hardly spare those men, but Alfred insisted his family go.
Ælswith was to wait in Scireburnan and, if news came that her husband was defeated and the Danes victorious, she was to flee south to
the coast and find a ship that would take her to Frankia. She was also
instructed to take with her whatever books she could find in Scireburnan, for Alfred reckoned the Danes would burn every book in Wessex
and so Ælswith was to rescue the gospel books and saints' lives and
church fathers and histories and philosophers and thus raise her son
Edward to become a learned king in exile.

Iseult was with the army, walking with Hild and with Eanflæd who
had upset Ælswith by insisting on following Leofric. The women led
packhorses, which carried the army's shields, food, and spare spears.
Nearly every woman was equipped with some kind of weapon. Even
Hild, a nun, wanted to take revenge on the Danes who had whored her
and so carried a long, narrow-bladed knife. "God help the Danes,"
Father Pyrlig had said when he saw the women gather, "if that lot get
among them."

He and I now trotted eastward. I had horsemen ringing the column, riding on every crest, staying in sight of one another, and ready
to signal if they saw any sign of the enemy, but there was none. We
rode or marched under a spring sky, through a land bright with flowers, and the priests and monks kept up their chanting, and sometimes
the men behind, who followed Alfred's two standard-bearers, would
start singing a battle song.

Father Pyrlig beat his hand in time to the singing, then gave me a
big grin. "I expect Iseult sings to you, doesn't she?"

"She does."

"We Britons love to sing! I must teach her some hymns." He saw
my sour look and laughed. "Don't worry, Uhtred, she's no Christian."

"She isn't?" I asked, surprised.

"Well, she is, for the moment. I'm sorry you didn't come to her
baptism. It was cold, that water! Fair froze me!"

"She's baptized," I said, "but you say she's no Christian?"

"She is and she isn't," Pyrlig said with a grin. "She is now, see, because she's among Christians. But she's still a shadow queen, and she won't forget it."

"You believe in shadow queens?"

"Of course I do! Good God, man! She is one!" He made the sign of the cross.

"Brother Asser called her a witch," I said, "a sorceress."

"Well, he would, wouldn't he? He's a monk! Monks don't marry. He's terrified of women, Brother Asser, unless they're very ugly, and then he bullies them. But show him a pretty young thing and he goes all addled. And of course he hates the power of women."

"Power?"

"Not just tits, I mean. God knows tits are powerful enough, but the real thing. Power! My mother had it. She was no shadow queen, mind you, but she was a healer and a scryer."

"She saw the future?"

He shook his head. "She knew what was happening far off. When my father died she suddenly screamed. Screamed fit to kill herself because she knew what had happened. She was right, too. The poor man was cut down by a Saxon. But she was best as a healer. Folk came to her from miles around. It didn't matter that she was born a Saxon, they'd walk for a week to fetch the touch of her hand. Me? I got it for free! She banged me about, she did, and I dare say I deserved it, but she was a rare healer. And, of course, the priests don't like that."

"Why not?"

"Because we priests tell folk that all power comes from God, and if it doesn't come from God then it must be evil, see? So when folk are ill the church wants them to pray and to give the priests money. Priests don't like it when they don't understand things, and they don't like folk going to the women to be healed. But what else are folk to do? My mother's hand, God rest her Saxon soul, was better than any prayer! Better than the touch of the sacraments! I wouldn't stop folk going to see a healer. I'd tell them to!" He stopped talking because I raised my hand. I had seen movement on a hillside to the north, but it was only a deer. I dropped my hand and kicked the horse on. "Now your Iseult," Pyrlig went on, "she's been raised with the power and she won't lose it."

"Didn't the baptism wash it out of her?"

"Not at all! It just made her a bit colder and cleaner. Nothing wrong with a scrub once or twice a year." He laughed. "But she was frightened back there in the swamp. You were gone and all around were Saxons and they were spitting that she was a pagan, so what did you think she would do? She wants to be one of them, she wants folk to stop spitting at her, so she said she'd be baptized. And maybe she really is a Christian? I'd praise God for that mercy, but I'd rather praise him for making her happy."

"You don't think she is?"

"Of course she isn't! She's in love with you!" He laughed. "And being in love with you means living among the Saxons, doesn't it? Poor girl. She's like a beautiful young hind that finds herself living among grunting pigs."

"What a gift for words you have," I said.

He laughed, delighted with his insult. "Win your war, Lord Uhtred," he said, "then take her away from us priests and give her lots of children. She'll be happy, and one day she'll be truly wise. That's the women's real gift, to be wise, and not many men have it."

And my gift was to be a warrior, though there was no fighting that day. We saw no Danes, though I was certain they had seen us and that by now Guthrum would have been told that Alfred had at last come from the swamp and was marching inland. We were giving him the opportunity to destroy us, to finish Wessex, and I knew that the Danes would be readying to march on us.

We spent that night in an earthen fort built by the old people, and next morning went north and east through a hungry land. I rode ahead, going into the hills to look for the enemy, but again the world seemed empty. Rooks flew, hares danced, and cuckoos called from the woods that were thick with bluebells, but there were no Danes. I rode along a high ridge, gazing northward, and saw nothing, and when the sun was at its height I turned east. There were ten of us in my band and our guide was a man from Wiltunscir who knew the country and he led us toward the valley of the Wilig where Egbert's Stone stands.

A mile or so short of the valley we saw horsemen, but they were to the south of us and we galloped across ungrazed pastureland to find it was Alfred, escorted by Leofric, five soldiers, and four priests. "Have you been to the stone?" Alfred called out eagerly as we closed on him.

"No, lord."

"Doubtless there are men there," he said, disappointed that I could not bring him news.

"I didn't see any Danes either, lord."

"It'll take them two days to organize," he said dismissively. "But they'll come! They'll come! And we shall beat them!" He twisted in his saddle to look at Father Beocca, who was one of the priests. "Are you sore, father?"

"Mightily sore, lord."

"You're no horseman, Beocca, no horseman, but it's not much farther. Not much farther, then you can rest!" Alfred was in a feverish mood. "Rest before we fight, eh! Rest and pray, father, then pray and fight. Pray and fight!" He kicked his horse into a gallop and we pounded after him through a pink-blossomed orchard and up a slope, then across a long hilltop where the bones of dead cattle lay in the new grass. White mayflower edged the woods at the foot of the hill and a hawk slanted away from us, sliding across the valley toward the charred remains of a barn.

"It's just across the crest, lord!" my guide shouted at me.

"What is?"

"Defereal, lord!"

Defereal was the name of the settlement in the valley of the river Wilig where Egbert's Stone waited, and Alfred now spurred his horse so that his blue cloak flapped behind him. We were all galloping, spread across the hilltop, racing to be the first over the crest to see the Saxon forces. Then Father Beocca's horse stumbled. He was, as Alfred said, a bad horseman, but that was no surprise for he was both lame and palsied, and when the horse tipped forward Beocca tumbled from the saddle. I saw him rolling in the grass and turned my horse back. "I'm not hurt," he shouted at me. "Not hurt! Not much. Go on, Uhtred, go on!"

I caught his horse. Beocca was on his feet now, limping as fast as he could to where Alfred and the other horsemen stood in a line gazing into the valley beyond. "We should have brought the banners," Beocca said as I gave him his reins.

"The banners?"

"So the fyrd knows their king has come," he said breathlessly. "They should see his banners on the skyline, Uhtred, and know he has come. The cross and the dragon, eh? *In hoc signo!* Alfred will be the new Constantine, Uhtred, a warrior of the cross! *In hoc signo*, God be praised, God be praised, God indeed be mightily praised."

I had no idea what he meant, nor did I care.

For I had reached the hilltop and could stare down into the long lovely valley of the Wilig.

Which was empty.

Not a man in sight. Just the river and the willows and the water meadows and the alders and a heron flying and the grass bending in the wind and the triple stone of Egbert on a slope above the Wilig where an army was supposed to gather. And there was not one man there. Not one single man in sight. The valley was empty.

The men we had brought from Æthelingæg straggled into the valley, and with them now was the fyrd of Sumorsæte. Together they numbered just over a thousand men, and about half were equipped to fight in the shield wall while the rest were only good for shoving the front ranks forward or dealing with the enemy wounded or, more likely, dying.

I could not face Alfred's disappointment. He said nothing about it, but his thin face was pale and set hard as he busied himself with deciding where the thousand men should camp and where our few horses should be pastured. I rode up a high hill that lay to the north of the encampment, taking a score of men including Leofric, Steapa, and Father Pyrlig. The hill was steep, though that had not stopped the old people from making one of their strange graves high on the slope. The grave was a long mound and Pyrlig made a wide detour rather than ride past it. "Full of dragons, it is," he explained to me.

"You've seen a dragon?" I asked.

"Would I be alive if I had? No one sees a dragon and lives!"

I turned in the saddle and stared at the mound. "I thought folk were buried there?"

"They are! And their treasures! So the dragon guards the hoard. That's what dragons do. Bury gold and you hatch a dragon, see?"

The horses had a struggle to climb the steep slope, but at its summit we were rewarded by a stretch of firm turf that offered views far to the north. I had climbed the hill to watch for Danes. Alfred might believe that it would be two or three days before we saw them, but I expected their scouts to be close and it was possible that a war band might try to harass the men camping by the Wilig.

Yet I saw no one. To the northeast were great downs, sheep hills, while straight ahead was the lower ground where the cloud shadows raced across fields and over blossoming mayflower and darkened the bright green new leaves.

"So what happens now?" Leofric asked me.

"You tell me."

"A thousand men? We can't fight the Danes with a thousand men."

I said nothing. Far off, on the northern horizon, there were dark clouds.

"We can't even stay here!" Leofric said. "So where do we go?"

"Back to the swamp?" Father Pyrlig suggested.

"The Danes will bring more ships," I said, "and eventually they'll capture the swamp. If they send a hundred ships up the rivers, then the swamp is theirs."

"Go to Defnascir," Steapa growled.

And the same thing would happen there, I thought. We would be safe for a time in Defnascir's tangle of hills and woods, but the Danes would come and there would be a succession of little fights and bit by bit Alfred would be bled to death. And once the Danes across the sea knew that Alfred was pinned into a corner of Wessex, they would bring more ships to take the good land that he could not hold. And that, I thought, was why he had been right to try to end the war in one blow, because he dared not let Wessex's weakness become known.

Except we were weak. We were a thousand men. We were pathetic. We were dreams fallen to earth, and suddenly I began to laugh. "What is it?" Leofric asked.

"I was thinking that Alfred insisted I learn to read," I said, "and for what?"

He smiled, remembering. It was one of Alfred's rules that every man who commanded sizable bodies of troops must be able to read, though it was a rule he had ignored when he made Leofric commander of his bodyguard. It seemed funny at that moment. All that effort so I could read his orders and he had never sent me one. Not one. "Reading is useful," Pyrlig said.

"What for?"

He thought about it. The wind gusted, flapping his hair and beard. "You can read all those good stories in the gospel book," he suggested brightly, "and the saints' lives! How about those, eh? They're full of lovely things, they are. There was Saint Donwen! Beautiful woman she was, and she gave her lover a drink that turned him into ice."

"Why did she do that?" Leofric asked.

"Didn't want to marry him, see?" Pyrlig said, trying to cheer us up, but no one wanted to hear more about the frigid Saint Donwen so he turned and stared northward. "That's where they'll come from, is it?" he asked.

"Probably," I said, and then I saw them, or thought I did. There was a movement on the far hills, something stirring in the cloud shadow, and I wished Iseult was on the hilltop for she had remarkable eyesight, but she would have needed a horse to climb to that summit and there were no horses to spare for women. The Danes had thousands of horses, all the beasts they had captured from Alfred at Cippanhamm, and all the animals they had stolen across Wessex, and now I was watching a group of horsemen on that far hill. Scouts, probably, and they would have seen us. Then they were gone. It had been a glimpse, no more, and so far away that I could not be sure of what I had seen. "Or perhaps they won't come at all," I went on. "Perhaps they'll march around us. Capture Wintanceaster and everywhere else."

"The bastards will come," Leofric said grimly, and I thought he was probably right. The Danes would know we were here, they would want to destroy us, and after that all would be easy for them.

Pyrlig turned his horse as if to ride back down to the valley, then paused. "So it's hopeless, is it?" he asked.

"They'll outnumber us four or five to one," I said.

"Then we must fight harder!"

I smiled. "Every Dane who comes to Britain, father," I explained, "is a warrior. The farmers stay in Denmark, but the wild men come here. And us? We're nearly all farmers and it takes three or four farmers to beat down a warrior."

"You're warriors," he said, "all of you! You're warriors! You all know how to fight! You can inspire men, and lead men, and kill your enemies. And God is on your side. With God on your side, who can beat you, eh? You want a sign?"

"Give me a sign," I said.

"Then look," he said, and pointed down to the Wilig, and I turned my horse and there, in the afternoon sun, was the miracle we had wanted. Men were coming. Men in their hundreds. Men from the east and men from the south, men streaming down from the hills, men of the West Saxon fyrd, coming at their king's command to save their country.

"Now it's only two farmers to one warrior!" Pyrlig said cheerfully.

"Up to our arseholes," Leofric said.

But we were not alone any longer. The fyrd was gathering.

TWELVE

ost men came in large groups, led by their thegns, while
others arrived in small bands, but together they swelled into
an army. Arnulf, Ealdorman of Suth Seaxa, brought close to four hundred men and apologized that it could not be more, but there were
Danish ships off his coast and he had been forced to leave some of his
fyrd to guard the shore. The men of Wiltunscir had been summoned
by Wulfhere to join Guthrum's army, but the reeve, a grim man named
Osric, had scoured the southern part of the shire and over eight hundred men had ignored their ealdorman's summons and came to Alfred
instead. More arrived from the distant parts of Sumorsæte to join
Wiglaf's fyrd that now numbered a thousand men, while half that
many came from Hamptonscir, including Burgweard's garrison in
which were Eadric and Cenwulf, crewmen from the *Heahengel*, and
both embraced me, and with them was Father Willibald, eager and
nervous. Almost every man came on foot, weary and hungry, with
their boots falling apart, but they had swords and axes and spears and
shields, and by midafternoon there were close to three thousand men
in the Wilig valley and more were still coming as I rode toward the distant hill where I thought I had seen the Danish scouts.

Alfred sent me and, at the last moment, Father Pyrlig had offered
to accompany me and Alfred had looked surprised, appeared to think

about it for a heartbeat, then nodded assent. "Bring Uhtred home safe, father," he had said stiffly.

I said nothing as we rode through the growing camp, but once we were on our own I gave Pyrlig a sour look. "That was all arranged," I said.

"What was?"

"You coming with me. He had your horse already saddled! So what does Alfred want?"

Pyrlig grinned. "He wants me to talk you into becoming a Christian, of course. The king has great faith in my powers of speech."

"I am a Christian," I said.

"Are you now?"

"I was baptized, wasn't I? Twice, as it happens."

"Twice! Doubly holy, eh? How come you got it twice?"

"Because my name was changed when I was a child and my stepmother thought heaven wouldn't recognize me under my old name."

He laughed. "So they washed the devil out of you the first time and slopped him back in the second?" I said nothing to that and Pyrlig rode in silence for a time. "Alfred wants me to make you a good Christian," he said after a while, "because he wants God's blessing."

"He thinks God will curse us because I'm fighting for him?"

Pyrlig shook his head. "He knows, Uhtred, that the enemy are pagans. If they win, then Christ is defeated. This isn't just a war over land, it's a war about God. And Alfred, poor man, is Christ's servant, so he will do all he can for his master, and that means trying to turn you into a pious example of Christian humility. If he can get you onto your knees, then it'll be easy to make the Danes grovel."

I laughed, as he had meant me to. "If it encourages Alfred," I said, "tell him I'm a good Christian."

"I planned to tell him that anyway," Pyrlig said, "just to cheer him up, but in truth I wanted to come with you."

"Why?"

"Because I miss this life. God, I miss it! I loved being a warrior. All that irresponsibility! I relished it. Kill and make widows, frighten children! I was good at it, and I miss it. And I was always good at scouting. We'd see you Saxons blundering away like swine and you never knew we were watching you. Don't worry, I'm not going to talk Christ into you, whatever the king wants."

Our job was to find the Danes, if they were near. Alfred had marched to the Wilig valley to block any advance Guthrum made into the heartland of Wessex, but he still feared that the Danes might resist the lure of destroying his small army and instead march around us to take southern Wessex, which would leave us stranded and surrounded by Danish garrisons. That uncertainty meant Alfred was desperate for news of the enemy, so Pyrlig and I rode north and east up the valley until we came to where a smaller river flowed south into the Wilig and we followed that lesser stream past a large village that had been reduced to ashes. The small river passed through good farmland, but there were no cattle, no sheep, and the fields were unplowed and thick with weeds. We went slowly, for the horses were tired and we were well north of our army now. The sun was low in the west, though it was early May so the days were lengthening. There were mayfly on the river and trout rising to them, and then a scuffling sound made us both pause, but it was only a pair of otter cubs scrabbling down to the water through the roots of a willow. Doves were nesting in the blackthorn, and warblers called from the riverbank, and somewhere a woodpecker drummed intermittently. We rode in silence for a while, turning away from the river to go into an orchard where wrynecks sang among the pink blossom.

Pyrlig curbed his horse under the trees and pointed at a muddy patch in the grass and I saw hoofprints sifted with fallen petals. The prints were fresh and there were a lot of them. "Bastards were here, weren't they?" he said. "And not so long ago."

I looked up the valley. There was no one in sight. The hills rose steeply on either side with thick woods on their lower slopes. I had the sudden uncomfortable feeling that we were being watched, that we were blundering and the wolves were close. "If I were a Dane,"

Pyrlig spoke softly, and I suspected he shared my discomfort, "I'd be over there." He jerked his head to the western trees.

"Why?"

"Because when you saw them, they saw us, and that's the way to where they saw us. Does that make sense?" He laughed wryly. "I don't know, Uhtred, I just think the bastards are over there."

So we went east. We rode slowly, as if we did not have a care in the world, but once we were in the woods we turned north. We both searched the ground for more hoofprints, but saw none, and the feeling of being watched had gone now, though we did wait for a long time to see if anyone was following us. There was only the wind in the trees. Yet I knew the Danes were near, just as a hall's hounds know when there are wolves in the nearby darkness. The hair on their necks stands up, they bare their teeth, they quiver.

We came to a place where the trees ended and we dismounted, tied the horses, and went to the wood's edge and just watched.

And at last we saw them.

Thirty or forty Danes were on the valley's farther side, above the woods, and they had plainly ridden to the top of the hills, looked southward, and were now coming back. They were scattered in a long line that was riding down into the woods. "Scouting party," Pyrlig said.

"They can't have seen much from that hilltop."

"They saw us," he said.

"I think so."

"But they didn't attack us?" He was puzzled. "Why not?"

"Look at me," I said.

"I get a treat every day."

"They thought I was a Dane," I said. I was not in mail and had no helmet, so my long hair fell free down my leather-clad back and my arms were bright with rings. "And they probably thought you were my performing bear," I added.

He laughed. "So shall we follow them?"

The only risk was crossing the valley, but if the enemy saw us, they would probably still assume I was a fellow Dane, so we cantered over the open ground, then rode up into the farther woods. We heard the Danes before we saw them. They were careless, talking and laughing, unaware that any Saxons were close. Pyrlig tucked his crucifix beneath his leather coat. Then we waited until we were sure the last of the Danes had passed before kicking the horses uphill to find their tracks and so follow them. The shadows were lengthening, and that made me think that the Danish army must be close for the scouting party would want to reach safety before dark, but as the hilly country flattened we saw that they had no intention of joining Guthrum's forces that evening. The patrolling Danes had their own camp, and as we approached it we were nearly caught by another group of mounted scouts who rode in from the east. We heard the newcomers and swerved aside into a thicket and watched a dozen men ride by, and then we dismounted and crept through the trees to see how many enemy were in the camp.

There were perhaps a hundred and fifty Danes in a small pasture. The first fires were being lit, suggesting they planned to spend the night where they were. "All scouting parties," Pyrlig suggested.

"Confident bastards," I said. These men had been sent ahead to explore the hills, and they felt safe to camp in the open countryside, sure that no Saxon would attack them. And they were right. The West Saxon army was a long way south and we had no war band in the area, and so the Danes would have a quiet night and, in the morning, their scouts would ride again to watch Alfred's movements.

"But if they're here," Pyrlig suggested, "then it means Guthrum is following them."

"Maybe," I said. Or perhaps Guthrum was marching well to the east or west and had sent these men to make sure that Alfred was ignorant of his movements.

"We should go back," Pyrlig said. "Be dark soon."

But I had heard voices and I held up a hand to silence him, and then went to my right, keeping to the places where the undergrowth was thickest, and heard what I thought I had heard. English. "They've got Saxons here," I said.

"Wulfhere's men?"

Which made sense. We were in Wiltunscir and Wulfhere's men would know this country, and who better to guide the Danes as they watched Alfred?

The Saxons were coming into the wood and we stayed behind some hawthorn bushes until we heard the sound of axes. They were cutting firewood. There seemed to be about a dozen of them. Most of the men who followed Wulfhere would probably be reluctant to fight Alfred, but some would have embraced their ealdorman's new cause and doubtless those were the men who had been dispatched to guide the Danish scouting parties. Wulfhere would only have sent men he could trust, fearing that less loyal men would desert to Alfred or just run away, so these Saxons were probably from the ealdorman's household troops, the warriors who would profit most from being on the winner's side in the war between the Danes and the West Saxons.

"We should get back to Alfred before it's dark," Pyrlig whispered.

But just then a voice sounded close and petulant. "I will go tomorrow," the voice said.

"You won't, lord," a man answered. There was the sound of splashing and I realized one of the two men had come to the bushes for a piss and the other had followed him. "You'll go nowhere tomorrow," the second man went on. "You'll stay here."

"I just want to see them!" the petulant voice pleaded.

"You'll see them soon enough. But not tomorrow. You'll stay here with the guards."

"You can't make me."

"I can do what I like with you, lord. You might command here, but you take my orders all the same." The man's voice was hard and deep. "And my orders are that you're staying here."

"I'll go if I want," the first voice insisted weakly and was ignored.

Very slowly, so that the blade made no noise against the scabbard's throat, I drew Serpent-Breath. Pyrlig watched me, puzzled. "Walk away," I whispered to him, "and make some noise." He frowned in puzzlement at that, but I jerked my head and he trusted me. He stood and walked toward our horses, whistling softly, and immediately the two men followed. The one with the deep voice led. He was an old warrior, scar-faced and bulky. "You!" he shouted. "Stop!" And just then I stepped out from behind the hawthorn and swung Serpent-Breath once and her blade cut under his beard and into his throat, and cut so deep that I felt her scrape against his spine and the blood, sudden and bright in the spring dusk, sprayed across the leaf mold. The man went down like a felled ox. The second man, the petulant man, was following close behind and he was too astonished and much too scared to run away and so I seized his arm and pulled him down behind the bushes.

"You can't," he began, and I placed the flat of Serpent-Breath's bloody blade against his mouth so that he whimpered with terror.

"Not a sound," I said to him, "or you're dead." Pyrlig came back then, sword drawn.

Pyrlig looked at the dead man whose breeches were still untied. He stooped to him and made the sign of the cross on his forehead. The man's death had been quick, and the capture of his companion had been quiet, and none of the woodcutters seemed to have taken alarm. Their axes went on thumping, the echoes rattling in the trees. "We're taking this one back to Alfred," I told Pyrlig. Then I moved Serpent-Breath to my captive's throat. "Make one sound," I said, pressing the blade into his skin, "and I'll gut you from your overused gullet to your overused crotch. Do you understand?"

He nodded.

"Because I'm doing you the favor I owe you," I explained, and smiled nicely.

Because my captive was Æthelwold, Alfred's nephew and the would-be king of the West Saxons.

The man I had killed was named Osbergh and he had been the commander of Wulfhere's household troops. His job on the day of his death was to make certain Æthelwold got into no trouble.

Æthelwold had a talent for misfortune. By rights he should have been the king of Wessex, though I daresay he would have been the last king for he was impetuous and foolish, and the twin solaces for having lost the throne to his uncle Alfred were ale and women. Yet he had ever wanted to be a warrior. Alfred had denied him the chance, for he dared not let Æthelwold make a name for himself on the battlefield. Æthelwold, the true king, had to be kept foolish so that no man saw in him a rival for Alfred's throne. It would have been far easier to have killed Æthelwold, but Alfred was sentimental about family. Or perhaps it was his Christian conscience. But for whatever reason, Æthelwold had been allowed to live and had rewarded his uncle's mercy by constantly making a fool of himself.

But in these last months he had been released from Alfred's leash and his thwarted ambition had been given encouragement. He dressed in mail and carried swords. He was a startling-looking man, handsome and tall, and he looked the part of the warrior, though he had no warrior's soul. He had pissed himself when I put Serpent-Breath to his throat, and now that he was my captive he showed no defiance. He was submissive, frightened, and glad to be led.

He told us how he had pestered Wulfhere to be allowed to fight, and when Osbergh had brought a score of men to guide the Danes in the hills, he had been given notional charge of them. "Wulfhere said I was in command," Æthelwold said sullenly, "but I still had to obey Osbergh."

"Wulfhere was a damned fool to let you go so far from him," I said.

"I think he was tired of me," Æthelwold admitted.

"Tired of you? You were humping his woman?"

"She's only a servant! But I wanted to join the scouting parties, and Wulfhere said I could learn a lot from Osbergh."

"You've just learned never to piss into a hawthorn bush," I said, "and that's worth knowing."

Æthelwold was riding Pyrlig's horse and the Welsh priest was leading the beast by its reins. I had tied Æthelwold's hands. There was still a hint of light in the western sky, just enough to make our journey down the smaller river easy. I explained to Pyrlig who Æthelwold was, and the priest grinned up at him. "So you're a prince of Wessex, eh?"

"I should be king," Æthelwold said sullenly.

"No you shouldn't," I said.

"My father was! And Guthrum promised to crown me."

"And if you believed him," I said, "you're a damned fool. You'd be king as long as he needed you, then you'd be dead."

"Now Alfred will kill me," he said miserably.

"He ought to," I said, "but I owe you a favor."

"You think you can persuade him to let me live?" he asked eagerly.

"You'll do the persuading," I said. "You'll kneel to him, and you're going to say that you've been waiting for a chance to escape the Danes, and at last you succeeded, and you got away, found us, and have come to offer him your sword."

Æthelwold just stared at me.

"I owe you a favor," I explained, "and so I'm giving you life. I'll untie your hands, you go to Alfred, and you say you're joining him because that's what you've wanted to do ever since Christmas. You understand that?"

Æthelwold frowned. "But he hates me!"

"Of course he does," I agreed, "but if you kneel to him and swear you never broke your allegiance to him, then what can he do? He'll embrace you, reward you, and be proud of you."

"Truly?"

"So long as you tell him where the Danes are," Pyrlig put in.

"I can do that," Æthelwold said. "They're coming south from Cippanhamm. They marched this morning."

"How many?"

"Five thousand."

"Coming here?"

"They're going to wherever Alfred is. They reckon they'll have a chance to destroy him, and after that it's just a summer of women and silver." He said the last three words plaintively and I knew he had been relishing the prospect of plundering Wessex. "So how many men does Alfred have?" he asked.

"Three thousand," I said.

"Sweet Jesus," he said fearfully.

"You always wanted to be a warrior," I said, "and what name can you make for yourself fighting a smaller army?"

"Jesus Christ."

The last of the light went. There was no moon, but by keeping the river on our left we knew we could not get lost and after a while we saw the glow of firelight showing over the loom of the hills and knew we were seeing Alfred's encampment. I twisted in the saddle then and thought I saw another such glow far to the north. Guthrum's army.

"If you let me go," Æthelwold asked sulkily, "what's to stop me going back to Guthrum?"

"Absolutely nothing," I said, "except the certainty that I'll hunt you down and kill you."

He thought about that for a short while. "You're sure my uncle will welcome me?"

Pyrlig answered for me. "With open arms!" he said. "It will be like the return of the prodigal son. You'll be welcomed by slaughtered calves and psalms of rejoicing. Just tell Alfred what you told us, about Guthrum marching toward us."

We reached the Wilig and the going was easy now, for the light of the campfires was much brighter. I cut Æthelwold free at the edge of the encampment, then gave him back his swords. He carried two, as I did, a long one and a short saxe. "Well, my prince," I said, "time to grovel, eh?"

We found Alfred at the camp's center. There was no pomp here. We did not have the animals to drag wagons loaded with tents or furniture, so Alfred was seated on a spread cloak between two fires. He looked dispirited and later I learned that he had assembled the army in the twilight and made them a speech, but the speech, even Beocca admitted, had been less than successful. "It was more a sermon than a speech," Beocca told me glumly. Alfred had invoked God, spoken of Saint Augustine's doctrine of a righteous war, and talked about Boethius and King David, and the words had flown over the heads of the tired, hungry troops. Now Alfred sat with the leading men of the army, all of them eating stale hard bread and smoked eel. Father Adelbert, the priest who had accompanied us to Cippanhamm, was playing a lament on a small harp. A bad choice of music, I thought. Then Alfred saw me and waved Adelbert to silence. "You have news?" he asked.

For answer I stood aside and bowed to Æthelwold, gesturing him toward the king. "Lord," I said to Alfred, "I bring you your nephew."

Alfred stood. He was taken aback, especially as Æthelwold was plainly no prisoner for he wore his swords. Æthelwold looked good; indeed, he looked more like a king than Alfred. He was well made and handsome, while Alfred was much too thin and his face was so haggard that he looked many years older than his twenty-nine years. And of the two it was Æthelwold who knew how to behave at that moment. He unbuckled his swords and threw them with a great clatter at his uncle's feet. Then he went to his knees and clasped his hands and looked up into the king's face. "I have found you!" he said with what sounded like utter joy and conviction.

Alfred, bemused, did not know what to say, so I stepped forward. "We discovered him, lord," I said, "in the hills. He was searching for you."

"I escaped Guthrum," Æthelwold said. "God be praised, I escaped the pagan." He pushed his swords to Alfred's feet. "My blades are yours, lord king."

This extravagant display of loyalty gave Alfred no choice except to raise his nephew and embrace him. The men around the fires applauded. Then Æthelwold gave his news, which was useful enough. Guthrum was on the march and Svein of the White Horse came with him. They knew where Alfred was and so they came, five thousand strong, to give him battle in the hills of Wiltunscir.

"When will they get here?" Alfred wanted to know.

"They should reach these hills tomorrow, lord," Æthelwold said.

So Æthelwold was seated beside the king and given water to drink, which was hardly a fit welcome for a prodigal prince and caused him to throw me a wry glance, and it was then that I saw Harald, shire reeve of Defnascir, among the king's companions. "You're here?" I asked, surprised.

"With five hundred men," he said proudly.

We had expected no men from either Defnascir or Thornsæta, but Harald, the shire reeve, had brought four hundred of his own fyrd and a hundred more from Thornsæta. "There's enough men left to protect the coast against the pagan fleet," he said, "and Odda insisted we help defeat Guthrum."

"How is Mildrith?"

"She prays for her son," Harald said, "and for all of us."

There were prayers after the meal. There were always prayers when Alfred was around, and I tried to escape them, but Pyrlig made me stay. "The king wants to talk with you," he said.

So I waited while Bishop Alewold droned, and afterward Alfred wanted to know whether Æthelwold had truly run away from the Danes.

"That's what he told me, lord," I said, "and I can only say we found him."

"He didn't run from us," Pyrlig offered, "and he could have run."

"So there's good in the boy," Alfred said.

"God be praised for that," Pyrlig said.

Alfred paused, gazing down into the glowing embers of a camp fire. "I spoke to the army tonight," he told us.

"I heard you did, lord," I said.

He looked up at me sharply. "What did you hear?"

"That you preached to them, lord."

He flinched at that, then seemed to accept the criticism. "What do they want to hear?" he asked.

"They want to hear," Pyrlig answered, "that you are ready to die for them."

"Die?"

"Men follow, kings lead," Pyrlig said. Alfred waited. "They don't care about Saint Augustine," Pyrlig went on. "They only care that their women and children are safe, that their lands are safe, and that they'll have a future of their own. They want to know that they'll win. They want to know the Danes are going to die. They want to hear that they'll be rich on plunder."

"Greed, revenge, and selfishness?" Alfred asked.

"If you had an army of angels, lord," Pyrlig went on, "then a rousing speech about God and Saint Augustine would doubtless fire their ardor, but you have to fight with mere men, and there's nothing quite like greed, revenge, and selfishness to inspire mortals."

Alfred frowned at that advice, but did not argue with it. "So I can trust my nephew?" he asked me.

"I don't know that you can trust him," I said, "but nor can Guthrum. And Æthelwold did seek you out, lord, so be content with that."

"I shall, I shall." He bade us good night, going to his hard bed.

The fires in the valley were dying. "Why didn't you tell Alfred the truth about Æthelwold?" I asked Pyrlig.

"I thought I would trust your judgment," he said.

"You're a good man."

"And that constantly astonishes me."

I went to find Iseult, then slept.

Next day the whole of the northern sky was dark with cloud, while over our army, and above the hills, was sunlight.

The West Saxon army, now almost three and a half thousand strong, marched up the Wilig, then followed the smaller river that Pyrlig and I had explored the previous evening. We could see the Danish scouts on the hills and knew they would be sending messengers back to Guthrum.

I led fifty men to one of the hilltops. We were all mounted, all armed, all with shields and helmets, and we rode ready to fight, but the Danish scouts yielded the ground. There were only a dozen of them and they rode off the hill long before we reached the summit where a host of blue butterflies flickered above the springy turf. I

gazed northward at the ominous dark sky and watched a spar-
rowhawk stoop. Down the bird went, and I followed its plunging fall
and suddenly saw, beneath the folded wings and reaching claws, our
enemy.

Guthrum's army was coming south.

The fear came then. The shield wall is a terrible place. It is where a
warrior makes his reputation, and reputation is dear to us. Reputation
is honor, but to gain that honor a man must stand in the shield wall
where death runs rampant. I had been in the shield wall at Cynuit and
I knew the smell of death, the stink of it, the uncertainty of survival,
the horror of the axes and swords and spears, and I feared it. And it was
coming.

I could see it coming, for in the lowlands north of the hills, in the
green ground stretching long and level toward distant Cippanhamm,
was an army. The great army, the Danes called it, the pagan warriors of
Guthrum and Svein, the wild horde of wild men from beyond the sea.

They were a dark smear on the landscape. They were coming
through the fields, band after band of horsemen, spread across the
country, and because their leading men were only just emerging into
the sunlight, it seemed as if their horde sprang from the shadowlands.
Spears and helmets and mail and metal reflected the light, a myriad
glints of broken sunlight that spread and multiplied as yet more men
came from beneath the clouds. They were nearly all mounted.

"Jesus, Mary, and Joseph," Leofric said.

Steapa said nothing. He just glowered at them.

Osric, the shire reeve of Wiltunscir, made the sign of the cross.
"Someone has to tell Alfred," he said.

"I'll go," Father Pyrlig offered.

"Tell him the pagans have crossed the Afen," Osric said. "Tell him
they're heading toward"—he paused, trying to judge where the horde
was going—"Ethandun," he finally said.

"Ethandun." Pyrlig repeated the name.

"And remind him there's a fort of the old people there," Osric
said. This was his shire, his country, and he knew its hills and fields,

and he sounded grim, doubtless wondering what would happen if the Danes found the old fortress and occupied it. "God help us," Osric said. "They'll be in the hills tomorrow morning, tell him."

"Tomorrow morning, at Ethandun," Pyrlig said, then turned his horse and spurred away.

"Where's the fort?" I asked.

Osric pointed. "You can see it." From this distance the ancient fastness looked like nothing more than green wrinkles on a far hilltop. All across Wessex there were such forts with their massive earthen walls, and this one was built at the top of the escarpment that climbed from the lowlands, a place guarding the sudden edge of the chalk downs. "Some of the bastards will get up there tonight," Osric said, "but most won't make it till morning. Let's just hope they ignore the fort."

We had all thought that Alfred would find a place where Guthrum must attack him, a slope made for defense, a place where our smaller numbers would be helped by the difficult ground, but the sight of that distant fort was a reminder that Guthrum might adopt the same tactics. He might find a place where it would be hard for us to attack him, and Alfred would have a grim choice then. To attack would be to court disaster, while to retreat would guarantee it. Our food would be exhausted in a day or two, and if we tried to withdraw south through the hills Guthrum would release a horde of horsemen against us. And even if the army of Wessex escaped unscathed, it would be a beaten army. If Alfred brought the fyrd together, then marched it away from the enemy, men would take it for a defeat and begin to slip away to protect their homes. We had to fight, because to decline battle was a defeat.

The army camped that evening to the north of the woods where I had found Æthelwold. He was in the king's entourage now and went with Alfred and his war leaders to the hilltop to watch the Danish army as it closed on the hills. Alfred looked a long time. "How far away are they?" he asked.

"From here," Osric answered, "four miles. From your army, six."

"Tomorrow, then," Alfred said, making the sign of the cross. The northern clouds were spreading, darkening the evening, but the slanting light reflected from spears and axes at the old people's fort. It seemed Guthrum had not ignored the place after all.

We went back down to the encampment to find yet more men arriving. Not many now, just small bands, but still they came, and one such band, travel weary and dusty, was mounted on horses and all sixteen men had chain mail and good helmets.

They were Mercians and they had ridden far to the east, crossed the Temes, then looped through Wessex, ever avoiding Danes, and so come to help Alfred. Their leader was a short young man, wide in the chest, round-faced, and with a pugnacious expression. He knelt to Alfred, then grinned at me, and I recognized my cousin, Æthelred.

My mother was a Mercian, though I never knew her, and her brother Æthelred was a power in the southern part of that country and I had spent a short time in his hall when I first fled from Northumbria. Back then I had quarreled with my cousin, called Æthelred like his father, but he seemed to have forgotten our youthful enmity and embraced me instead. The top of his head just came up to my collarbone. "We've come to fight," he told me, his voice muffled by my chest.

"You'll have a fight," I promised him.

"Lord." He let go of me and turned back to Alfred. "My father would have sent more men, but he must protect his land."

"He must," Alfred said.

"But he sent the best he has," Æthelred went on. He was young and bumptious, a little strut of a youth, but his confidence pleased Alfred, as did the gleaming silver crucifix hanging over Æthelred's chain mail. "Allow me to present Tatwine," my cousin went on, " the chief of my father's household troops."

I remembered Tatwine, a barrel of a man and a real fighter, whose arms were smothered in blotchy black marks, each made with a needle and ink and representing a man killed in battle. He gave me a crooked smile. "Still alive, lord?"

"Still alive, Tatwine."

"Be good to fight alongside you again."

"Good to have you here," I said, and it was. Few men are natural-born warriors, and a man like Tatwine was worth a dozen others.

Alfred had ordered the army to assemble again. He did it partly so the men could see their own numbers and take heart from that, and he did it, too, because he knew his speech the previous night had left men confused and uninspired. He would try again. "I wish he wouldn't," Leofric grumbled. "He can make sermons, but he can't make speeches."

We gathered at the foot of a small hill. The light was fading. Alfred had planted his two banners, the dragon and the cross, on the summit of the hill, but there was small wind so the flags stirred rather than flew. He climbed to stand between them. He was alone, dressed in a mail coat over which he wore the faded blue cloak. A group of priests began to follow him, but he waved them back to the hill's foot. Then he just stared at us huddled in the meadow beneath him and for a time he said nothing and I sensed the discomfort in the ranks. They wanted fire put into their souls and expected holy water instead.

"Tomorrow!" he said suddenly. His voice was high, but it carried clearly enough. "Tomorrow we fight! Tomorrow! The Feast of Saint John the Apostle!"

"Oh God," Leofric grumbled next to me, "up to our arseholes in more saints."

"John the Apostle was condemned to death!" Alfred said. "He was condemned to be boiled in oil! Yet he survived the ordeal! He was plunged into the boiling oil and he lived! He came from the cauldron a stronger man! And we shall do the same." He paused, watching us, and no one responded. We all just gazed at him, and he must have known that his homily on Saint John was not working, for he made an abrupt gesture with his right hand as if he was sweeping all the saints aside. "And tomorrow," he went on, "is also a day for warriors. A day to kill your enemies. A day to make the pagans wish they had never heard of Wessex!"

He paused again, and this time there were some murmurs of agreement.

"This is our land! We fight for our homes! For our wives! For our children! We fight for Wessex!"

"We do," someone shouted.

"And not just Wessex!" Alfred's voice was stronger now. "We have men from Mercia, men from Northumbria, men from East Anglia!" I knew of none from East Anglia and only Beocca and I were from Northumbria, but no one seemed to care. "We are the men of England," Alfred shouted, "and we fight for all Saxons."

Silence again. The men liked what they heard, but the idea of England was in Alfred's head, not theirs. He had a dream of one country, but it was too big a dream for the army in the meadow. "And why are the Danes here?" Alfred asked. "They want your wives for their pleasure, your children for their slaves, and your homes for their own, but they do not know us!" He said the last six words slowly, spacing them out, shouting each one distinctly. "They do not know our swords," he went on. "They do not know our axes, our spears, our fierceness! Tomorrow we teach them! Tomorrow we kill them! Tomorrow we hack them into pieces! Tomorrow we make the ground red with their blood and make them whimper! Tomorrow we shall make them call for our mercy!"

"None!" a man called out.

"No mercy!" Alfred shouted, and I knew he did not mean it. He would have offered every mercy to the Danes, he would have offered them the love of God and tried to reason with them, but in the last few minutes he had at last learned how to talk to warriors.

"Tomorrow," he shouted, "you do not fight for me! I fight for you! I fight for Wessex! I fight for your wives, for your children, and your homes! Tomorrow we fight and, I swear to you on my father's grave and on my children's lives, tomorrow we shall win!"

And that started the cheering. It was not, in all honesty, a great battle speech, but it was the best Alfred ever gave and it worked. Men stamped the ground and those who carried their shields beat them with swords or spears so that the twilight was filled with a rhythmic thumping as men shouted, "No mercy!" The sound echoed back from the hills. "No mercy, no mercy."

We were ready. And the Danes were ready.

That night it clouded over. The stars vanished one by one, and the thin moon was swallowed in the darkness. Sleep came hard. I sat with Iseult who was cleaning my mail while I sharpened both swords. "You will win tomorrow," Iseult said in a small voice.

"You dreamed that?"

She shook her head. "The dreams don't come since I was baptized."

"So you made it up?"

"I have to believe it," she said.

The stone scraped down the blades. All around me other men were sharpening weapons. "When this is over," I said, "you and I will go away. We shall make a house."

"When this is over," she said, "you will go north. Ever north. Back to your home."

"Then you'll come with me."

"Perhaps." She heaved the mail coat to start on a new patch, scrubbing it with a scrap of fleece to make the links shine. "I can't see my own future. It's all dark."

"You shall be the lady of Bebbanburg," I said, "and I shall dress you in furs and crown you with bright silver."

She smiled, but I saw there were tears on her face. I took it for fear. There was plenty of that in the camp that night, especially when men noticed the glow of light showing where the Danes had lit their fires in the nearby hills. We did sleep, but I was woken long before dawn by a small rain. No one slept through it, but all stirred and pulled on war gear.

We marched in the gray light. The rain came and went, spiteful and sharp, but always at our backs. Most of us walked, using our few horses to carry shields. Osric and his men went first, for they knew the shire. Alfred had said that the men of Wiltunscir would be on the

right of the battle line, and with them would be the men of Suth
Seaxa. Alfred was next, leading his bodyguard that was made of all
the men who had come to him in Æthelingæg, and with him was
Harald and the men of Defnascir and Thornsæta. Burgweard and the
men from Hamptonscir would also fight with Alfred, as would my
cousin Æthelred from Mercia, while on the left would be the strong
fyrd of Sumorsæte under Wiglaf. Three and a half thousand men. The
women came with us. Some carried their men's weapons; others had
their own.

No one spoke much. It was cold that morning, and the rain made
the grass slippery. Men were hungry and tired. We were all fearful.

Alfred had told me to collect fifty or so men to lead, but Leofric
was unwilling to lose that many from his ranks, so I took them from
Burgweard instead. I took the men who had fought with me in the
Heahengel when she had been the *Fyrdraca*, and twenty-six of those
men had come from Hamtun. Steapa was with us, for he had taken a
perverse liking to me, and I had Father Pyrlig who was dressed as a
warrior, not a priest. We were fewer than thirty men, but as we climbed
past a green-mounded grave of the old folk, Æthelwold came to us.
"Alfred said I could fight with you," he said.

"He said that?"

"He said I'm not to leave your side."

I smiled at that. If I wanted a man by my side it would be Eadric
or Cenwulf, Steapa or Pyrlig, men I could trust to keep their shields
firm. "You're not to leave my back," I said to Æthelwold.

"Your back?"

"And in the shield wall you stay close behind me. Ready to take
my place."

He took that as an insult. "I want to be in the front," he insisted.

"Have you ever fought in a shield wall?"

"You know I haven't."

"Then you don't want to be in the front," I said, "and, besides, if
Alfred dies, who'll be king?"

"Ah." He half smiled. "So I stay behind you?"

"You stay behind me."

Iseult and Hild were leading my horse. "If we lose," I told them, "you both get in the saddle and ride."

"Ride where?"

"Just ride. Take the money," I said. My silver and treasures, all I possessed, were in the horse's saddlebags. "Take it and ride with Hild."

Hild smiled at that. She looked pale and her fair hair was plastered tight to her scalp by the rain. She had no hat, and was dressed in a white shift belted with rope. I was surprised that she had come with the army, thinking she would have preferred to find a convent, but she had insisted on coming. "I want to see them dead," she told me flatly. "And the one called Erik I want to kill myself." She patted the long, narrow-bladed knife hanging from her belt.

"Erik is the one who—" I began, then hesitated.

"The one who whored me," she said.

"So he wasn't the one we killed that night?"

She shook her head. "That was the steersman of Erik's ship. But I'll find Erik, and I won't go back to a convent till I see him screaming in his own blood."

"Full of hate, she is," Father Pyrlig told me as we followed Hild and Iseult up the hill.

"Isn't that bad in a Christian?"

Pyrlig laughed. "Being alive is bad in a Christian! We say people are saints if they're good, but how few of us become saints? We're all bad! Some of us just try to be good."

I glanced at Hild. "She's wasted as a nun," I said.

"You do like them thin, don't you?" Pyrlig said, amused. "Now I like them meaty as well-fed heifers! Give me a nice dark Briton with hips like a pair of ale barrels and I'm a happy priest. Poor Hild. Thin as a ray of sunlight, she is, but I pity a Dane who crosses her path today."

Osric's scouts came back to Alfred. They had ridden ahead and seen the Danes. The enemy was waiting, they reported, at the edge of the escarpment, where the hills were highest and where the old people's fort stood. Their banners, the scouts said, were numberless. They had also seen Danish scouts, so Guthrum and Svein must have known we were coming.

On we went, ever higher, climbing into the chalk downs, and the rain stopped, but no sun appeared for the whole sky was a turmoil of gray and black. The wind gusted from the west. We passed whole rows of graves from the ancient days and I wondered if they contained warriors who had gone to battle as we did, and I wondered if in the thousands of years to come other men would toil up these hills with swords and shields. Of warfare there is no end, and I looked into the dark sky for a sign from Thor or Odin, hoping to see a raven fly, but there were no birds. Just clouds.

And then I saw Osric's men slanting away to the right. We were in a fold of the hills and they were going around the right-hand hill and, as we reached the saddle between the two low slopes, I saw the level ground and there, ahead of me, was the enemy.

I love the Danes. There are no better men to fight with, drink with, laugh with, or live with. Yet that day, as on so many others of my life, they were the enemy and they waited for me in a gigantic shield wall arrayed across the down. There were thousands of Danes, spear Danes and sword Danes, Danes who had come to make this land theirs, and we had come to keep it ours. "God give us strength," Father Pyrlig said when he saw the enemy who had begun shouting as we appeared. They clashed spears and swords against limewood shields, making a thunder on the hilltop. The ancient fort was the right wing of their army, and men were thick on the green turf walls. Many of those men had black shields and above them was a black banner, so that was where Guthrum was, while their left wing, which faced our right, was strung out on the open down and it was there I could see a triangular banner, supported by a small cross staff, showing a white horse. So Svein commanded their left, while to the Danish right, our left, the escarpment dropped to the river plains. It was a steep drop, a tumbling hill. We could not hope to outflank the Danes on that side, for no one could fight on such a slope. We had to attack straight ahead, directly into the shield wall and against the earthen ramparts and onto the spears and the swords and the war axes of our outnumbering enemy.

I looked for Ragnar's eagle-wing banner and thought I saw it in the fort, but it was hard to be certain, for every crew of Danes flew their standard, and the small flags were crowded together and the rain had started to fall again, obscuring the symbols, but off to my right, outside the fort and close to the bigger standard of the white horse, was a Saxon flag. It was a green flag with an eagle and a cross, which meant Wulfhere was there with that part of the Wiltunscir fyrd that had followed him. There were other Saxon banners in the enemy horde. Not many, maybe a score, and I guessed that the Danes had brought men from Mercia to fight for them. All the Saxon banners were in the open ground; none was inside the fort.

We were still a long way apart, much farther than a man could shoot an arrow, and none of us could hear what the Danes were shouting. Osric's men were making our right wing as Wiglaf led his Sumorsæte fyrd off to the left. We were making a line to oppose their line, but ours would inevitably be shorter. The odds were not quite two Danes to one Saxon, but it was close. "God help us," Pyrlig said, touching his crucifix.

Alfred summoned his commanders, gathering them under the rain-sodden banner of the dragon. The Danish thunder went on, the clattering of thousands of weapons against shields, as the king asked his army's leaders for advice.

Arnulf of Suth Seaxa, a wiry man with a short beard and a perpetual scowl, advised attack. "Just attack," he said, waving at the fort. "We'll lose some men on the walls, but we'll lose men anyway."

"We'll lose a lot of men," my cousin Æthelred warned. He only led a small band, but his status as the son of a Mercian ealdorman meant he had to be included in Alfred's council of war.

"We do better defending," Osric growled. "Give a man land to defend and he stands, so let the bastards come to us." Harald nodded agreement.

Alfred cast a courteous eye on Wiglaf of Sumorsæte who looked surprised to be consulted. "We shall do our duty, lord," he said, "do our duty whatever you decide." Leofric and I were present, but the king did not invite our opinion so we kept silent.

Alfred gazed at the enemy, then turned back to us. "In my experience," he said, "the enemy expect something of us." He spoke pedantically, in the same tone he used when he was discussing theology with his priests. "They want us to do certain things. What are those things?"

Wiglaf shrugged, while Arnulf and Osric looked bemused. They had both expected something fiercer from Alfred. Battle, for most of us, was a hammering rage, nothing clever, a killing orgy, but Alfred saw it as competition of wisdom, or perhaps as a game of tafl that took cleverness to win. That, I am sure, was how he saw our two armies, as tafl pieces on their checkered board.

"Well?" he asked.

"They expect us to attack!" Osric said uncertainly.

"They expect us to attack Wulfhere," I said.

Alfred rewarded me with a smile. "Why Wulfhere?"

"Because he's a traitor and a bastard and a piece of whore-begotten goat shit," I said.

"Because we do not believe," Alfred corrected me, "that Wulfhere's men will fight with the same passion as the Danes. And we're right, they won't. His men will pull back from killing fellow Saxons."

"But Svein is there," I said.

"Which tells us?" he asked.

The others stared at him. He knew the answer, but he could never resist being a teacher, and so he waited for a response. "It tells us," I supplied it again, "that they want us to attack their left, but they don't want their left to break. That's why Svein is there. He'll hold us and they'll launch an assault out of the fort to hit the flank of our attack. That breaks the right of our army and then the whole damned lot come and kill the rest of us."

Alfred did not respond, but looked worried, suggesting that he agreed with me. The other men turned and looked at the Danes, as if some magical answer might suggest itself, but none did.

"So do as Lord Arnulf suggests," Harald said. "Attack the fort."

"The walls are steep," Wiglaf warned. The Ealdorman of Sumor-sæte was a man of sunny disposition, frequent laughter, and casual generosity, but now, with his men arrayed opposite the fort's green ramparts, he was downcast.

"Guthrum would dearly like us to assail the fort," the king observed.

This caused some confusion for it seemed, according to Alfred, that the Danes wanted us to attack their right just as much as they wanted us to attack their left. The Danes, meanwhile, were jeering us for not attacking at all. One or two ran toward our lines and screamed insults, and their whole shield wall was still banging weapons in a steady, threatening rhythm. The rain made the colors of the shields darker. The colors were black and red and blue and brown and dirty yellow.

"So what do we do?" Æthelred asked plaintively.

There was silence and I realized that Alfred, though he under-stood the problem, had no answer to it. Guthrum wanted us to attack and probably did not care whether we went against Svein's seasoned warriors on the left of the enemy line or against the steep, slippery ditches in front of the fort's walls. And Guthrum must also have known that we dared not retreat because his men would pursue and break us like a horde of wolves savaging a frightened flock.

"Attack their left," I said.

Alfred nodded as though he had already come to that conclusion. "And?" he invited me.

"Attack it with every man we've got," I said. There were probably two thousand men outside the fort and at least half of those were Sax-ons. I thought we should assault them in one violent rush, and over-whelm them by numbers. Then the weakness of the Danish position would be revealed, for they were on the very lip of the escarpment and once they were forced over the edge they had nowhere to go but down the long, precipitous slope. We could have destroyed those two thou-sand men, then re-formed our lines for the harder task of attacking the three thousand inside the fort.

"Employ all our men?" Alfred asked. "But then Guthrum will attack our flank with every man he has."

"Guthrum won't," I said. "He'll send some men to attack our flank, but he'll keep most of his troops inside the fort. He's cautious. He won't abandon the fort, and he won't risk much to save Svein. They don't like each other."

Alfred thought about it, but I could see he did not like the gamble. He feared that while we attacked Svein the other Danes would charge from the fort and overwhelm our left. I still think he should have taken my advice, but fate is inexorable and he decided to imitate Guthrum by being cautious. "We will attack on our right," he said, "and drive off Wulfhere's men, but we must be ready for their counterstroke and so our left stays where it is."

So it was decided. Osric and Arnulf, with the men of Wiltunscir and Suth Seaxa, would give battle to Svein and Wulfhere on the open land to the east of the fort, but we suspected that some Danes would come from behind the ramparts to attack Osric's flank and so Alfred would take his own bodyguard to be a bulwark against that assault. Wigulf, meanwhile, would stay where he was, which meant a third of our men were doing nothing. "If we can defeat them," Alfred said, "then their remnant will retreat into the fort and we can besiege it. They have no water there, do they?"

"None," Osric confirmed.

"So they're trapped," Alfred said as though the whole problem was neatly resolved and the battle as good as won. He turned to Bishop Alewold. "A prayer, bishop, if you would be so kind."

Alewold prayed, the rain fell, the Danes went on jeering, and I knew the awful moment, the clash of the shield walls, was close. I touched Thor's hammer, then Serpent-Breath's hilt, for death was stalking us. God help me, I thought, touching the hammer again, Thor help us all, for I did not think we could win.

THIRTEEN

The Danes made their battle thunder and we prayed. Alewold harangued God for a long time, mostly begging him to send angels with flaming swords, and those angels would have been useful, though none appeared. It would be up to us to do the job.

We readied for battle. I took my shield and helmet from the horse that Iseult led, but first I teased out a thick hank of her black hair. "Trust me," I said to her, because she was nervous, and I used a small knife to cut the tress. I tied one end of the hair to Serpent-Breath's hilt and made a loop with the other end. Iseult watched. "Why?" she asked.

"I can put the loop over my wrist." I showed her. "Then I can't lose the sword. And your hair will bring me luck."

Bishop Alewold was angrily demanding that the women go back. Iseult stood on tiptoe to buckle my wolf-crested helmet in place. Then she pulled my head down and kissed me through the gap in the faceplate. "I shall pray for you," she said.

"So will I," Hild said.

"Pray to Odin and Thor," I urged them, then watched as they led the horse away. The women would hold the horses a quarter mile behind our shield wall and Alfred insisted they went that far back so that no man was tempted to make a sudden dash for a horse and gallop away.

It was time to make the shield wall, and that is a cumbersome business. Some men offer to be in the front rank, but most try to be behind, and Osric and his battle leaders were shoving and shouting as they tried to settle the men. "God is with us!" Alfred was shouting at them. He was still mounted and rode down Osric's slowly forming shield wall to encourage the fyrd. "God is with us!" he shouted again. "We cannot lose! God is with us!" The rain fell harder. Priests were walking down the lines offering blessings and adding to the rain by throwing handfuls of holy water at the shields. Osric's fyrd was mostly five ranks thick, and behind them was a scatter of men with spears. Their job, as the two sides met, was to hurl the spears over their comrades' heads, and the Danes would have similar spear-throwers readying their own weapons. "God is with us!" Alfred shouted. "He is on our side! Heaven watches over us! The holy saints pray for us! The angels guard us! God is with us!" His voice was already hoarse. Men touched amulets for luck, closed their eyes in silent prayer, and tugged at buckles. In the front rank they obsessively touched their shields against their neighbors' shields. The right-hand edge of every man's shield was supposed to overlap the next shield so that the Danes were confronted with a solid wall of iron-reinforced limewood. The Danes would make the same wall, but they were still jeering us, daring us to attack. A young man stumbled from the back of Osric's fyrd and vomited. Two dogs ran to eat the vomit. A spear-thrower was on his knees, shaking and praying.

Father Beocca stood beside Alfred's standards with his hands raised in prayer. I was in front of the standards with Steapa to my right and Pyrlig to my left. "Bring fire on them, oh most holy Lord!" Beocca wailed. "Bring fire on them and strike them down! Punish them for their iniquities." His eyes were closed tight and his face raised to the rain so that he did not see Alfred gallop back to us and push through our ranks. The king would stay mounted so he could see what happened, and Leofric and a dozen other men were also on horseback so that their shields could protect Alfred from thrown spears and axes.

"Forward!" Alfred shouted.

"Forward!" Leofric repeated the order because the king's voice was so hoarse.

No one moved. It was up to Osric and his men to begin the advance, but men are ever reluctant to go against an enemy shield wall. It helps to be drunk. I have been in battles where both sides struggled in a reeking daze of birch wine and ale, but we had little of either and our courage had to be summoned out of sober hearts and there was not much to be found on that cold wet morning.

"Forward!" Leofric shouted again, and this time Osric and his commanders took up the shout and the men of Wiltunscir shuffled a few paces forward and the Danish shields clattered into the wall and locked together and the sight of that *skjaldborg* checked the advance. That is what the Danes call their shield wall, the *skjaldborg* or shield fort. The Danes roared mockery, and two of their younger warriors strutted out of their line to taunt us and invite a duel. "Stay in the wall!" Leofric roared.

"Ignore them!" Osric shouted.

Horsemen rode from the fort, perhaps a hundred of them, and they trotted behind the *skjaldborg* that was formed of Svein's warriors and Wulfhere's Saxons. Svein joined the horsemen. I could see his white horse, the white cloak, and the white horsetail plume. The presence of the horsemen told me that Svein expected our line to break and he wanted to ride our fugitives down just as his riders had slaughtered Peredur's broken Britons at Dreyndynas. The Danes were full of confidence, and so they should have been, for they outnumbered us and they were all warriors, while our ranks were filled with men more used to the plow than the sword.

"Forward!" Osric shouted. His line quivered, but did not advance more than a yard.

Rain dripped from the rim of my helmet. It ran down inside the faceplate, worked itself inside my mail coat, and ran in shivers down my chest and belly.

"Strike them hard, lord!" Beocca shouted. "Slaughter them without mercy! Break them in pieces!"

Pyrlig was praying, at least I think he was praying, for he was speaking in his own tongue, but I heard the word *duw* repeated over and over and I knew, from Iseult, that *duw* was the Britons' word for god. Æthelwold was behind Pyrlig. He was supposed to be behind me, but Eadric had insisted on being at my back, so Æthelwold would protect Pyrlig instead. He was chattering incessantly, trying to cover his nervousness, and I turned on him. "Keep your shield up," I told him.

"I know, I know."

"You protect Pyrlig's head, understand?"

"I know!" He was irritated that I had given him the advice. "I know," he repeated petulantly.

""Forward! Forward!" Osric called. Like Alfred he was on horseback and he went up and down behind his line, sword drawn, and I thought he would jab the blade at his men to goad them onward. They went a few paces and the Danish shields came up again and the limewood made a knocking sound as the *skjaldborg* was made and once again our line faltered. Svein and his horsemen were now at the very far flank, but Osric had placed a group of picked warriors there, ready to guard the open end of his line.

"For God! For Wiltunscir!" Osric shouted. "Forward!"

Alfred's men were on the left of Osric's fyrd where we were bent slightly back, ready to receive the expected flank attack from the fort. We went forward readily enough, but then we were mostly warriors and knew we could not advance in front of Osric's more nervous troops. I almost stepped into a scrap of the ground where, astonishingly, three leverets lay low and quivering. I stared at them and hoped that the men behind me would avoid the little beasts and knew they would not. I do not know why hares leave their young in the open, but they do and there they lay, three sleek leverets in a hollow of the downs, doubtless the first things to die in that day of wind and rain.

"Shout at them!" Osric called. "Tell them they're bastards! Call them sons of whores! Say they're shit from the north! Shout at them!" He knew that was one way to get men moving. The Danes were screaming at us, calling us women, saying we had no courage, and no one in our ranks was shouting back, but Osric's men started now and the wet sky was filled with the noise of weapons banging on shields and men calling insults.

I had hung Serpent-Breath on my back. In the crush of battle a sword is easier to draw over the shoulder than from the hip, and the first stroke can then be a vicious downward hack. I carried Wasp-Sting in my right hand. Wasp-Sting was a saxe, a short sword, a stout blade for stabbing, and in the press of men heaving against an enemy shield wall a short sword can do more damage than a long blade. My shield, iron-rimmed, was held on my left forearm by two leather loops. The shield had a metal boss the size of a man's head, a weapon in itself. Steapa, to my right, had a long sword, not as long as the one with which he had fought me at Cippanhamm, but still a hefty blade, though in his big hand it looked almost puny. Pyrlig carried a boar spear, short and stout and with a wide blade. He was saying the same phrase over and over. *"Ein tad, yr hwn wyt yn y nefoedd, sancteiddier dy enw."* I learned later it was the prayer Jesus had taught his disciples. Steapa was muttering that the Danes were bastards. "Bastards," he said, then, "God help me, bastards." He kept saying it. Over and over. "Bastards, God help me, bastards." My mouth was suddenly too dry to speak, and my stomach felt sour and my bowels loose.

"Forward! Forward!" Osric called and we shuffled on, shields touching, and we could see the enemies' faces now. We could see men's unkempt beards and yellow-toothed snarls, see their scarred cheeks, pocked skin, and broken noses. My faceplate meant I could only see directly ahead. Sometimes it is better to fight without a face-plate, to see the attacks coming from the side, but in the clash of shield walls a faceplate is useful. The helmet was lined with leather. I was sweating. Arrows flicked from the Danish line. They did not have many bowmen and the arrows were scattered, but we raised shields to protect our faces. None came near me, but we were bent back from the line to watch the fort's green walls that were rimmed with men, thick with sword Danes, and I could see Ragnar's eagle-wing banner there and I wondered what would happen if I found myself face-to-face with him. I could see the axes and spears and swords, the blades that sought our souls. Rain drummed on helmets and shields.

The line paused again. Osric's shield wall and Svein's *skjaldborg* were only twenty paces apart now and men could see their immediate foes, could see the face of the man they must kill or the man who would kill them. Both sides were screaming, spitting anger and insults, and the spear-throwers had their first missiles hefted.

"Keep close!" someone shouted.

"Shields touching!"

"God is with us!" Beocca called.

"Forward!" Another two paces, more of a shuffle forward than stepping.

"Bastards," Steapa said. "God help me, bastards."

"Now!" Osric screamed. "Now! Forward and kill them! Forward and kill them! Go! Go! Go!" And the men of Wiltunscir went. They let out a great war shout, as much to hearten themselves as to frighten the enemy, and suddenly, after so long, the shield wall went forward fast, men screaming, and the spears came over the Danish line and our own spears were hurled back and then came the clash, the real battle thunder as shield wall met *skjaldborg*. The shock of the collision shook our whole line so that even my troops, who were not yet engaged, staggered. I heard the first screams, the clangor of blades, the thump of metal driving into shield wood, the grunting of men, and then I saw the Danes coming over the green ramparts, a flood of Danes charging us, intent on hacking into the flank of our attack, but that was why Alfred had put us on the left of Osric's force.

"Shields!" Leofric roared.

I hoisted my shield, touched Steapa's and Pyrlig's shields, then crouched to receive the charge. Head down, body covered by wood, legs braced, Wasp-Sting ready. Behind us and to our right Osric's men fought. I could smell blood and shit. Those are the smells of battle. Then I forgot Osric's fight for the rain was in my face, and the Danes were coming at a run, no shield wall formed, just a frenzied charge intent on winning the battle in one furious assault. There were hundreds of them, and then our spear-throwers let their missiles go.

"Now!" I shouted, and we stepped one pace forward to meet the charge and my left arm was crushed into my chest as a Dane hit me, shield against shield, and he slammed an ax down and I rammed Wasp-Sting forward, past his shield, into his flank, and his ax buried itself in Eadric's shield that was above my head. I twisted Wasp-Sting's blade, pulled her free, and stabbed again. I could smell ale on the Dane's sour breath. His face was a grimace. He yanked his ax free. I stabbed again and twisted the saxe's tip into mail or bone, I could not tell which. "Your mother was a piece of pig shit," I told the Dane, and he screamed in rage and tried to bring the ax down onto my helmet, but I ducked and shoved forward, and Eadric protected me with his shield, and Wasp-Sting was red now, warm and sticky with blood, and I ripped her upward.

Steapa was screaming incoherently, his sword slashing left and right, and the Danes avoided him. My enemy stumbled, went down onto his knees, and I hit him with the shield boss, breaking his nose and teeth, then shoved Wasp-Sting into his bloody mouth. Another man immediately took his place, but Pyrlig buried his boar spear in the newcomer's belly.

"Shields!" I shouted, and Steapa and Pyrlig instinctively lined their shields with mine. I had no idea what happened elsewhere on the hilltop. I only knew what happened within Wasp-Sting's reach.

"Back one! Back one!" Pyrlig called, and we stepped back one pace so that the next Danes, taking the place of the men we had wounded or killed, would trip over the fallen bodies of their comrades, and then we stepped forward as they came so that we met them when they were off balance. That was how to do it, the way of the warrior, and we in Alfred's immediate force were his best soldiers. The Danes had charged us wildly, not bothering to lock shields in the belief that their fury alone would overwhelm us. They had been drawn, too, by the sight of Alfred's banners and the knowledge that should those twin flags topple, then the battle was as good as won, but their assault hit our shield wall like an ocean wave striking a cliff, and it shattered there. It left men on the turf and blood on the grass, and now the Danes at last made a proper shield wall and came at us more steadily.

I heard the enemy shields touching, saw the Danes' wild eyes over the round rims, saw their grimaces as they gathered their strength. Then they shouted and came to kill us.

"Now!" I shouted and we thrust forward to meet them.

The shield walls crashed together. Eadric was at my back, pressing me forward, and the art of fighting now was to keep a space between my body and my shield with a strong left arm, and then to stab under the shield with Wasp-Sting. Eadric could fight over my shoulder with his sword. I had space to my right for Steapa was left-handed, which meant his shield was on his right arm, and he kept moving it away from me to give his long sword room to strike. That gap, no wider than a man's foot is long, was an invitation to the Danes, but they were scared of Steapa and none tried to burst through the small space. His height alone made him distinctive, and his skull-tight face made him fearsome. He was bellowing like a calf being gelded, half shriek and half belligerence, inviting the Danes to come and be killed. They refused. They had learned the danger of Pyrlig, Steapa, and me, and they were cautious. Elsewhere along Alfred's shield wall there were men dying and screaming, swords and axes clanging like bells, but in front of me the Danes hung back and merely jabbed with spears to keep us at bay. I shouted that they were cowards, but that did not goad them onto Wasp-Sting, and I glanced left and right and saw that all along Alfred's line we were holding them. Our shield wall was strong. All that practice in Æthelingæg was proving itself, and for the Danes the fight grew ever more difficult for they were attacking us, and to reach us they had to step over the bodies of their own dead and wounded. A man does not see where he treads in battle for he is watching the enemy, and some Danes stumbled, and others slipped on the rain-slicked grass, and when they were off balance we struck hard, spears and swords like snake tongues, making more bodies to trip the enemy.

We of Alfred's household troops were good. We were steady. We were beating the Danes, but behind us, in Osric's larger force, Wessex was dying.

Because Osric's shield wall unraveled.

Wulfhere's men did it. They did not break Osric's shield wall by fighting it, but by trying to join it. Few of them wanted to fight for the Danes and, now that the battle was joined, they shouted at their countrymen that they were no enemy and wanted to change sides, and the shield wall opened to let them through, and Svein's men went for the gaps like wildcats. One after the other those gaps widened as sword Danes burst through. They cut Wulfhere's men down from behind; they prized open Osric's ranks and spread death like a plague. Svein's Vikings were warriors among farmers, hawks among pigeons, and all of Alfred's right wing shattered. Arnulf saved the men of Suth Seaxa by leading them to the rear of our ranks, and they were safe enough there, but Osric's fyrd was broken, harried, and driven away east and south.

The rain had stopped and a cold damp wind scoured the edge of the downs now. Alfred's men, reinforced by Arnulf's four hundred and a dozen or so of Osric's fugitives, stood alone as the Wiltunscir fyrd retreated. They were being driven away from us, and Svein and his horsemen were panicking them. The fyrd had been eight hundred strong, ranked firm, and now they were shattered into small groups that huddled together for protection and tried to fend off the galloping horsemen who thrust with their long spears. Bodies lay all across the turf. Some of Osric's men were wounded and crawled south as if there might be safety where the women and horses were gathered around a mounded grave of the old folk, but the horsemen turned and speared them, and the unmounted Danes were making new shield walls to attack the fugitives. We could do nothing to help, for we were still fighting Guthrum's men who had come from the fort and, though we were winning that fight, we could not turn our backs on the enemy. So we thrust and hacked and pushed, and slowly they went backward, and then they realized that they were dying man by man, and I heard the Danish shouts to go back to the fort, and we let them go. They retreated from us, walking backward, and when they saw we would not follow, they turned and ran to the green walls. They left a tide line of corpses, sixty or seventy Danes on the turf, and we had lost no more than twenty men. I took a silver chain off one corpse, two arm rings from another, and a fine bone-handled knife with a knob of amber in its hilt from a third.

"Back!" Alfred called.

It was not till we retreated to where we had begun the fight that I realized the disaster on our right. We had been the center of Alfred's army, but now we were its right wing, and what had been our strong right flank was splintered chaos. Many of Osric's men had retreated to where the women and horses waited, and they made a shield wall there, which served to protect them, but most of the fyrd had fled farther east and was being carved into smaller and smaller groups.

Svein at last hauled his men back from the pursuit, but by then nearly all our right wing was gone. Many of those men lived, but they had been driven from the field and would be reluctant to come back and take more punishment. Osric himself had survived, and he brought the two hundred men who had retreated to the women and horses back to Alfred, but that was all he had left. Svein formed his men again, facing us, and I could see him haranguing them. "They're coming for us," I said.

"God will protect us," Pyrlig said. He had blood on his face. A sword or ax had pierced his helmet and cut open his scalp so that blood was crusted thick on his left cheek.

"Where was your shield?" I demanded of Æthelwold.

"I've got it," he said. He looked pale and frightened.

"You're supposed to protect Pyrlig's head," I snarled at him.

"It's nothing." Pyrlig tried to calm my anger.

Æthelwold looked as if he would protest, then suddenly jerked forward and vomited. I turned away from him. I was angry, but I was also disappointed. The bowel-loosening fear was gone, but the fighting had seemed halfhearted and ineffective. We had seen off the Danes who had attacked us, but we had not hurt them so badly that they would abandon the fight. I wanted to feel the battle rage, the screaming joy of killing, and instead all seemed ponderous and difficult.

I had looked for Ragnar during the fight, fearing having to fight my friend, and when the Danes had gone back to the fort I saw he had been engaged farther down the line. I could see him now, on the rampart, staring at us. Then I looked right, expecting to see Svein lead his men in an assault on us, but instead I saw Svein galloping to the fort and I suspected he went to demand reinforcements from Guthrum.

The battle was less than an hour old, yet now it paused. Some women brought us water and moldy bread while the wounded sought what help they could find. I wrapped a rag around Eadric's left arm where an ax blade had gone through the leather of his sleeve. "It was aimed at you, lord," he said, grinning at me toothlessly.

I tied the rag into place. "Does it hurt?"

"Bit of an ache," he said, "but not bad. Not bad." He flexed his arm, found it worked, and picked up his shield. I looked again at Svein's men, but they seemed in no hurry to resume their attack. I saw a man tip a skin of water or ale to his mouth. Just ahead of us, among the line of dead, a man suddenly sat up. He was Danish and had plaited black hair that had been tied in knots and decorated with ribbons. I had thought he was dead, but he sat up and stared at us with a look of indignation and then, seemingly, yawned. He was looking straight at me, his mouth open, and then a flood of blood rimmed and spilled over his lower lip to soak his beard. His eyes rolled white and he fell backward. Svein's men were still not moving. There were some eight hundred of them arrayed in their line. They were still the left wing of Guthrum's army, but that wing was much smaller now that it had been shorn of Wulfhere's men, and so I turned and pushed through our ranks to find Alfred. "Lord!" I called, getting his attention. "Attack those men!" I pointed to Svein's troops. They were a good two hundred paces from the fort and, for the moment at least, without their leader because Svein was still inside the ramparts. Alfred looked down on me from his saddle and I urged him to attack with every man in the center division of our army. The Danes had the escarpment at their back and I reckoned we could tip them down that treacherous slope. Alfred listened to me, looked at Svein's men, then shook his head dumbly. Beocca was on his knees, hands spread wide and face screwed tight in an intensity of prayer.

"We can drive them off, lord," I insisted.

"They'll come from the fort," Alfred said, meaning that Guthrum's Danes would come to help Svein's men.

Some would, but I doubted enough would come. "But we want them out of the fort," I insisted. "They're easier to kill in open ground, lord."

Alfred just shook his head again. I think, at that moment, he was almost paralyzed by the fear of doing the wrong thing, and so he chose to do nothing. He wore a plain helmet with a nasal, no other protection for his face, and he looked sickly pale. He could not see an obvious opportunity, and so he would let the enemy make the next decision.

It was Svein who made it. He brought more Danes out of the fort, three or four hundred of them. Most of Guthrum's men stayed behind the ramparts, but those men who had made the first attack on Alfred's bodyguard now streamed onto the open downland where they joined Svein's troops and made their shield wall. I could see Ragnar's banner among them.

"They're going to attack, aren't they?" Pyrlig said. The rain had washed much of the blood from his face, but the split in his helmet looked gory. "I'm all right," he said, seeing me glance at the damage. "I've had worse from a row with the wife. But those bastards are coming, aren't they? They want to keep killing us from our right."

"We can beat them, lord," I called back to Alfred. "Put all our men against them. All of them!"

He seemed not to hear.

"Bring Wiglaf's fyrd across, lord!" I appealed to him.

"We can't move Wiglaf," he said indignantly.

He feared that if he moved the Sumorsæte fyrd from its place in front of the fort, then Guthrum would lead all his men out to assault our left flank, but I knew Guthrum was far too cautious to do any such thing. He felt safe behind the turf ramparts and he wanted to stay safe while Svein won the battle for him. Guthrum would not move until our army was broken; then he would launch an assault. But Alfred would not listen. He was a clever man, perhaps as clever as any man born, but he did not understand battle. He did not understand that battle is not just about numbers, it is not about moving tafl pieces, and it is not even about who has the advantage in ground, but about passion and madness and a screaming, ungovernable rage.

And so far I had felt none of those things. We in Alfred's household troops had fought well enough, but we had merely defended ourselves. We had not carried slaughter to the enemy, and it is only when you attack that you win. Now, it seemed, we were to defend ourselves again, and Alfred stirred himself to order me and my men to the right of his line. "Leave the standards with me," he said, "and make sure our flank is safe."

There was honor in that. The right end of the line was where the enemy might try to wrap around us and Alfred needed good men to hold that open flank, and so we formed a tight knot there. Far off across the down I could see the remnants of Osric's fyrd. They were watching us. Some of them, I thought, would return if they thought we were winning, but for the moment they were too full of fear to rejoin Alfred's army.

Svein rode his white horse up and down the face of his shield wall. He was shouting at his troops, encouraging them. Telling them we were weaklings who needed only one push to topple.

" 'And I looked,' " Pyrlig said to me, " 'and I saw a pale horse, and the rider's name was death.' " I just stared in astonishment. "It's in the gospel book, " he explained sheepishly, "and it just came to mind."

"Then put it out of your mind," I said harshly, "because our job is to kill him, not fear him." I turned to tell Æthelwold to make certain he kept his shield up, but saw he had taken a new place in the rear rank. He was better there, I decided, so let him alone. Svein was shouting that we were lambs waiting to be slaughtered, and his men had begun beating weapons against their shields. There were just over a thousand men in Svein's ranks now, and they would be assaulting Alfred's division, which numbered about the same, but the Danes still had the advantage, for every man in their shield wall was a warrior, while over half our men were from the fyrds of Defnascir, Thornsæta, and Hamptonscir. If we had brought Wiglaf's fyrd to join us we could have overwhelmed Svein, but by the same token he could have swamped us if Guthrum had the courage to leave the fort. Both sides were being cautious. Neither was willing to throw everything into the battle for fear of losing everything.

Svein's horsemen were on his left flank, opposite my men. He wanted us to feel threatened by the riders, but a horse will not charge into a shield wall. It will sheer away, and I would rather face horsemen than foot soldiers. One horse was tossing its head and I could see blood on its neck. Another horse was lying dead out where the corpses lay in the cold wind, which was bringing the first ravens from the north. Black wings in a dull sky. Odin's birds.

"Come and die!" Steapa suddenly shouted. "Come and die, you bastards! Come on!"

His shout prompted others along our line to call insults to the Danes. Svein turned, apparently surprised by our sudden defiance. His men had started forward, but stopped again, and I realized, with surprise, that they were just as fearful as we were. I had always held the Danes in awe, reckoning them the greatest fighting men under the sky. Alfred, in a moment of gloom, once told me it took four Saxons to beat one Dane, and there was truth in that, but it was not a binding truth, and it was not true that day, for there was no passion in Svein's men. There was unhappiness there, a reluctance to advance, and I reckoned that Guthrum and Svein had quarreled. Or perhaps the cold, damp wind had quelled everyone's ardor. "We're going to win this battle!" I shouted, and surprised myself by shouting it.

Men looked at me, wondering if I had been sent a vision by my god.

"We're going to win!" I was hardly aware of speaking. I had not meant to make a speech, but I made one anyway. "They're frightened of us!" I called out. "They're scared! Most of them are skulking in the fort because they daren't come out to face Saxon blades! And those men," I gestured at Svein's ranks with Wasp-Sting, "know they're going to die! They're going to die!" I took a few paces forward and spread my arms to get the Danes' attention. I held my shield out to the left and Wasp-Sting to the right. "You're going to die!" I shouted it in Danish, loud as I could, then in English. "You're going to die!"

And all Alfred's men took up that shout. "You're going to die! You're going to die!"

Something odd happened then. Beocca and Pyrlig claimed that the spirit of God wafted through our army, and maybe that did happen, or else we suddenly began to believe in ourselves. We believed we could win and as the chant was shouted at the enemy, we began to go forward, step by step, beating swords against shields and shouting that the enemy would die. I was ahead of my men, taunting the enemy, screaming at them, dancing as I went, and Alfred called me back to the ranks. Later, when all was done, Beocca told me that Alfred called me repeatedly, but I was capering and shouting, out ahead on the grass where the corpses lay, and I did not hear him. And Alfred's men were following me and he did not call them back though he had not ordered them forward.

"You bastards!" I screamed, "you goat turds! You fight like girls!" I do not know what insults I shouted that day, only that I shouted them and that I went ahead, on my own, asking just one of them to come and fight me man-to-man.

Alfred never approved of those duels between the shield walls. Perhaps, sensibly, he disapproved because he knew he could not have fought one himself, but he also saw them as dangerous. When a man invites an enemy champion to a fight, man on man, he invites his own death, and if he dies he takes the heart from his own side and gives courage to the enemy, and so Alfred ever forbade us to accept Danish challenges, but on that cold wet day one man did accept my challenge.

It was Svein himself. Svein of the White Horse, and he turned the white horse and spurred toward me with his sword in his right hand. I could hear the hooves thumping, see the clods of wet turf flying behind, see the stallion's mane tossing, and I could see Svein's boar-masked helmet above the rim of his shield. Man and horse coming for me, and the Danes were jeering and just then Pyrlig shouted at me. "Uhtred! Uhtred!"

I did not turn to look at him. I was too busy sheathing Wasp-Sting and was about to pull Serpent-Breath from her scabbard, but just then Pyrlig's thick-shafted boar spear skidded beside me in the wet grass, and I understood what he was trying to tell me. I left Serpent-Breath on my shoulder and snatched up the Briton's spear just as Svein closed on me. All I could hear was the thunder of hooves, see the white cloak spreading, the bright shine of the lofted blade, the tossing horsehair plume, white eyes on the horse, teeth bared, and then Svein twitched the stallion to his left and cut the sword at me. His eyes were glitters behind the eyepieces of his helmet as he leaned to kill me, but as his sword came I threw myself into his horse and rammed the spear into the beast's guts. I had to do it one-handed, for I had my shield on my left arm, but the wide blade pierced hide and muscle, and I was screaming, trying to drive it deeper, and then Svein's sword struck my lifted shield like a hammer blow and his right knee struck my helmet so that I was thrown hard back to sprawl on the grass. I had let go of the spear, but it was well buried in the horse's belly and the animal was screaming and shaking, bucking and tossing, and thick blood was pouring down the spear's shaft that banged and bounced along the grass.

The horse bolted. Svein somehow stayed in the saddle. There was blood on the beast's belly. I had not hurt Svein, I had not touched him, but he was fleeing from me, or rather his white horse was bolting in pain, and it ran straight at Svein's own shield wall. A horse will instinctively swerve away from a shield wall, but this horse was blinded by pain, and then, just short of the Danish shields, it half fell. It slid on the wet grass and skidded hard into the *skjaldborg*, breaking it open. Men scattered from the animal. Svein tumbled from the saddle, and then the horse somehow managed to get back on its feet, and it reared and screamed. Blood was flying from its belly, and its hooves were flailing at the Danes, and now we were charging them at the run. I was on my feet, Serpent-Breath in my right hand, and the horse was thrashing and twisting, and the Danes backed away from it, and that opened their shield wall as we hit them.

Svein was just getting to his feet as Alfred's men arrived. I did not see it, but men said Steapa's sword took Svein's head off in one blow, a blow so hard that the helmeted head flew into the air. Perhaps that was true, but what was certain was that the passion was on us now: the blinding, seething passion of battle. The blood lust, the killing rage, and the horse was doing the work for us, breaking the Danish shield wall apart so all we had to do was ram into the gaps and kill.

And so we killed. Alfred had not meant this to happen. He had expected to wait for the Danish attack and hoped we would resist it, but instead we had thrown off his leash and were doing his work, and he had the wit to send Arnulf's men out to the right because my men were among the enemy. The horsemen had tried to come around our rear, but the men of Suth Seaxa saw them off with shields and swords, then guarded the open flank as all Alfred's men from Æthelingæg and all Harald's men from Defnascir and Thornsæta joined the slaughter. My cousin was there, with his Mercians, and he was a stout fighter. I watched him parry, stab, put down a man, take on another, kill him, and go on steadily. We were making the hilltop rich with Danish blood because we had the fury and they did not, and the men who had fled the field, Osric's men, were coming back to join the fight.

The horsemen went. I did not see them go, though their tale will be told. I was fighting, screaming, shouting at Danes to come and be killed, and Pyrlig was beside me, holding a sword now, and the whole left-hand side of Svein's shield wall had broken and its survivors were making small groups, and we attacked them. I charged one group with the shield, using its boss to slam a man back, and stabbing with Serpent-Breath, feeling her break through mail and leather. Leofric appeared from somewhere, ax swinging, and Pyrlig was ramming his sword's tip into a man's face, and for every Dane there were two Saxons and the enemy stood no chance. One man shouted for mercy and Leofric broke his helmet apart with the ax so that blood and brains oozed onto the jagged metal. I kicked the man aside and plunged Serpent-Breath into a man's groin so that he screamed like a woman in childbirth. The poets often sing of that battle, and for once they get something right when they tell of the sword joy, the blade song, the slaughter. We tore Svein's men to bloody ruin, and we did it with passion, skill, and savagery. The battle calm was on me at last and I could do no wrong. Serpent-Breath had her own life and she stole it from the Danes who tried to oppose me, but those Danes were broken and running and all the left wing of Svein's vaunted troops was defeated.

And there was suddenly no enemy near me except for the dead and injured. Alfred's nephew, Æthelwold, was jabbing his sword at one of the wounded Danes. "Either kill him," I snarled, "or let him live." The man had a broken leg and had an eye hanging down his bloody cheek and he was no danger to anyone.

"I have to kill one pagan," Æthelwold said. He prodded the man with the sword tip and I kicked his blade aside and would have helped the wounded man except it was then that I saw Haesten.

He was at the hill's edge, a fugitive, and I shouted his name. He turned and saw me, or saw a blood-drenched warrior in mail and a wolf-crested helmet, and he stared at me. Then perhaps he recognized the helmet, for he fled. "Coward!" I shouted at him. "You treacherous, bastard coward! You swore me an oath! I made you rich! I saved your rotten life!"

He turned then, half grinned at me, and waved his left arm on which hung the splintered remnants of a shield. Then he ran to what remained of the right-hand side of Svein's shield wall, and that was still in good order, its shields locked tight. There were five or six hundred men there, and they had swung back, then retreated toward the fort, but now they checked because Alfred's men, having no one left

to kill, were turning on them. Haesten joined the Danish ranks, pushing through the shields, and I saw the eagle-wing banner above them
and knew that Ragnar, my friend, was leading those survivors.

I paused. Leofric was shouting at men to form a shield wall and I
knew this attack had lost its fury, but we had damaged them. We had
killed Svein and a good number of his men, and the Danes were now
penned back against the fort. I went to the hill's edge, following a trail
of blood on the wet grass, and saw that the white horse had bolted
over the down's lip and now lay, its legs grotesquely cocked in the air
and its white pelt spattered with blood, a few yards down the slope.

"That was a good horse," Pyrlig said. He had joined me on the
edge of the hill. I had thought this crest was the top of the escarpment,
but the land was tangled here, as though a giant had kicked the hillside with a massive boot. The ground fell away to make a steep valley
that suddenly climbed to a farther crest that was the real edge of the
downs. The steep valley sloped up to the fort's eastern corner, and I
wondered whether it would offer a way into the fastness. Pyrlig was
still staring at the dead horse. "You know what we say at home?" he
asked me. "We say that a good horse is worth two good women, that a
good woman is worth two good hounds, and that a good hound is
worth two good horses."

"You say what?"

"Never mind." He touched my shoulder. "For a Saxon, Uhtred,
you fight well. Like a Briton."

I decided the valley offered no advantage over a direct assault and
turned away to see that Ragnar was retreating step by step toward the
fort. I knew this was the moment to attack him, to keep the battle
anger alive and the slaughter fresh, but our men were plundering the
dead and the dying and none had the energy to renew the assault,
and that meant we would have the harder task of killing Danes protected by a rampart. I thought of my father, killed in an attack on a
wall. He had not shown much liking for me, probably because I had
been a small child when he died, and now I would have to follow
him into the death trap of a well-protected wall. Fate is inexorable.

Svein's banner of the white horse had been captured and a man was waving it toward the Danes. Another had Svein's helmet on the tip of a spear, and at first I thought it was Svein's head, then I saw it was only the helmet. The white horsetail plume was pink now. Father Willibald was holding his hands to heaven, saying a prayer of thanks, and that was premature, I thought, for all we had done was break Svein's men; Guthrum's troops still waited for us behind their walls. And Ragnar was there, too, safe in the fort. Its walls made a semicircle jutting into the downs, ending at the escarpment's lip. They were high walls, protected by a ditch. "It'll be a bastard crossing those ramparts," I said.

"Maybe we won't have to," Pyrlig answered.

"Of course we have to."

"Not if Alfred can talk them out of there," Pyrlig said, and he pointed and I saw that the king, accompanied by two priests and by Osric and Harald, was approaching the fort. "He's going to let them surrender," Pyrlig said.

I could not believe this was the time to talk. This was the killing time, not a place for negotiations. "They won't surrender," I said. "Of course they won't! They still think they can beat us."

"Alfred will try to persuade them," Pyrlig said.

"No." I shook my head. "He'll offer them a truce." I spoke angrily. "He'll offer to take hostages. He'll preach to them. It's what he always does." I thought about going to join him, if for nothing else than to add some sourness to his reasonable suggestions, but I could not summon the effort. Three Danes had gone to talk to him, but I knew they would not accept his offer. They were not beaten; far from it. They still had more men than we did and they had the walls of the fort, and the battle was still theirs to win.

Then I heard the shouts. Shouts of anger and screams of pain, and I turned and saw that the Danish horsemen had reached our women, and the women were screaming and there was nothing we could do.

The Danish horsemen had expected to slaughter the broken remnants of Alfred's shield wall, but instead it had been Svein's men who
had been broken and the riders, out on Svein's left flank, had
retreated into the downs. They must have thought to circle about our
army and rejoin Guthrum from the west, and on the way they had
seen our women and horses and smelled easy plunder.

Yet our women had weapons, and there were a few wounded men
there, and together they had resisted the horsemen. There was a brief
flurry of killing. Then the Danish riders, with nothing to show for
their attack, rode away westward. It had taken a few moments, nothing more, but Hild had snatched up a spear and run at a horseman,
screaming hate for the horrors the Danes had inflicted on her in Cippanhamm, and Eanflæd, who saw it all, said that Hild sank the spear
in a Dane's leg and the man had chopped down with his sword, and
Iseult, who had gone to help Hild, had parried the blow with another
sword, and a second Dane caught her from behind with an ax, and
then a rush of screaming women drove the Danes away. Hild lived, but
Iseult's skull had been broken open and her head almost split into two.
She was dead.

"She has gone to God," Pyrlig told me when Leofric brought us
the news. I was weeping, but I did not know whether it was sorrow or
anger that consumed me. I could say nothing. Pyrlig held my shoulders. "She is with God, Uhtred."

"Then the men who sent her there must go to hell," I said. "Any
hell. Freeze or burn, the bastards!"

I pulled away from Pyrlig and strode toward Alfred. I saw Wulfhere
then. He was a prisoner, guarded by two of Alfred's bodyguards, and he
brightened when he saw me as though he thought I was a friend, but
I just spat at him and walked on past.

Alfred frowned when I joined him. He was escorted by Osric and
Harald, and by Father Beocca and Bishop Alewold, none of whom
spoke Danish, but one of the Danes was an English speaker. There
were three of them, all strangers to me, but Beocca told me their
spokesman was called Hrothgar Ericson and I knew he was one of

Guthrum's chieftains. "They attacked the women," I told Alfred. The king just stared at me, perhaps not understanding what I had said. "They attacked the women!" I repeated.

"He's whimpering," the Danish interpreter said to his two companions, "that the women were attacked."

"If I whimper," I turned on the man in fury, "then you will scream." I spoke in Danish. "I shall pull your guts out of your arse-hole, wrap them around your filthy neck, and feed your eyeballs to my hounds. Now if you want to translate, you shriveled bastard, translate properly, or else go back to your vomit."

The man blinked but said nothing. Hrothgar, resplendent in mail and silvered helmet, half smiled. "Tell your king," he said, "that we might agree to withdraw to Cippanhamm, but we shall want hostages."

I turned on Alfred. "How many men does Guthrum still have?"

He was still unhappy that I had joined him, but he took the question seriously. "Enough," he said.

"Enough to hold Cippanhamm and a half dozen other towns. We break them now."

"You are welcome to try," Hrothgar said when my words were translated.

I turned back to him. "I killed Ubba," I said, "and I put Svein down, and next I shall cut Guthrum's throat and send him to his whore-mother. We'll try."

"Uhtred." Alfred did not know what I had said, but he had heard my tone and he tried to calm me.

"There's work to be done, lord," I said. It was anger speaking in me, a fury at the Danes and an equal fury at Alfred who was once again offering the enemy terms. He had done it so often. He would beat them in battle and immediately make a truce because he believed they would become Christians and live in brotherly peace. That was his desire, to live in a Christian Britain devoted to piety, but on that day I was right. Guthrum was not beaten, he still outnumbered us, and he had to be destroyed.

"Tell them," Alfred said, "that they can surrender to us now. Tell them they can lay down their weapons and come out of the fort."

Hrothgar treated that proposal with the scorn it deserved. Most of Guthrum's men had yet to fight. They were far from defeated, and the green walls were high and the ditches were deep, and it was the sight of those ramparts that had prompted Alfred to speak with the enemy. He knew men must die, many men, and that was the price he had been unwilling to pay a year before when Guthrum had been trapped in Exanceaster, but it was a price that had to be paid. It was the price of Wessex.

Hrothgar had nothing more to say, so he turned away. "Tell Earl Ragnar," I called after him, "that I am still his brother."

"He will doubtless see you in Valhalla one day," Hrothgar called back, then waved a negligent hand to me. I suspected that the Danes had never intended to negotiate a truce, let alone a surrender, but when Alfred offered to talk they had accepted because it gave them time to organize their defenses.

Alfred scowled at me. He was plainly annoyed that I had intervened, but before he could say anything Beocca spoke. "What happened to the women?" he asked.

"They fought the bastards off," I said, "but Iseult died."

"Iseult," Alfred said, and then he saw the tears in my eyes and did not know what to say. He flinched, stuttered incoherently, then closed his eyes as if in prayer. "I am glad," he said after he had collected his thoughts, "that she died a Christian."

"Amen," Beocca said.

"I would rather she was a live pagan," I snarled, and then we went back to our army and Alfred again summoned his commanders.

There was really no choice. We had to assault the fort. Alfred talked for a time about establishing a siege, but that was not practical. We would have to sustain an army on the summit of the downs and, though Osric insisted the enemy had no springs inside the fort, neither did we have springs close by. Both armies would be thirsty, and we did not have enough men to stop Danes going down the steep embankment at night to fetch water. And if the siege lasted longer than a week, then men of the fyrd would begin to slip home to look after their fields, and Alfred would be tempted to mercy, especially if Guthrum promised to convert to Christianity.

So we urged an assault on Alfred. There could be nothing clever. Shield walls must be made and men sent against the ramparts, and Alfred knew that every man in the army must join the attack. Wiglaf and the men of Sumorsæte would attack on the left, Alfred's men in the center, while Osric, whose fyrd had gathered again and was now reinforced by the men who had deserted from Guthrum's army, would assault on the right. "You know how to do it," Alfred said, though without any enthusiasm for he knew he was ordering us into a feast of death. "Put your best men in the center, let them lead, and make the others press behind and on either side."

No one said anything. Alfred offered a bitter smile. "God has smiled on us so far," he said, "and he will not desert us."

Yet he had deserted Iseult. Poor, fragile Iseult, shadow queen and lost soul, and I pushed into the front rank because the only thing I could do for her now was to take revenge. Steapa, as smothered in blood as I was, pushed into the rank beside me. Leofric was to my left and Pyrlig was now behind me. "Spears and long swords," Pyrlig advised us, "not those short things."

"Why not?" Leofric asked.

"You climb that steep wall," he said, "and all you can do is go for their ankles. Bring them down. I've done it before. You need a long reach and a good shield."

"Jesus help us," Leofric said. We were all fearful, for there is little in warfare as daunting as an assault on a fortress. If I had been in my senses I would have been reluctant to make that attack, but I was filled with a keening sorrow for Iseult and nothing except revenge filled my mind. "Let's go," I said, "let's go."

But we could not go. Men were collecting spears thrown in the earlier fighting, and the bowmen were being brought forward. Whenever we attacked we wanted a shower of spears to precede us and a plague of stinging arrows to annoy the enemy, but it took time to array the spearmen and archers behind the men who would make the assault.

Then, ominously for our archers, it began to rain again. Their bows would still work, but water weakened the strings. The sky became darker as a great belly of black cloud settled over the down and the rain started to drum on helmets. The Danes were lining the ramparts, clashing their weapons against shields, as our army curled about their fastness.

"Forward!" Alfred shouted, and we went toward the ramparts, but stopped just out of bowshot. Rain beaded the rim of my shield. There was a new, bright scar in the iron there, a blade strike, but I had not been aware of the blow. The Danes mocked us. They knew what was coming, and they probably welcomed it. Ever since Guthrum had climbed the escarpment and discovered the fort, he had probably imagined Alfred's men assaulting its walls and his men cutting the enemy down as we struggled up the steep banks. This was Guthrum's battle now. He had placed his rival, Svein, and his Saxon ally, Wulfhere, outside the fort, and doubtless he had hoped they could destroy a good part of our army before the assault on the ramparts, but it would not matter much to Guthrum that those men had been destroyed themselves. Now his own men would fight the battle he had always envisaged.

"In the name of God!" Alfred called, then said no more for suddenly a clap of thunder crashed, a vast sound that consumed the heavens and was so loud that some of us flinched. A crack of lightning splintered white inside the fort. The rain pelted now, a cloudburst that hammered and soaked us, and more thunder rolled away in the distance. Perhaps we thought that noise and savage light was a message from God for suddenly the whole army started forward. No one had given a command unless Alfred's invocation was an order. We just went.

Men were shouting as they advanced. They were not calling insults, but just making a noise to give themselves courage. We did not run, but walked, because the shields had to be kept close. Then another bellow of thunder deafened us, and the rain seemed to have a new and vicious intensity. It seethed on the dead and the living, and we were close now, very close, yet the rain was so thick it was hard to see the waiting Danes. Then I saw the ditch, already flooding, and the bows sounded and the spears flew and we were splashing down the ditch's side and Danish spears were thumping into us. One stuck in my shield, fell away, and I stumbled on its shaft, half sprawled in the water, then recovered and began the climb.

Not all the army tried to cross the ditch. Many men's courage faltered at the brink, but a dozen or more groups went into the attack. We were what the Danes call the *svinfylkjas*, the swine wedges, the elite warriors who try to pierce the *skjaldborg* like a boar trying to gouge the hunter with its tusks. But this time we not only had to gouge the *skjaldborg*, but cross the rain-flooded ditch and clamber up the bank.

We held our shields over our heads as we splashed through the ditch. Then we climbed, but the wet bank was so slippery that we constantly fell back, and the Danish spears kept coming, and someone pushed me from behind and I was crawling up the bank on my knees, the shield over my head, and Pyrlig's shield was covering my spine and I heard a thumping above me and thought it was thunder. Except the shield kept banging against my helmet and I knew a Dane was hacking at me, trying to break through the limewood to drive his ax or sword into my spine, and I crawled again, lifted the shield's lower edge, and saw boots. I lunged with Serpent-Breath, tried to stand, felt a blow on my leg, and fell again. Steapa was roaring beside me. There was mud in my mouth, and the rain hammered at us. I could hear the crash of blades sinking in shields and I knew we had failed, but I tried to stand again and lunged with Serpent-Breath and on my left Leofric gave a shrill cry and I saw blood streaming into the grass. The blood was instantly washed away by the rain, and another peal of thunder crashed overhead as I slithered back to the ditch.

The bank was scarred where we had tried to climb; the grass had been gouged down to the white chalk. We had failed utterly and the Danes were screaming defiance. Then another rush of men splashed through the ditch and the banging of blades and shields began again. I climbed a second time, trying to dig my boots into the chalk, and my shield was raised so I did not see the Danes coming down to meet me, and the first I knew was when an ax struck the shield so hard that the boards splintered, and a second ax gave me a glancing blow on the helmet and I fell backward and would have lost Serpent-Breath if it had not been for the loop of Iseult's hair about my wrist. Steapa managed to seize a Danish spear and pulled its owner down the bank where a half dozen Saxons hacked and stabbed in fury so that the ditch was churning with water, blood, and blades, and someone shouted for us to go again, and I saw it was Alfred, dismounted, coming to cross the ditch, and I roared for my men to protect him.

Pyrlig and I managed to get in front of the king and we stayed there, protecting him as we tried to climb that blood-fouled bank a third time. Pyrlig was screaming in his native tongue, I was cursing in Danish, and somehow we got halfway up and stayed on our feet. Someone, perhaps it was Alfred, was pushing me from behind. Rain hammered us, soaked us. A peal of thunder shook the heavens and I swung Serpent-Breath, trying to hack the Danish shields aside, then swung again, and the shock of the blade striking a shield boss jarred up my arm. A Dane, all beard and wide eyes, lunged a spear at me. I lunged back with the sword, shouted Iseult's name, tried to climb, and the spear Dane slammed his spear forward again. The blade struck my helmet's forehead and my head snapped back, and another Dane hit me on the side of the head and all the world went drunken and dark. My feet slid and I was half aware of falling down into the ditch water. Someone pulled me clear and dragged me back to the ditch's far side, and there I tried to stand, but fell again.

The king. The king. He had to be protected and he had been in the ditch when I had last seen him, and I knew Alfred was no warrior. He was brave, but he did not love the slaughter as a warrior loves it. I tried to stand again, and this time succeeded, but blood squelched in my right boot and flowed over the boot top when I put my weight on that leg. The ditch bottom was thick with dead and dying men, half drowned by the flood, but the living had fled from the ditch and the Danes were laughing at us. "To me!" I shouted. There had to be one last effort. Steapa and Pyrlig closed on me, and Eadric was there, and I was groggy and my head was filled with a ringing sound and my arm seemed feeble, but we had to make that last effort. "Where's the king?" I asked.

"I threw him out of the ditch," Pyrlig said.

"Is he safe?"

"I told the priests to hold him down. Told them to hit him if he tried to go again."

"One more attack," I said. I did not want to make it. I did not want to clamber over the bodies in the ditch and try to climb that impossible wall, and I knew it was stupid, knew I would probably die if I went again, but we were warriors and warriors will not be beaten. It is reputation. It is pride. It is the madness of battle. I began beating Serpent-Breath against my half-broken shield, and other men took up the rhythm, and the Danes, so close, were inviting us to come and be killed, and I shouted that we were coming. "God help us," Steapa said.

"God help us," Pyrlig echoed.

I did not want to go. I was frightened, but I feared being called a coward more than I feared the ramparts, and so I screamed at my men to slaughter the bastards, and then I ran. I jumped over the corpses in the ditch, lost my footing on the far side, fell on my shield, and rolled aside so that no Dane could plunge a spear into my unprotected back. I hauled myself up and my helmet had skewed in the fall so that the faceplate half-blinded me, and I fumbled it straight with my sword hand as I began to climb and Steapa was there, and Pyrlig was with me, and I waited for the first hard Danish blow.

It did not come. I struggled up the bank, the shield over my head, and I expected the death blow, but there was silence and I lifted the shield and thought I must have died for all I saw was the rain-filled sky. The Danes had gone. One moment they had been sneering at us, calling us women and cowards, and boasting how they would slice open our bellies and feed our guts to the ravens, and now they were gone. I clambered to the top of the wall and saw a second ditch and second wall beyond, and the Danes were scrambling up that inner rampart and I supposed that they intended to make a defense there, but instead they vanished over its top and Pyrlig grabbed my arm and pulled me on. "They're running!" he shouted. "By God, the bastards are running!" He had to shout to make himself heard over the rain.

"On! On!" someone shouted and we ran into the second flooded ditch and up over the undefended inner bank and I saw Osric's men, the fyrd of Wiltunscir that had been defeated in the opening moments of the fight, had managed to cross the fort's walls. We learned later that they had gone into the valley where the white horse lay dead, and in the blinding rain they had made it to the fort's eastern corner, which, because Guthrum thought it unapproachable, was only lightly defended. The rampart was lower there, hardly more than a grassy ridge on the valley's slope, and they had flooded over the wall and so got behind the other defenders.

Who now ran. If they had stayed, then they would have been slaughtered to a man, so they fled across the fort's wide interior, and some were slow to realize that the battle was lost and those we trapped. I just wanted to kill for Iseult's sake, and I put two fugitives down, hacking them with Serpent-Breath with such fury that she cut through mail, leather, and flesh to bite as deep as an ax. I was screaming my anger, wanting more victims, but we were too many and the trapped Danes were too few. The rain kept falling and the thunder bellowed as I looked about for enemies to kill, and then I saw one last group of them, back-to-back, fighting off a swarm of Saxons, and I ran toward them and suddenly saw their banner. The eagle's wing. It was Ragnar.

His men, outnumbered and overwhelmed, were dying. "Let him live!" I shouted. "Let him live!" And three Saxons turned toward me and they saw my long hair and my arm rings bright on my mailed sleeves, and they must have thought I was a Dane for they ran at me, and I fended off the first with Serpent-Breath. The second hammered my shield with his ax, and the third circled behind me and I turned fast, scything Serpent-Breath, shouted that I was a Saxon, but they did not hear me. Then Steapa slammed into them and they scattered, and Pyrlig grabbed my arm, but I shook him off and ran toward Ragnar who was snarling at the ring of enemies, inviting any one of them to try to kill him. His banner had fallen and his crewmen were dead, but he looked like a war god in his shining mail and with his splintered shield and his long sword and his defiant face, and then the ring began to close. I ran, shouting, and he turned toward me, thinking I had come to kill him, and he raised his sword and I brushed it aside with my shield, threw my arms around him, and drove him to the turf.

Steapa and Pyrlig guarded us. They fended off the Saxons, telling them to look for other victims, and I rolled away from Ragnar who sat up and looked at me with astonishment. I saw that his shield hand was bloody. A blade, cutting through the limewood, had sliced into his palm, hacking down between the fingers so that it looked as though he had two small hands instead of one. "I must bind that wound," I said.

"Uhtred," he just said, as if he did not really believe it was me.

"I looked for you," I told him, "because I did not want to fight you."

He flinched as he shook the shattered remnants of the shield away from his wounded hand. I could see Bishop Alewold running across the fort in mud-spattered robes, waving his arms and shouting that God had delivered the pagans into our hands. "I told Guthrum to fight outside the fort," Ragnar said. "We would have killed you all."

"You would," I agreed. By staying in the fort Guthrum had let us defeat his army piece by piece, but even so it was a miracle that the day was ours.

"You're bleeding," Ragnar said. I had taken a spear blade in the back of my right thigh. I have the scar to this day.

Pyrlig cut a strip of cloth from a dead man's jerkin and used it to bind Ragnar's hand. He wanted to bandage my thigh, but the bleeding had lessened and I managed to stand, though the pain, which I had not felt ever since the wound had been given, suddenly struck me. I touched Thor's hammer. We had won. "They killed my woman," I told Ragnar. He said nothing, but just stood beside me and, because my thigh was agony and I suddenly felt weak, I put an arm about his shoulders. "Iseult, she was called," I said, "and my son is dead, too." I was glad it was raining or else the tears on my face would have shown. "Where's Brida?"

"I sent her down the hill," Ragnar told me. We were limping together toward the fort's northern ramparts.

"And you stayed?"

"Someone had to stay as a rear guard," he said bleakly. I think he was crying, too, because of the shame of the defeat. It was a battle Guthrum could not lose, yet he had.

Pyrlig and Steapa were still with me, and I could see Eadric stripping a dead Dane of his mail, but there was no sign of Leofric. I asked Pyrlig where he was, and Pyrlig gave me a pained look and shook his head.

"Dead?" I asked.

"An ax," he said, "in the spine." I was numb, too numb to speak, for it did not seem possible that the indestructible Leofric was dead, but he was, and I wished I could give him a Danish funeral, a fire funeral, so that the smoke of his corpse would rise to the halls of the gods. "I'm sorry," Pyrlig said.

"The price of Wessex," I said, and then we climbed the northern ramparts that were crowded with Alfred's soldiers.

The rain was lessening, though it still fell in great swaths across the plain below. It was as if we stood on the rim of the world, and ahead of us was an immensity of cloud and rain, while beneath us, on the long steep slope, hundreds of Danes scrambled to the foot of the escarpment where their horses had been left. "Guthrum," Ragnar said bitterly.

"He lives?"

"He was the first to run," he said. "Svein told him we should fight outside the walls," he went on, "but Guthrum feared defeat more than he ever wanted victory."

A cheer sounded as Alfred's banners were carried across the captured fort to the northern ramparts. Alfred, mounted again, and with a bronze circlet about his helmet, rode with the flags. Beocca was on his knees giving thanks, while Alfred had a dazed smile and a look of disbelief, and I swear he wept as his standards were rammed into the turf at the world's edge. The dragon and the cross flew above his kingdom that had almost been lost, but had been saved so that there was still one Saxon king in England.

But Leofric was dead and Iseult was a corpse and a hard rain fell across the land we had rescued.

Wessex.

HISTORICAL NOTE

The Westbury white horse is cut into the chalk of the escarpment beneath Bratton Camp on the edge of the Wiltshire Downs. From the north it can be seen for miles. The present horse, a handsome beast, is over a hundred feet long and almost two hundred feet high and was cut in the 1770s, making it the oldest of Wiltshire's ten white horses, but local legend says that it replaced a much older horse that was blazoned into the chalk hillside after the battle of Ethandun in 878.

I should like to think that legend is true, but no historian can be certain of the location of the battle of Ethandun where Alfred met Guthrum's Danes, though Bratton Camp, above the village of Edington, is the prime candidate. Bratton Camp is an Iron Age fortress that still stands just above the Westbury white horse. John Peddie, in his useful book, *Alfred, Warrior King*, places Ethandun at Bratton Camp, and Edgar's Stone at Kingston Deverill in the Wylye valley, and I am persuaded by his reasoning.

There is no debate about the location of Æthelingæg. That is now Athelney, in the Somerset Levels, near Taunton, and if Bratton Camp is substantially unaltered since 878, the levels are changed utterly. Today, mostly thanks to the medieval monks who diked and drained the land, they make a wide, fertile plain, but in the ninth century they were a vast swamp mingled with tidal flats, an almost impenetrable marsh into which Alfred retreated after the disasters at Chippenham.

That disaster was the result of his generosity in agreeing to the truce that allowed Guthrum to leave Exeter and retreat to Gloucester in Danish-held Mercia. That truce was secured by Danish hostages, but Guthrum, just as he had broken the truce arranged at Wareham in 876, again proved untrustworthy and, immediately after Twelfth Night, attacked and captured Chippenham, thus precipitating the greatest crisis of Alfred's long reign. The king was defeated and most of his country taken by the Danes. Some great nobles, Wulfhere, the Ealdorman of Wiltshire, among them, defected to the enemy, and Alfred's kingdom was reduced to the watery wastes of the Somerset Levels. Yet in the spring, just four months after the disaster at Chippenham, Alfred assembled an army, led it to Ethandun, and there defeated Guthrum. All that happened. What, sadly, did not probably happen is the burning of the cakes. That story, how a peasant woman struck Alfred after he allowed her cakes to burn, is the most famous folktale attached to Alfred, but its source is very late and thus very unreliable.

Alfred, Ælswith, Wulfhere, Æthelwold, and Brother (later Bishop) Asser all existed, as did Guthrum. Svein is a fictional character. The great Danish enemies before Guthrum had been the three Lothbrok brothers, and the defeat of the last of them at the battle of Cynuit occurred while Alfred was at Athelney. For fictional reasons I moved that Saxon victory forward a year, and it forms the ending of *The Last Kingdom*, the novel that precedes *The Pale Horseman*, which meant I had to invent a character, Svein, and a skirmish, the burning of Svein's ships, to replace Cynuit.

The two primary sources for Alfred's reign are the Anglo-Saxon Chronicle and Bishop Asser's life of the king, and neither, alas, tells us much about how Alfred defeated Guthrum at Ethandun. Both armies, by later standards, were small, and it is almost certain that Guthrum considerably outnumbered Alfred. The West Saxon fyrd that won Ethandun was mostly drawn from Somerset, Wiltshire, and western Hampshire, suggesting that all eastern Wessex, and most of the north of the country, had been subdued by the Danes. We know the fyrd of Devonshire was intact (it had won the victory at Cynuit), as was the fyrd of Dorset, yet neither is mentioned as part of Alfred's army, suggesting that they were held back to deter a seaborne attack. The lack of the fyrds from those two powerful shires, if indeed they were absent, only confirms what a remarkable victory Alfred won.

The Saxons had been in Britain since the fifth century. By the ninth century they ruled almost all of what is now England, but then the Danes came and the Saxon kingdoms crumbled. *The Last Kingdom* tells of the defeat of Northumbria, Mercia, and East Anglia, and *The Pale Horseman* describes how Wessex almost followed those northern neighbors into history's oblivion. For a few months in early 878 the idea of England, its culture and language, were reduced to a few square miles of swamp. One more defeat and there would probably never have been a political entity called England. We might have had a Daneland instead, and this novel would probably have been written in Danish. Yet Alfred survived, he won, and that is why history awarded him the honorific "the Great." His successors were to finish his work. They were to take back the three northern kingdoms and so, for the first time, unite the Saxon lands into one kingdom called England, but that work was begun by Alfred the Great.

Yet in 878, even after the victory at Ethandun, that must have seemed an impossible dream. It is a long way from Ethandun's white horse to the bleak moors north of Hadrian's Wall, so Uhtred and his companions must campaign again.